PERMACULTURE

Principles & Pathways Beyond Sustainability

David Holmgren was born in Fremantle, Western Australia, in 1955. As a child of working class political activists he was greatly influenced by the social revolution of the late 60s and early 70s. While travelling around Australia in 1973 he fell in love with the Tasmanian landscape and joined the innovative Environmental Design School in Hobart. An intense working relationship with his mentor Bill Mollison, over the following three years, led to the permaculture concept and set the course of his later life. As the young co-author of Permaculture One in 1978, David shunned the limelight and focused on further developing his practical and design skills for a self-reliant lifestyle.

Since then he has written several more books, developed three properties using permaculture principles, conducted workshops and courses in Australia, New Zealand, Israel and Europe. For the last 17 years he has lived and worked in Hepburn Springs, central Victoria. As a consultant designer he has developed a great expertise in the temperate landscapes of south-eastern Australia with a strong bioregional focus on his home territory. With his partner Su Dennett and their son Oliver he maintains their property, Melliodora as one of Australia's best known permaculture demonstration sites. Over the last seven years he has been a driving force in the design and development of the Fryer's Forest eco-village. Within the international permaculture movement, David is respected for his commitment to presenting permaculture ideas through practical projects. He teaches by personal example that a sustainable lifestyle is a realistic, attractive and powerful alternative to dependant consumerism. This book is the distillation of a life lived by the principles of permaculture.

To the memory of H.T Odum (died 11 September 2002).

PERMACULTURE

Principles & Pathways Beyond Sustainability

DAVID HOLMGREN
Co-Originator of the Permaculture Concept

Holmgren
Design Services

Published December 2002, reprinted December 2004.

by Holmgren Design Services
16 Fourteenth Street
Hepburn, Victoria 3461
Australia
www.holmgren.com.au

The Process

Over 25 years since the first conceptualisation of permaculture, the pattern of work and practice behind this book reflects the principles within.
This book was written over a four year period at Melliodora on a second hand Power Mac 9600 computer primarily using Clarisworks 5. During that time the author was supported by the production from the property and occasional permaculture courses, tours and consultancy.
The major proof reading drafts were printed on second hand paper on a 10 year old laser printer. The editing, graphics and book production was done by other self-employed people in the local community.
The book was printed and bound at Maryborough, a nearby regional centre by McPhersons, an Australian owned company on a Timson T32 offset press.
The paper is Australian made, primarily from plantation and recycled sources.
The cover is Australian made, from fully recycled sources.
No grants, subsidies or work for corporations or governments contributed to this book.

Credits

Editing: Janet Mackenzie, Ian Lillington
Proof reading: Su Dennett, Catherine Jones
Design and book production: Rob & Terttu Mancini
Graphics: Luke Mancini from originals by David Holmgren
Principle icons: Richard Telford
Back cover photo: Christian Wild
Printing and binding: McPhersons Printing Group
Typeset in Novarese Book

Holmgren, David
Permaculture: principles and pathways beyond sustainability

Bibliography.
Includes index.
ISBN 0 646 41844 0.

1. Permaculture. I. Holmgren Design Services. II. Title.

631.58

Foreword

If the 'Permaculture Principles' that David Holmgren discusses in this extremely important book were applied to all that we do, we would be well on the road to sustainability, and beyond. Furthermore, we would be liberated from the lurking feelings of guilt that most of us feel when we reflect on what we are currently passing on to future generations.

Permaculture is about values and visions, and designs and systems of management that are based on holistic understanding, especially on our bio-ecological and psychosocial knowledge and wisdom. It is particularly about our relationships with, and the design and redesign of, natural resource management systems, so that they may support the health and well-being of all present and future generations. What is particularly puzzling is that whereas all engineers — people who work primarily with non-living materials — learn about design principles, nearly all agriculturists, and others working with living systems, are still able to graduate without ever discussing principles of design, let alone having any unit devoted to this critical competence. It is the persistent lack of recognition of the importance of design, of the importance of mutualistic relationships and high biodiversity within sustainable ecosystems, and of the need to design managed ecosystems based on this awareness that is responsible for so many of the problems we currently face in natural resource management.

Permaculture may be described in a diverse range of complementary ways. It is one expression of a next step in the evolution of natural resource management, particularly as it relates to agriculture, most of which is still stuck at an earlier evolutionary stage, characterized by deceptively simple designs based on specialisation, monocultures and simple rotations. These designs, the problems they produce, and the disruptive solutions commonly used to address them, have led to losses of topsoil, moisture holding capacity, fertility, productivity, resilience, wildlife habitat, biodiversity, including natural control organisms, and the gene pool upon which the system depends. To a permaculturist, agriculture's growing dependence on resource inputs to compensate for this progressive degradation of its resource base and associated need to control pests and diseases, its increasingly negative energy budget, and growing waste production problems and environmental impacts, are all obviously predictable. This situation is particularly distressing because it can be largely avoided by applying the 'Permaculture Principles' outlined in the following pages. Instead of repeatedly wasting expertise, time, energy and resources in efforts to address such problems, at the 'back-end' of the system, permaculture enables us to avoid and minimise them by focusing on 'front-end' imaginative design and redesign initiatives. My particular experience of doing this has been focused mainly on pest control and soil management.

Permaculture also reflects the ongoing evolution of our knowledge systems. These are currently being driven by challenges from post-modernists and post-structuralists, feminists and eco-feminists, social ecologists, deep ecologists and eco-psychologists, and those interested in post-normal science, holism, sense-of-place, sustainability, communalism, spirituality and indigenous knowledge systems.

Many factors have contributed to the development of permaculture. Key among them are:

- synchronicity and collaboration across difference (the chance association between David Holmgren — the modest, reflective, thorough, follow-through person — and Bill Mollison — the wild ideas man with the public persona);
- the visioning of permaculture as an international movement;
- the requirement for teachers to have extensive training and field experience and to maintain ongoing practice in order to teach courses; and
- the integration of ethical and design principles into all aspects of theory and practice.

This comprehensive quality, and its associated heavy demands on the process of holistic planning and action, has also been a major barrier to many who would benefit from permaculture. Just as most people tend to opt for the aspirin rather than get their life in order, most farmers and gardeners remain similarly dependent on chemicals to 'fix the headaches' in their maldesigned and mismanaged production systems. Those who have crossed this barrier however, and found permanent design solutions to problems — that only need to be discovered once — are never willing to go back to the dependence, inefficiency and illusion of 'magic-bullet solutions'.

David Holmgren has provided in the following chapters a reasoned, systematic and documented account, based particularly on his extensive experience, of the key principles for developing the intellectual competence to practise Permaculture. This must be matched with parallel experience in the field. Ideally this might include both work as an apprentice with a mentor, such as David, and also opportunities to experiment freely and boldly alone, without supervision. This latter work should focus on what I call 'small, meaningful initiatives that you can guarantee to carry through to completion'. Such initiatives minimise the chances of negative impacts from inappropriate designs, and feelings of discouragement from failure to follow-through on mega-projects.

As a holographic thinker — being open to the idea that anything one observes anywhere is likely to have parallel expressions everywhere — I am led to go beyond the usual boundaries that are put around permaculture. In fact, when I lived in North America I used to run workshops for permaculturists entitled 'Permaculture of the Inner Landscape'. I did this because I had observed that many of these designers were being limited, not by their knowledge of external systems, but by their woundedness and need to 'heal and redesign' their internal systems. I encourage you to similarly try applying these Permaculture Principles to any area that might benefit from such holistic design theory and practice. Areas that immediately come to mind include human settlements and business enterprises, political and economic systems, and the health field, child rearing and learning environments.

This is the most advanced presentation of Permaculture concepts that I am aware of. The 12 principles have been extensively tested, not only by the author, who is the co-originator of permaculture, but also by thousands of permaculturists around the world.

If Permaculture is new to you, this volume will provide you with an outstanding introduction to this holistic approach to landscape design. If you are a long time practitioner or teacher of Permaculture, it is likely that this is the book that you have been waiting for — to challenge and hone your ideas, and to use as the core text in your Permaculture courses. I hope you enjoy reading and referring to this extremely valuable book as much as I have.

Professor Stuart B. Hill
Foundation Chair of Social Ecology
University of Western Sydney
NSW, Australia

Permaculture Design Principles

1 Observe and Interact

Beauty is in the eye of the beholder

2 Catch and Store Energy

Make hay while the sun shines

3 Obtain a Yield

You can't work on an empty stomach

4 Apply Self-regulation and Accept Feedback

The sins of the fathers are visited on the children unto the seventh generation

5 Use and Value Renewable Resources and Services

Let nature take it's course

6 Produce No Waste

A stitch in time saves nine
Waste not, want not

7 Design from Patterns to Details

Can't see the wood for the trees

8 Integrate Rather than Segregate

Many hands make light work

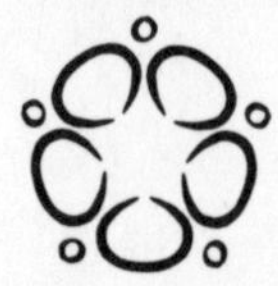

9 Use Small and Slow Solutions

The bigger they are, the harder they fall
Slow and steady wins the race

10 Use and Value Diversity

Don't put all your eggs in one basket

11 Use Edges and Value the Marginal

Don't think you are on the right track just because it is a well-beaten path

12 Creatively Use and Respond to Change

Vision is not seeing things as they are but as they will be

Contents

List of figures

Purpose of this Book

Permaculture is much more than a form of organic gardening. My aim in writing this book is to explain permaculture to a wider audience that may not be attracted by the organic gardening label. It is especially aimed at activists, designers, teachers, researchers, students and others grappling with the vexed issues of sustainability within a wide range of fields.

Permaculture One[1] was written over 25 years ago when I was 20 years old. Most of my more recent publications have been case studies with a practical focus that only hint at a deeper framework which guides that work. With this book, I want to build on the strengths and successes of 25 years of permaculture thinking and action around the world to provide a more evolved picture of the principles that inform permaculture design and action. In the process, I hope to invigorate the intellectual debate within the permaculture movement and address some of the real and perceived weaknesses of the concept.

I know from 25 years experience of applying, writing and teaching permaculture that people will use what they find relevant and meaningful and leave the rest. The quest by some for a completely consistent and logical picture of permaculture may not be useful. Rather than seeking to define or control permaculture, I write about it as simply one more contribution to understanding, meaning and action in a world full of uncertainty.

Evolution of the Project

This project began with a suggestion from permaculture colleague, Ian Lillington, that we publish an edited collection of my writings over the last two decades. The aim was to provide those interested in permaculture with access to more cerebral writing that illustrated permaculture thinking in a wide range of contexts. In the process it would show an important lineage in the ongoing evolution of permaculture ideas and applications by its lesser-known author.

Towards the end of the process, Ian (as editor) suggested that what the collection really needed was an article that directly explained permaculture principles, as taught on our residential Permaculture Design Courses. The moment he suggested this, I knew he was right; but my heart sank, for I knew that the task was not as simple as it seemed. The idea of quickly getting my ideas out to a permaculture readership receded.

Three years later, the project was again transformed by input from professional editor and permaculture activist Janet Mackenzie. The manuscript had grown into a new, hopefully profound reinterpretation of permaculture principles while the *Collected Writings* remained references for further reading best able to stand alone. That collection has since been published as a CD and on the Holmgren Design Services web site.

Format of this Book

A chapter on the ethical principles is followed by one on each of the 12 design principles, roughly in accordance with the structure of our residential Permaculture Design Courses over the last five years.

Each design principle takes the form of a brief action statement with an associated icon and a proverb or saying which exemplify the principle. While the action statements emphasise the positive aspect of permaculture based on the abundance of nature, the proverbs provide a cautionary warning about the constraints and limits of nature,

Each principle is explained in terms of design observable both in the wider world of nature and in the use of land and natural resources by traditional pre-industrial societies.

Then I discuss the ways in which our high-energy industrial society has transformed, or ignored, or apparently overturned the principle, especially where this is relevant to the claim of universality for these design principles.

Included in each chapter are examples of the application of the principle towards creating an ecological culture. The applications of the principle start with examples from gardening, land use and the built environment as the most concrete and widely understood; but they also include the more vexed and complex issues of personal behaviour and social and economic organisation.

I use our own property, documented in the book *Melliodora (Hepburn Permaculture Gardens)*[2] to illustrate each principle. References to the various articles published in the CD *David Holmgren: Collected Writings* 1978-2000[3] further explain or illustrate aspects of each principle. Where possible, I have made reference to published and other sources for the great diversity of concepts and ideas which I condense under each principle.

As always when trying to use the inevitably linear logic of writing to convey wholistic concepts, the division between the issues and perspectives covered under each principle is arbitrary. My choices, and therefore even the principles themselves, are simply tools to help us with multiple perspectives on whole-systems thinking. Cross-references to the other principles point out a selection of the more important linkages. In this sense, each principle can be thought of as a door into the labyrinth of whole-systems thinking.

Acknowledgements

My thanks to Ian Lillington for his strategic thinking, and his gentle but persistent encouragement and follow-up which have kept this project moving (even when it changed tracks) over many years. Other permaculture colleagues who have given inspiration, encouragement and feedback include Jason Alexandra, Stephen Bright, Andrea Furness, Stuart Hill, Sholto Maud, Kale Sniderman and Terry White . I particularly acknowledge Janet Mackenzie for the generous donation of her professional skills and time and for instilling the confidence to keep going when I thought I was finished; Richard Telford for his original ideas and artistic rendering of the principle icons; Luke Mancini for his artistic reworking of my graphics; and Rob and Terttu Mancini for their graphic and production skills in making this book a local business venture. I also want to acknowledge the permaculture students and colleagues who have patiently waited for the publication of this book over the last three years.

Su Dennett

It is traditional for authors, especially men, to acknowledge the patience and support that their partners have provided through the long and at times difficult process of authorship. All of that is appropriate in my acknowledgment of Su Dennett, my partner in life and livelihood over two decades.

In the early days of our relationship it was frustrating to us that, because Su was not prominent in the "important" activities of permaculture teaching, writing and public speaking, she was often seen as a follower implementing my ideas. Ironically, this view often came from women who were both permaculturalists and feminists. Su's commitment and energy in pursuing a low-impact, simple lifestyle have been a constant source of renewal of my own commitment to voluntary frugality as an empowering pathway adapted to ecological realities.

Su's contribution to the ideas in this book has come not as a heavyweight intellectual sparring partner but in helping me get past the limitations of my own overly intellectual and rational approach. Early in life I came to rely on a rational consideration of all factors and perspectives as a constraint on acting like a bull at a gate. But I have found that this adaptation has, in turn, often led me into dogged persistence in the certainty that I had considered all the factors or, alternatively, "paralysis by analysis". It has been through my partnership with Su that I have gradually overcome my deep suspicion of my intuitive capacity and been able to use it as a pathway toward more wholistic understanding and action.

At the practical level, Su has performed a major role facilitating and managing this do-it-yourself publication.

Oliver Holmgren

From his birth at home to his work experience on organic farms in Italy at 15, Oliver has been immersed in the permaculture lifestyle. Like any teenager, his views and behaviour are a challenge to his parents, but over the years of writing this book, Oliver's thinking and action have been an inspiration in refining my conception of permaculture. He has confirmed for me that it takes more than a single generation to create a new ecological culture; some difficult aspects that I have grappled with, he has easily digested and integrated.

Gerard Holmgren

My brother has often reminded me of the political dimension to permaculture, not only through his passion, intellect and action. The hard road of his experience has been a constant reminder that the pathway to a better world will not necessarily be as fortunate and positive as mine has been.

Venie Holmgren

Having got this far in family acknowledgements, an anecdote about my mother is apt. To the public exclamation from a permaculture enthusiast "So you're David Holmgren's mother!", she replied "No, he is my son."

1 B. Mollison & D. Holmgren, *Permaculture One* Corgi 1978 and since published in 5 languages (now out of print)

2 David Holmgren, *Melliodora (Hepburn Permaculture Gardens): Ten Years of Sustainable Living* Holmgren Design Services 1996.

3 *David Holmgren: Collected Writings* 1978-2000 CD, Holmgren Design Services 2002. The project to publish my collected writings was the starting-point of this book, and the articles in the collection complement this book in several ways. Firstly, they provide further examples of the application of the principles. Secondly, the content of some of the articles will be of particular interest to some readers and they provide references to further sources. Finally, the articles provide a historical lineage of my personal evolution, application and explanation of permaculture concepts to a wide range of audiences since 1978, the year *Permaculture One* was published.

Preface

Permaculture in an Uncertain Age

Uncertainty is one of the defining characteristics of our age. Contributions to this state of affairs come from diverse sources.

- Theoretical science has elevated uncertainty from simply a result of inadequate information to something which is inherent in everything.
- The clash between the world's multifarious cultural traditions and modernity leaves most people unsure of their values and their role in society.
- The avalanche of evidence and information about the impermanence of almost every aspect of modern society and economy, especially due to looming environmental threats, undermines any sense of certainty about the continuity of everyday life.
- At the same time, accelerating technology and the emergence of endless new ideas, ways of seeing and being, movements, spiritual pathways and subcultures have expanded possibilities, hopes and fears beyond previously imaginable horizons.

The permaculture concept and movement are part of this global cultural reality, which some call post-modernism, where all meaning is relative and contingent.

The permaculture concept was a product of an intense but relatively brief working relationship between Bill Mollison and myself in the mid 1970s. It was a response to the environmental crisis facing modern society. The publication of *Permaculture One* in 1978 was the culmination of that initial work and a starting point for both the evolution of the concept and the emergence of the worldwide permaculture movement.

Bill Mollison has described permaculture as a "positivistic"[1] response to environmental crisis. That means it is about what we want to do and can do, rather than what we oppose and want others to change. This response is both ethical and pragmatic, philosophical and technical.

Like all ideas, permaculture is founded on some fundamental assumptions that are critical to both understanding and evaluating it. The assumptions on which permaculture was originally based were implied in *Permaculture One* and are worth repeating.

- The environmental crisis is real and of a magnitude that will certainly transform modern global industrial society beyond recognition. In the process, the well-being and even survival of the world's expanding population is directly threatened.
- The ongoing and future impacts of global industrial society and human numbers on the world's wondrous biodiversity are assumed to be far greater than the massive changes of the last few hundred years.

- Humans, although unusual within the natural world, are subject to the same scientific (energy) laws that govern the material universe, including the evolution of life.
- The tapping of fossil fuels during the industrial era was seen as the primary cause of the spectacular explosion in human numbers, technology, and every other novel feature of modern society.
- Despite the inevitably unique nature of future realities, the inevitable depletion of fossil fuels within a few generations will see a return to the general patterns observable in nature and pre-industrial societies dependent on renewable energy and resources.

The conceptual underpinning of these assumptions arises from many sources, but I recognise a clear and special debt to the published work of American ecologist Howard Odum.[2] The ongoing influence of Odum's work on the evolution of my own ideas will become clear through the numerous references in this book, as well as the articles referred to in *David Holmgren: Collected Writings* 1978-2000.[3]

Some of the predictions of resource decline and economic collapse made in the 1970s have proven to be mistaken, at least in their timing. However, the evidence that natural resources are already constraining human development, after approximately 300 years of growth and 50 years of super-accelerated growth, is strong and increasing. The evidence that the current oil crisis reflects the permanent end to cheap energy is compelling.[4] Models from natural systems suggest a collapse back to low energy and resource use (mostly renewable), and a decline in world population is likely. Within this broad scenario, an almost infinite array of pathways and local possibilities can be considered, from the benign to the horrific.

On the other hand, technological and economic optimists argue that we are at the beginning of a new industrial/biological revolution which will lead to a golden age of material well-being. Again some of the evidence is compelling. Amory Lovins' ideas of natural capitalism and the dramatic examples of science and industry doing more with fewer resources and less energy are perhaps the most credible.[5]

Although a future of much diminished use of energy and resources seems inevitable, the nature of that world (and its various parts) is uncertain, to say the least. In the emerging energy transition, ideas and models such as those of Lovins have had considerable influence because they can be applied by business within a capitalist market economy without waiting for fundamental changes either in the political and cultural realm or in the personal behaviour and habits of citizens.

Permaculture is a creative design response to a world of declining energy and resource availability, with many similarities and overlaps with Lovins' emphasis on design processes drawn from nature. For many, the permaculture focus on land and natural resource management is complementary to the industrial focus of the "green tech" optimists, but there are also differences.

Permaculture:

- gives priority to using existing wealth to rebuilding natural capital, especially trees and forests, as a proven storage of wealth to sustain humanity into a future with less fossil fuel
- emphasises bottom-up "redesign" processes, starting with the individual and household as the drivers for change at the market, community and cultural level

- more fundamentally, was predicated on the likelihood of some degree of collapse and breakdown in technology, economics and even society, which is not envisaged or designed for by the "green tech" optimists but is a current reality for many people around the world
- sees pre-industrial sustainable societies as providing models that reflect the more general system design principles observable in nature, and relevant to post-industrial systems.

Insofar as permaculture is an effective response to the limitations on use of energy and natural resources, it will move from its current status as "alternative response to environmental crisis" to the social and economic mainstream of the post-industrial era. Whether it will be called permaculture or not is a secondary matter.

The permaculture concept and movement have already changed the lives of thousands of people and affected perhaps millions in a myriad of ways.[6] All this has occurred without any substantial support from powerful institutions, corporations or governments. Some would attribute its influence solely to the tireless energy, intellect and charisma of Bill Mollison. Although his role in permaculture's initial global spread is unquestioned, its persistence, evolution and influence must be attributed to its relevance to people's lives and situations.

Having pinned the relevance of permaculture to a future with less energy, what might be its relevance in some brave new world of abundant energy and resources (nuclear, genetic engineering, space colonies, or any of the other hoped-for or feared possibilities)? I suspect that the impact of permaculture would contract to influence the lives of relatively isolated individuals and groups who hold to minimal energy and resource use for ethical reasons.

The question of defining what permaculture is and isn't troubles some people. Its multi-faceted character has allowed its progressive evolution into a catholic integration of "ecological alternatives". I have contributed[7] to this expansive evolution, but I also recognise there are dangers in attempts to develop "a theory of everything", and in being "a jack of all trades and master of none" and in "reinventing the wheel". Nevertheless, I see the progressive evolution of permaculture as a strength in influencing the patchy and pulsing nature of social change.

Third Wave Environmentalism

The emergence of environmental awareness and innovation in the last quarter of the 20th century can be seen as clusters of intense activity followed by longer, slower phases of consolidation. These phases of new activity tend to coincide with recession in the main-stream economy.[8] Permaculture was one of the environmental alternatives which emerged from the first great wave of modern environmental awareness, following the Club of Rome report in 1972 and the oil shocks of 1973 and 1975.

After the economic growth of the Reagan–Thatcher revolution in affluent nations during the 1980s, the public awareness of the greenhouse effect in the late 1980s triggered a second wave of environmentalism, which accelerated interest in permaculture. In the 1990s, as new technology and the global economy diverted attention, there was another phase of consolidation. By 1999 the signs were in place for a third wave of environmentalism. In this new phase we can expect public interest to lead to the mainstreaming of many of the innovations of the second wave.

But past experience suggests that each new phase also throws up new insights and innovations that challenge the assumptions of the previous wave. This book is my contribution to the third wave.

1 This description should not be confused with the philosophy of "logical positivism" but rather follows the common use of the terms positive and negative as descriptions of personal attitudes.

2 H.T. Odum, *Environment, Power & Society* John Wiley 1971 was a book which influenced many key environmental thinkers in the 1970s and was the first listed reference in *Permaculture One*. Odum's prodigious published output over the three decades since as well as the work of his students and colleagues has continued to inform my work.

3 *David Holmgren: Collected Writings* 1978-2000.Article 10 "The Development of The Permaculture Concept" and Article 22 "Energy and EMERGY: Revaluing Our World" are especially relevant in explaining the influence of Howard Odum's work on my development of permaculture. See Holmgren Design Services website http://www.holmgren.com.au

4 For one of the most authoritative books on the energy crisis see C. Campbell, *The Coming Oil Crisis* Multi-Science Publishing 1997. The following web site provides many other sources on the subject http://www.hubbertpeak.com/

5 See P. Hawken, A. Lovins & H Lovins, *Natural Capitalism: Creating the Next Industrial Revolution* Rocky Mountain Institute 1999 including the case studies of business achieving factor 4 and factor 10 improvements in energy and resources use per unit of value or profit.

6 See *Permaculture International Journal* (Australia), no longer in print but many libraries have back issues, *Permaculture Magazine* (UK) and *The Permaculture Activist* (USA) for an indication of the scope.

7 For a review of the influence of permaculture on the whole field of ecological alternatives see Article 10 "Development of the Permaculture Concept" in *David Holmgren: Collected Writings* 1978-2000.

8 I have noticed that sustainable agricultural innovation in Australia tends to be clustered in periods of economic recession in the 1880-1890s, the 1930s & 40s and since the 1970s (a more or less continuous rural recession in Australia) with very little of enduring significance during economic booms especially in the 1950s & 60s. More fundamentally, I see the ecological and social movements of these periods as well as the more recent decades being part of a continuous lineage of counterculture in the modern world. See Article 27 "The Counterculture As Dynamic Margin" in *David Holmgren: Collected Writings* 1978-2000.

Introduction

What is Permaculture?

Having said in the preface that I don't want to define or control permaculture, I now must do at least the first of these for the purpose of clarifying the subject of this book.

The vision

The word permaculture was coined by Bill Mollison and myself in the mid-1970s to describe an "integrated, evolving system of perennial or self-perpetuating plant and animal species useful to man".[1]

A more current definition of permaculture, which reflects the expansion of focus implicit in *Permaculture One*, is "Consciously designed landscapes which mimic the patterns and relationships found in nature, while yielding an abundance of food, fibre and energy for provision of local needs." People, their buildings and the ways they organise themselves are central to permaculture. Thus the permaculture vision of permanent (sustainable) agriculture has evolved to one of permanent (sustainable) culture.

The design system

For many people, myself included, the above conception of permaculture is so global in its scope that its usefulness is reduced. More precisely, I see permaculture as *the use of systems thinking and design principles that provide the organising framework for implementing the above vision*. It draws together the diverse ideas, skills and ways of living which need to be rediscovered and developed in order to empower us to move from being dependent consumers to becoming responsible and productive citizens.

In this more limited, but important sense, permaculture is not the landscape, or even the skills of organic gardening, sustainable farming, energy efficient building or eco-village development as such. But it can be used to design, establish, manage and improve these and all other efforts made by individuals, households and communities towards a sustainable future.

The Permaculture Flower (Figure 1) shows the key domains that require transformation to create a sustainable culture. Historically, permaculture has focused on land and nature stewardship as both a source for and an application of ethical and design principles. Those principles are now being applied to other domains dealing with physical and energetic resources as well as human organisation[2] (often called invisible structures in permaculture teaching). Some of the specific fields, design systems and solutions that have been associated with this wider view of permaculture are shown around the periphery of the flower. The spiral evolutionary path beginning with ethics and principles suggests a knitting together of these domains, initially at the personal and the local level and proceeding to the collective and global level. The spidery nature of that spiral suggests the uncertain and variable nature of that process of integration.

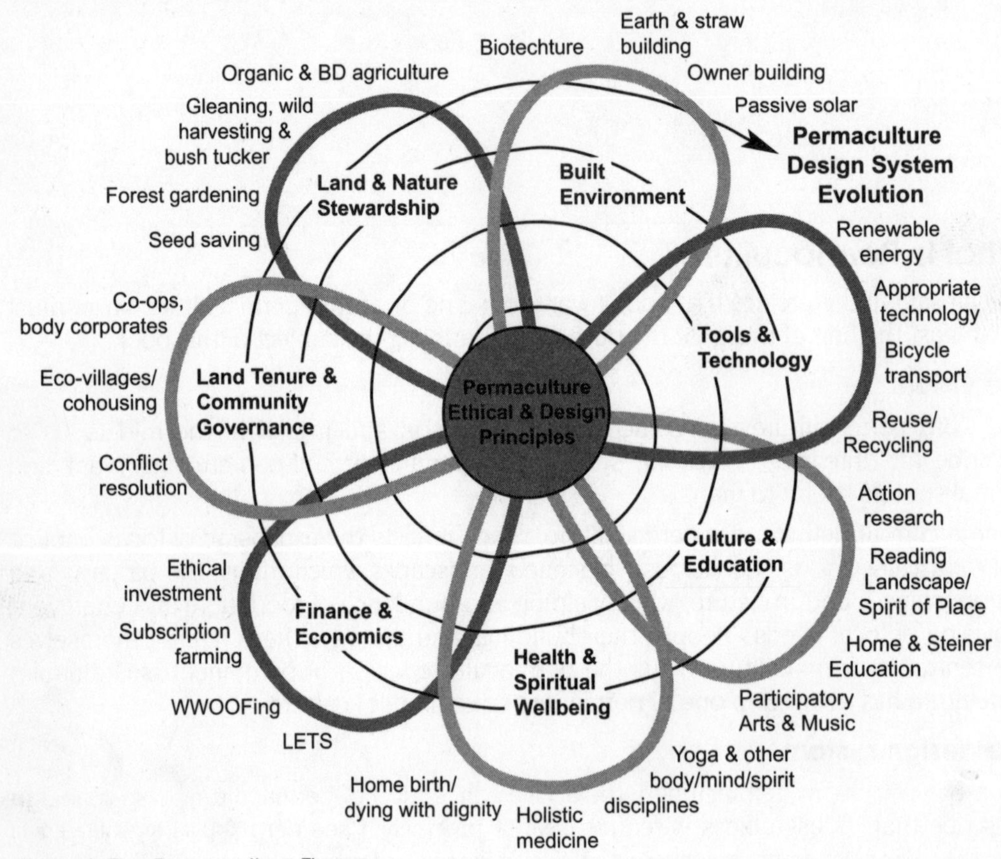

Figure 1: The Permaculture Flower

The network

Permaculture is also a worldwide network and movement of individuals and groups who are working in both rich and poor countries on all continents to demonstrate and spread permaculture design solutions. Largely unsupported by government or business, these people are contributing to a more sustainable future by reorganising their lives and work around permaculture design principles. In this way they are creating small local changes but ones which are directly and indirectly influencing action in the wider environment, organic agriculture, appropriate technology, communities and other movements for a more sustainable world. After 20 years permaculture may rank as one of Australia's most significant "intellectual exports".

The Permaculture Design Course

Most of the people involved in this movement have completed a Permaculture Design Course, which for over 15 years has been the prime vehicle for permaculture inspiration and training world wide. A curriculum was codified in 1984, but divergent evolution of both the form and the content of these courses, as presented by different permaculture teachers, has produced very varied and localised experiences and understandings of permaculture.

In the early 1990s when I began to teach Design Courses regularly, I used the curriculum as a foundation but I freely adapted the format to emphasise my own understandings, experience and priorities. I also contributed to the discussion and debate within the movement about how permaculture education should develop.[3]

The debate about the content of Permaculture Design Courses has become more intense in recent years. Bill Mollison and others[4] have asserted that a failure to adhere to the curriculum, the inclusion of religious beliefs outside the scope of "design science", and a lack of attention to principles and theory is diluting and devaluing some permaculture education. Although I might agree with some of these claims about some courses, I have always found the perception of dilution has to be balanced by the value of diversity even when, like weeds, it comes in forms we don't particularly like.

Popular conceptions of permaculture

In many countries, only the relatively small numbers of people who have done Permaculture Design Courses, or been closely associated with permaculture projects, are familiar with the concept. In Australia, however, a longer history of permaculture activism and influence within large related environmental movements,[5] as well as extensive media exposure, have resulted in broader public appreciation of permaculture.

Popular conceptions of permaculture as a system of gardening or as a counter-cultural lifestyle are both strengths and weaknesses, which provide a context for understanding and appreciation of the ideas in this book.

Permaculture as gardening

As a system of gardening, or perhaps even as commonsense, environmentally aware living, permaculture has been accepted by many people in Australian society as a benign activity. The effects of television gardening programs, easy do-it-yourself books[6] and videos, local school projects, community gardens, LETSystems, and its inclusion as an option in horticultural and other tertiary courses, have all contributed to enthusiasm for permaculture.

The process of providing for people's needs in more sustainable ways requires a cultural revolution, but to propose such a step as a prerequisite can alienate people and inhibit productive steps toward personal and social change. Permaculture has avoided some of the obstacles and opposition that revolutionary ideas encounter.

The permaculture movement and the rudimentary public understanding of permaculture show that it is possible for complex, abstract and revolutionary ideas to exercise influence through positive grassroots processes. This example provides an alternative to the reception of some sustainability concepts, which have been mired in the largely unsuccessful attempts to shift culture by top-down policy processes, exemplified by the Rio Earth Summit.

Permaculture as counterculture

The perception of permaculture as a phenomenon of the countercultural lifestyle, with regular gatherings, its own magazines and newsletters and local groups, has also had positive aspects. As such, permaculture has provided a wholistic framework for reorganising the lives and values of a small minority ready for more fundamental change. This has been particularly so for the minority of young people disillusioned with the conservative consumer youth culture of the late twentieth century.[7] For others, permaculture has provided a message of hope in the struggles against environmental and social evils.[8] The Permaculture Design Course, especially in its two-week residential format, has been particularly effective in galvanising fundamental change and new focus in the lives of participants, and in providing a sense of belonging. This subcultural or countercultural aspect of permaculture has facilitated the experimentation and pioneering of lifestyle models directed by the ecological imperative.[9]

Academic, professional and official reaction

The reaction of academics, professionals and decision-makers has been more varied than that from the wider public.

Among the small number of professionals and academics attempting to integrate the ethical, pragmatic, philosophical and technical aspects of ecological thinking in the late 1970s, *Permaculture One* produced some enthusiastic reactions, such as "Permaculture provides a valuable conceptual framework for future thought on sane, sustainable societies" from Earle Barnhart of the New Alchemy Institute.[10] On the other hand, Bill Mollison noted more generally, "The professional community was outraged because we were combining architecture with biology, agriculture with forestry and forestry with animal husbandry, [so] that almost everybody who considered themselves to be a specialist felt a bit offended."[11]

Permaculture itself was conceived within academia. Many who are involved in large-scale agriculture and land use policy saw it as theoretical, utopian and impractical because it was difficult to apply within the prevailing social, market and policy environment.[12]

Since the growth of the permaculture movement, permaculture has itself become a subject of academic study, with emphasis ranging from the sociological, political and educational aspects to the ecological and agricultural.[13] Some university teachers use permaculture texts and other resources; in 1992, a whole unit on permaculture, which I wrote, was included in the first Australian postgraduate course in sustainable agriculture.[14]

Elsewhere in the world, academics have been at the forefront of development and promotion of permaculture,[15] while the work of Stuart Hill has placed permaculture within a spectrum of concepts and ideas about sustainability.[16] Hill's "Deep Sustainability" perspective on the dilemmas of sustainable agriculture, reinforces the personal and bottom-up strategies for change of permaculture.

> Thus my analysis of the situation is primarily psychosocial, rather than just political, and that is exactly what makes such a proposition so difficult to accept, because for me this requires that I first recognise and act on my responsibilities and change myself before pointing fingers at others, or at least while concurrently doing this. This is not to deny the inequities and oppressions that exist and that need to be addressed within our societies, but rather to acknowledge that each of these can be traced to collective and individual patterns of behaviour, which if not changed will continue to wreak havoc with our precious planet, our societies and our individual wellbeing. Furthermore, I believe that the more empowered, aware, informed, competent and clear about our values that each of us is, then the more effective we are likely to be in bringing about the structural and institutional changes that are required. Trying to do the latter without addressing the former can only ever result in initiatives that will fail

> to address the causes of our problems and that at best can only slightly reduce the levels of unsustainability and degradation.[17]

Hill's views are grounded in his experience in ecological agricultural and entomological research at McGill University in Canada and in knowledge of the organic and other ecological agriculture practices around the world.

Permaculture has been granted a degree of recognition, and has generated strong interest from some students through inclusion as an option in formal horticultural and other tertiary courses. However, a perception of lack of intellectual rigour, and the populist image of permaculture, has continued to inhibit the concept being taken seriously in academia.

Some of the caution about permaculture in academia can be attributed to Bill Mollison, or at least to his personality. His charisma, his ego, and his abrasive and confrontationist manner have made him the perfect media subject. The image of Bill as someone's eccentric uncle with irreverent but neat ideas has struck a chord with many Australians, and he has become a guru to the permaculture faithful. Dr John Wamsley,[18] another environmental iconoclast, had a similar media role and public reaction in recent years. This media portrayal produces an automatic suspicion and rejection by many people, even when they accept such characteristics in mainstream leaders. In addition, Bill Mollison's occasional use of outrageous statements to break down people's preconceptions and get a point across has enraged many in the scientific and academic communities who are already suspicious of wholistic approaches and who resent generalists delving into their own fields.

Over-promotion

Sometimes permaculture-designed solutions have proved, at least in retrospect, naive, misguided or counterproductive. More commonly, a lack of finance, information and skill has often seen good ideas flounder and be discarded. Robert Gilman, editor of *In Context* magazine, has spoken[19] of the community being "inoculated against good ideas" by too early and strong a promotion before the ideas are well tested and proven. Occasionally, superficial and cynical use of permaculture design has given credibility to large projects with little or no environmental or ethical basis.

The fact that permaculture was catapulted into the popular domain so quickly may have had the effect of muddling, and perhaps short-circuiting, the further intellectual development of the concept. This process can be compared and contrasted to the Sustainable Development concept, which was muddled and discredited by its rapid projection into the world of intergovernmental policy and corporate spin doctors.

Whatever path they follow, ideas have to get dirty in one of many "real worlds" outside academia if they are to have life and utility.

Permaculture Principles

In *Permaculture One* (1978), Bill Mollison and I outlined the theory and some initial applications of permaculture design without explicitly listing a clear set of permaculture principles. The permaculture tree[20] presented the concept as analogous to the germinating tree seed, giving rise to interdependent root and aerial structures. The germination of the idea generates both the physical reality of ecological human support systems and the wholistic conceptual framework of knowledge.

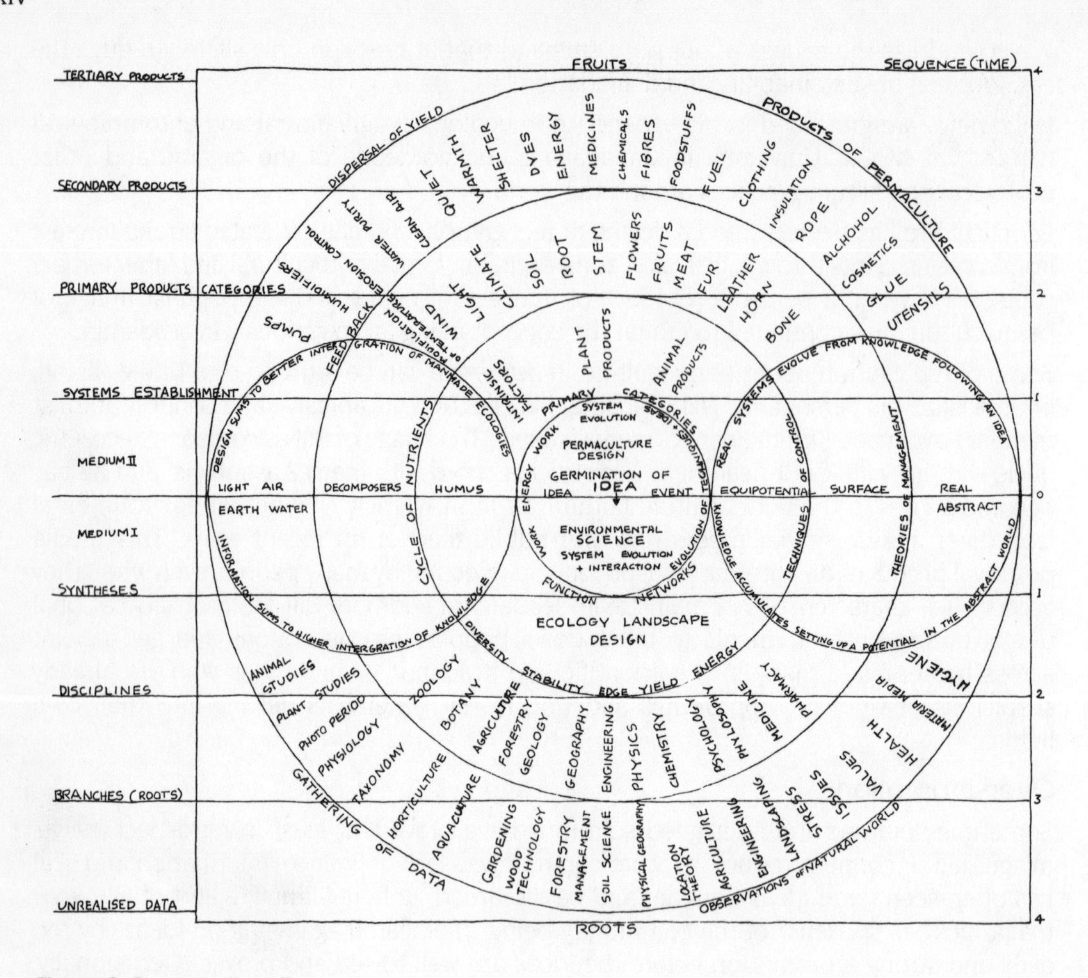

Figure 2: Permaculture Tree

In *Permaculture: A Designers' Manual* (1988), Bill Mollison provided an encyclopaedic coverage of the scope and possibilities of permaculture design as well as an enlargement of the theory and design principles which underlie the applications. Chapters 2 and 3, which deal with these conceptual foundations, are multifaceted and insightful, but they hardly provide a clear list of principles.

In *Introduction to Permaculture* (1991), Bill Mollison and Reny Slay presented design principles in a much simpler format, attributed to American permaculture teacher John Quinney, which has since been widely used or adapted by many permaculture teachers.

The value and use of principles

The idea behind permaculture principles is that generalised principles can be derived from the study of both the natural world and pre-industrial sustainable societies, and that these will be universally applicable to fast-track the post-industrial development of sustainable use of land and resources.

The process of providing for people's needs within ecological limits requires a cultural revolution. Inevitably such a revolution is fraught with many confusions, false leads, risks and inefficiencies. We appear to have little time to achieve this revolution. In this historical context, the idea of a simple set of guiding principles which have wide, even universal application is attractive.

Permaculture principles are brief statements or slogans which can be remembered as a checklist when considering the inevitably complex options for design and evolution of ecological support systems. These principles are seen as universal, although the methods which express them will vary greatly according to place and situation. By still developing extension, these principles are also applicable to our personal, economic, social and political reorganisation, as illustrated in the Permaculture Flower.

These principles can be divided into ethical principles[21] and design principles.

Permaculture ethical principles were distilled from "research of community ethics as adopted by older religious and cooperative groups".[22] Since the emergence of permaculture, ethics — especially environmental ethics — has become a very active field of academic and wider study. This move recognises the ethical problems which lie at the heart of the manifold crisis facing humanity at the end of the second Christian millennium. Permaculture itself has become a subject for study within the field of environmental ethics.[23]

Moral philosophers may argue that such a simple foundation, without extensive reference to the broad field of environmental ethics, let alone philosophy in general, is fraught with ethical and practical dangers. While I agree that ignorance of history condemns us to repeat it, I believe it is hard for us to proceed very far with ethical frameworks without at the same time acting in the real world to develop ourselves as whole persons. The dangers of isolation of philosophical thought from an integrated existence are as great as the dangers of ignorance of the history of philosophy and ethics.

In the modern world of complete uncertainty and questioning, the very simple permaculture ethics, inevitably, are being interpreted in many ways. My own understanding of permaculture ethics is informed by a range of sources both before and after *Permaculture One*. Others versed in environmental ethics may put these ideas in a wider context.

In seeking to be more explicit about ethical principles and their application in this book, I walk through another philosophical minefield with only limited awareness of the details of the hazards. Many academics[24] will see the use of energetics and systems theory to understand and inform ethical concepts, — implicit in permaculture and made explicit in this book — as a dangerously deterministic view of reality. Even within the permaculture movement, some feel uncomfortable with some of my interpretations of ethical and design principles.

I offer my thoughts in the belief that discomfort, especially ethical discomfort, is a healthy alternative to ideological certainty.

Design principles

The scientific foundation for permaculture design principles lies generally within the modern science of ecology, and more particularly within the branch of ecology called systems ecology. Other intellectual disciplines, most particularly landscape geography and ethnobiology, have contributed concepts that have been adapted to design principles.

Fundamentally, permaculture design principles arise from a way of perceiving the world which is often described as "systems thinking" and "design thinking" (see **Principle 1**: *Observe and Interact*). Other examples of systems and design thinking including:

- the Whole Earth Review, and its better-known offshoot the Whole Earth Catalogue, edited by Stewart Brand, did much to publicise systems and design thinking as a central tool in the cultural revolution to which permaculture is a contribution
- the widely known and applied ideas of Edward De Bono[25] fall under the broad rubric of systems and design thinking
- as the academic discipline of cybernetics,[26] systems thinking has been an esoteric and difficult subject, closely associated with the emergence of computing and communication networks and many other technological applications

Despite the powerful applications of systems thinking in the modern world, it has been slow to change the fundamental aspects of our everyday thinking patterns. Recently I was quizzed about my approach to teaching systems thinking in a permaculture context by an American academic who was nearing retirement after a career spanning labouring, mining engineering and academia. In teaching systems thinking as it applied to organisations, he had begun to wonder whether such thinking was an almost innate ability, to which teaching could contribute little. His experience was that most people seem stuck in simplistic understandings and reactive responses to complexity, unable to grasp the broader understandings of systemic thinking, which he saw as both enlightening and empowering.

Apart from the ecological energetics of Howard Odum, the influence of systems thinking in my development of permaculture and its design principles has not come through extensive study of the literature, but more through an osmotic absorption of ideas in the "cultural ether" which strike a chord with my own experience in permaculture design. Further, I believe many of the insights of systems thinking that are difficult to grasp as abstractions are truths that are embodied in the stories and myths of indigenous cultures.

Permaculture principles, both ethical and design, may be observed operating all around us. I argue that their absence or apparent contradiction by modern industrial culture does not invalidate their universal relevance to the descent into a low-energy future

Although the idea of a simple set of ethical and design principles has been central to permaculture teaching, any review of texts, teaching and websites about permaculture shows a diversity of approaches, and even confusion about ethical and design principles and their application. Permaculture-inspired projects and processes frequently illustrate a difficulty in using principles except in an illustrative and literal way.

It could be argued that permaculture has contributed to the spread of some innovative design solutions that illustrate permaculture principles, but that it has been less effective in spreading the systems and design thinking which underlies those solutions.

Inevitably, any set of principles that has been found useful needs to be constantly questioned and further articulated to help us to recognise the creative solutions more clearly. This book represents a culmination of my own efforts to understand and explain the thinking behind permaculture solutions over the last 25 years.

I organise the diversity of permaculture thinking under 12 design principles. My set of design principles varies significantly from those used by most other permaculture teachers. Some of this is simply a matter of emphasis and organisation; in a few cases it may indicate difference of substance. This is not surprising, given the new and still emerging nature of permaculture.

Permaculture Zones and Sectors

Among the great diversity of design concepts taught in Permaculture Design Courses, the zone and sector concepts are by far the most widely understood and applied to make sense of the myriad choices in permaculture site design (see **Principle 7**: *Design from Patterns to Details*). For those familiar with these concepts, Figure 3 provides a meta-analysis of the zones and sectors of permaculture, in which zones of influence and direct power start with the personal and extend to the global.

- The zones, like the permaculture site design zones, are partly physical and geographic and partly conceptual. They work from a core of integration and strength to a wider domain of uncertainty and flexibility. The particular strategies and methods that work in one zone will not necessarily be effective in another.
- The sectors of external energetic forces and material flows inform, support, constrain, influence and damage our meta-system. We can focus, amplify and/or ameliorate these forces and flows, both by spatial and conceptual design responses. At the same time, we need to accept that our influence on their large-scale dynamics may be minimal.

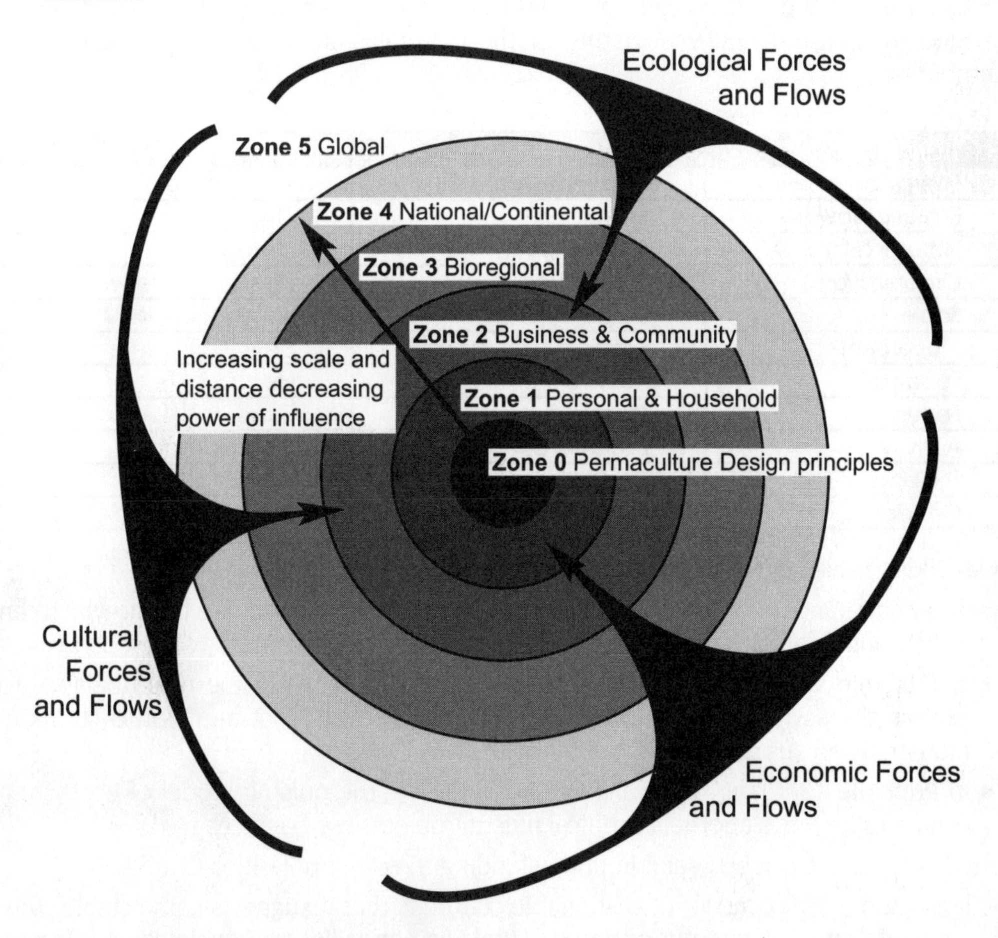

Figure 3: Zone and sector analysis of permaculture

In the same way that permaculture site design has empowered people to make sense of their site and improve their design decisions, this meta-analysis may be useful in empowering people to better understand their world and to act both for themselves and for the future.

Sustainable Culture?

Permaculture principles have an immediacy and relevance to day-to-day life for people in many different situations and cultures, without the need for a particularly unified view of the future. As I explained in the preface, the uncertainty of the future and the ambiguity of concepts of sustainability are unavoidable. However, we cannot live with change or think about sustainability without considering large-scale perspectives that provide some context for the past, present and future. In many ways this book is as much about making sense of our world as it is about what we should do.

One way to view sustainability is as a set of coherent system priorities. The following table gives a snapshot of a contrast between prevailing industrial culture, which is currently reaching a global climax, and a sustainable culture, which reflects long-range ecological realities. This set of polarised characteristics is inevitably artificial, but it quickly identifies the fundamental and universal nature of the cultural shift to which permaculture is contributing.

Characteristic	Industrial culture	Sustainable culture
Energy base	Non-renewable	Renewable
Material flows	Linear	Cyclical
Natural assets	Consumption	Storage
Organization	Centralised	Distributed Network
Scale	Large	Small
Movement	Fast	Slow
Feedback	Positive	Negative
Focus	Centre	Edge
Activity	Episodic change	Rythmic stability
Thinking	Reductionist	Wholistic
Gender	Masculine	Feminine

Table: Characteristics of two cultural systems

The dynamic balance between these polarised pairs of characteristics is a theme which can be found running through my explanation of permaculture principles.

- In **Principle** 9: *Use Small and Slow Solutions*, I use the image of the beam balance to explore the asymmetrical and dynamic balance between fast and slow systems in industrial and sustainable cultures.
- In **Principle** 12: *Creatively Use and Respond to Change*, the pulsing model of ecosystem dynamics provides another graphical illustration of these balances.

Both those models can be useful in understanding the other polarities listed above.

The limitation of this concept of sustainable culture is that it suggests some stable state that we might arrive at sometime soon (by applying permaculture principles). A future in which much smaller human populations are in balance with their renewable resource base may be hundreds of years ahead, but this is no longer than the lifespan of an old tree, a well-built and well-maintained building, or some universities. Paradoxically, it is easier to characterise that low-energy sustainable culture than to explain how we get there.

This process is best visualised using the graphs of dynamic change that have been recorded and predicted for self-organising systems across many scales, from populations of microbes to economies and galaxies. Figure 4 shows such a graph of civilisational growth and predicted[27] decline. Industrial culture and permaculture are stable only in their direction of energy use. The current cultural and economic dynamic of globalisation is one of chaotic climax[28] and transition from growth in population and energy use to decline. The philosophical and artistic concepts of modernism and post-modernism can be loosely linked to these energetic and ecological realities. We have trouble visualising decline as positive, but this simply reflects the dominance of our prior culture of growth. Permaculture is a whole-hearted adaptation to the ecological realities of decline, which are as natural and creative as those of growth. The proverb "what goes up, must come down" reminds us that, in our hearts, we know this to be true. The real issue of our age is how we make a graceful and ethical descent.

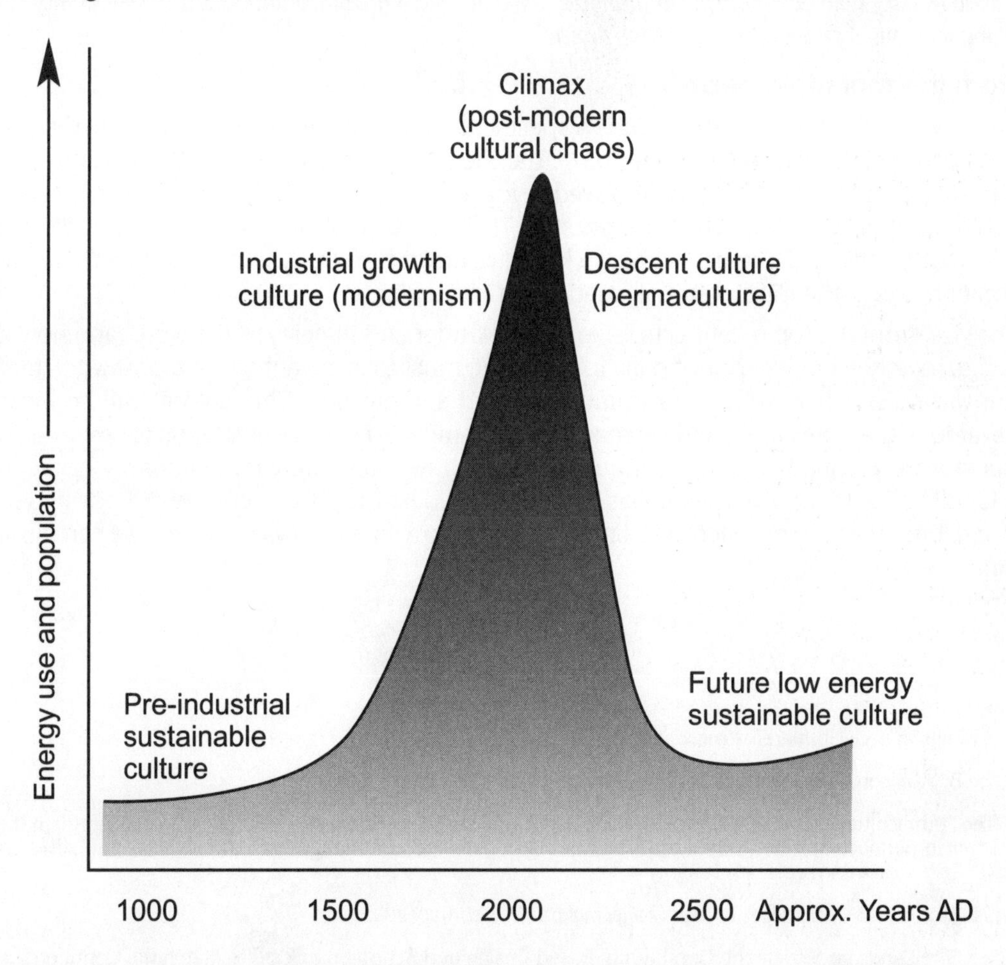

Figure 4: Large-scale cultural dynamics based on fossil energy pulse

Beyond Sustainability

The lack of any reasonable definition of sustainability has left it open to inevitable appropriation by the corporate spin doctors. But even the most genuine and useful sustainability concepts including permaculture contain an ambiguity about sustainability as a state or a process. Once we accept the reality and magnitude of energy descent, we begin to ask what "sustainability", "sustainable systems" or "sustainable system design" might mean. Even the idea of permanence at the heart of permaculture is problematic to say the least.

For any human culture to be considered sustainable it must have the capacity (proven only with historical hindsight) to reproduce itself down the generations while providing human material needs without cataclysmic and long-term breakdown. If it is energetically impossible for high energy society to be anything more than a pulse in the long run of human history, then it cannot, by this definition, be sustainable, no matter how much we shuffle the technological deckchairs. *In articulating Permaculture as the Principles and Pathways Beyond Sustainability, I am suggesting that we need to get over our naive and simplistic notions of sustainability as a likely reality for ourselves or even our grandchildren and instead accept that our task is use our familiarity with continuous change to adapt to energy descent.*

From the mountain peak

When we picture the energy climax as a spectacular but dangerous mountain peak that we (humanity) have succeeded in climbing, the idea of descent to safety is a sensible and attractive proposition. The climb involved heroic effort, great sacrifice, but also exhilaration and new views and possibilities at every step. There are several false peaks, but when we see the whole world laid out around us we know we are at the top. Some argue that there are higher peaks in the mists, but the weather is threatening.

The view from the top reconnects us with the wonder and majesty of the world and how it all fits together, but we cannot dally for long. We must take advantage of the view to chart our way down while we have favourable weather and daylight. The descent will be more hazardous than the climb, and we may have to camp on a series of plateaus to rest and sit out storms. Having been on the mountain so long, we can barely remember the home in a far-off valley that we fled as it was progressively destroyed by forces we did not understand. But we know that each step brings us closer to a sheltered valley where we can make a new home.

1 B. Mollison & D. Holmgren, *Permaculture One* Corgi 1978.

2 See B. Mollison, *Permaculture: A Designers' Manual* Tagari 1988, chapter 14.

3 "The Permaculture Movement and Education: Searching For Ways Forward" 1993, was circulated through the Australian permaculture network of teachers, and later published in three parts in *Permaculture and Landcarers* (later *Green Connections*) vols 3-5, 1995.

4 Lisa Mollison in *Permaculture International Journal* no. 73, February 2000.

5 See "The Landcare Movement: Community-Based Design and Action on a Scale to Match the Continent" in *David Holmgren: Collected Writings 1978-2000*.

6 The following show some of the range of Australian books: B. Mollison and R. Slay, *Introduction to Permaculture* Tagari 1991; R. & J. Mars, *The Basics of Permaculture Design* Candlelight Trust 1994; L. Woodrow, *The Permaculture Home Garden* Viking 1996.

7 Graham Burnett in *Permaculture: A Beginners Guide* Land & Liberty 2000, provides a recent British example of the flavour of this countercultural view of permaculture, in his case complete with connections to veganism and anarchism.

8 G. Holmgren, *Beyond The Nuclear Age* (self-published) 1985, a small book that circulated through environmental and political activist networks in the 1980s, outlined the dangers of genetic engineering and promoted permaculture as part of an integrated alternative perspective.

9 I expand on the historical lineage of the counterculture as a context for permaculture in Article 27 "The Counterculture as Dynamic Margin" in *David Holmgren: Collected Writings 1978-2000*.

10 Earl Barnhart, Introduction to the American edition of *Permaculture One* International Tree Crops Institute 1981.

11 Preface to *Introduction to Permaculture*.

12 For example: Peter King, agricultural and environmental consultant, speaking on "Science Bookshop" ABC Radio, July 1978.

13 For example: Caroline Smith, *The Getting of Hope: personal empowerment through learning permaculture* PhD thesis, Faculty of Education, The University of Melbourne. 2000; Adam Nelson, *Permaculture: Against the New Orthodoxy of Sustainable Development* BA(Hons) thesis, School of Science and Technology Studies, University of New South Wales, 1994; Mona Loofs, *Permaculture, Ecology and Agriculture: an investigation into permaculture theory and practise using two case studies in northern New South Wales* BSc (Hons) thesis, Human Ecology Program, Dept of Geography, Australian National University.

14 UNE Orange Agriculture College 1992, Article 7 "Gardening as Agriculture" in *David Holmgren: Collected Writings 1978-2000* was written as part of the collected reading materials of that course.

15 Most notably, Declan and Margret Kennedy, professors of architecture strongly identified with the Global Eco-village Network, have been pivotal in the development of permaculture in Germany and throughout Europe since the early 1980s.

16 S. Hill, "Ecological and Psychological Prerequisites for the Establishment of Sustainable Agricultural Communities" in J. Martin (ed.), *Alternative Futures for Prairie Agricultural Communities* University of Alberta 1991.

17 S. Hill, "Redesigning Agroecosystems for Environmental Sustainability: A Deep Systems Approach", in *System Research and Behavioural Science* no. 15 John Wiley & Sons 1998.

18 Founder of a series of native animal sanctuaries, including Warrawong in South Australia, and critic of government policies for wildlife protection.

19 Address to Australian Permaculture Convergence, Roseworthy Campus, University of South Australia, Feb 1995.

20 Published in *Permaculture One* and reproduced in *Introduction to Permaculture* (1991).

21 Generally called ethics in most permaculture literature.

22 B. Mollison, *Permaculture: A Designers' Manual* p. 2.

23 Freya Mathews La Trobe University personal comm.

24 Sholto Maud, pers. comm.

25 Best known for coining the term "lateral thinking".

26 Norbet Wiener, *Cybernetics: Control and Communication in the Animal and the Machine*, 1948, is the foundation text. John Gall, *General Systematics* Harper & Row 1977, provides an accessible and useful guide for permaculture designers.

27 This diagram is not a prediction generated by a model, but a simple tool to visualise the change process.

28 The word *climax* is used in its common sense rather than the classic ecological meaning of a sustained mature ecosystem.

Ethical Principles of Permaculture

Ethics are the moral principles that are used to guide action toward good and right outcomes and away from bad and wrong outcomes.

Ethics act as constraints on survival instincts and the other personal and social constructs of self-interest that drive human behaviour in any society. They are culturally evolved mechanisms for more enlightened self-interest, a more inclusive view of who and what constitutes "us", and a longer-term understanding of good and bad outcomes. The greater the power of human civilisation (due to energy availability) and the greater the concentration and scale of power within society, the more critical ethics become in ensuring long-term cultural — and even biological — survival. This ecologically functional view of ethics makes them central in the development of a culture for energy descent.

Like design principles, ethical principles were not explicitly listed in early permaculture literature. Since the development of the Permaculture Design Course, ethics have generally been covered by three broad maxims or principles:

Care for the earth

Care for people

Set limits to consumption and reproduction, and redistribute surplus.

These principles were distilled from research into community ethics, as adopted by older religious and cooperative groups. The third principle, and even the second, can be seen as derived from the first.

The ethical principles have been taught and used as simple and relatively unquestioned ethical foundations for permaculture design within the movement, and within the wider "global nation" of like-minded people. More broadly, these principles can be seen as common to all indigenous tribal peoples, although their conception of "people" may have been more limited than the notion that has emerged in the last two millennia.[1] This focus in permaculture on learning from indigenous tribal cultures is based on the evidence that these cultures have existed in relative balance with their environment and survived for longer than any of our more recent experiments in civilisation.

Of course, in our attempt to live an ethical life we should not ignore the teachings of the great spiritual and philosophical traditions of literate civilisations or the great thinkers of the European Enlightenment and since. But in the long transition to a sustainable low energy culture we need to consider, and attempt to understand, a broader canvas of values and concepts than those delivered to us by recent cultural history.

Philosophical frameworks

Most philosophers acknowledge that ideas and values are never free of their ecological, economic and cultural context. However, few seem to prepared to accept the degree to which the energetically novel conditions of recent centuries are primary factors in the creation and spread of much of what we hold precious in human thinking and culture. In particular, we need to be suspicious of seeing the philosophy of individualism as the source, rather than outcome, of material well-being. Further, we should expect that the beliefs and values that have developed with a rising energy base are likely to be dysfunctional — even destructive — in a world of limited and declining energy.

Most philosophers reject the primacy of energetic and ecological forces. I see this rejection as a continuing expression of the "Cartesian dualism" that separates mind and body, humanity and nature, thought and action, subject and object. Reductionist science, which seeks the fundamental causes of material and living complexity in simple constituents, arises from the same philosophical base. Reductionism both explained the physical reality of the industrial world and reflected its fundamental ideology. Despite the substantial philosophical critiques and alternatives[2] to the world-view created by reductionism, it has held sway partly because it is powerful in dealing with the increasingly disintegrated[3] world created by the rising energy base. Through it we have achieved much. But, despite the hubris about new contributions to humanity's well-being, I believe reductionist science is now an impediment to human survival.

Some productive alternatives have emerged within the culture of science, such as dialectical materialism, systems theory and design science. The development of a truly wholistic science is important; otherwise we will see a wholesale rejection of the culture of science in the new millennium as it increasingly fails to explain and predict the novel phenomena of energy descent. The fundamentalist adherence to reductionism and rationality that characterises much of the scientific, economic and political establishment will increase the likelihood of cultural revolutions similar to those that have already occurred in several Islamic countries. Arguably, the centre of scientific rational power, the United States, is the most likely candidate for this type of revolution in response to the major decline in the well-being and security of people brought on by the energy crisis and related phenomena.

Permaculture as design science

Bill Mollison has described permaculture as integrated design science. This brief definition places permaculture firmly within the culture of science. Permaculture is applied science in that it is essentially concerned with improving the long-term material well-being of people. In drawing together strategies and techniques from modern and traditional cultures, permaculture seeks a wholistic integration of utilitarian values. By using an ecological perspective, permaculture sees a much broader canvas of utility than the more reductionist perspectives, especially the econometric ones, that dominate modern society.

Spiritual dimensions

Although permaculture can be reasonably seen as essentially materialist and scientific, it depends on an ecological perspective. Spiritual beliefs about a higher purpose in nature have been universal and defining features of all cultures before scientific rationalism. We ignore this aspect of sustainable cultures at our peril.

Robert Theobold[4] and others have expounded the idea that the very success of science and materialism has led us to a state of disharmony and discontent which is almost without historical precedent, and they argue that a shift to a more spiritually based value system is essential if we are to survive.

The more we understand the world through the lens of system thinking and ecology, the more we see the wisdom in spiritual perspectives and traditions. The same process has happened in the field of psychology, especially Jungian psychology. Many thinkers and writers have suggested that the most progressive aspects of science are moving towards a union with the universal aspects of spiritual belief. Rudolf Steiner's spiritual science, although generally ignored, is a past attempt at union that has since borne some practical fruits in the fields of education (Waldorf schools) and agriculture (biodynamics).

Permaculture attracts many people raised in a culture of scientific rationalism because its wholism does not depend on a spiritual dimension. For others, permaculture reinforces their spiritual beliefs, even if these are simply a basic animism that recognises the earth as alive and, in some unknowable way, conscious. For most people on the planet, the spiritual and rational still coexist in some fashion. Can we really imagine a sustainable world without spiritual life in some form?

For myself, I am proud of my atheist upbringing, in which humanist values defined an ethical framework for a rational world; but I also accept that, through the project of permaculture, my life is by small increments being drawn towards some sort of spiritual awareness and perspective that is not yet clear. To deny this, based on the evidence, would be irrational. However, for the present, my own interpretation of the ethical principles of permaculture rests firmly on rational and humanist foundations.

The deliberate design of a new spirituality that reflects ecological realities may be an unrealistic and dangerous extension of the permaculture agenda. However, an organic growth of spirituality from ecological foundations promises more hope for the world than the increasingly strident clashes between religious and scientific fundamentalism. While I baulk at the idea of designing this spiritual union, I can't help but use my systems thinking framework to help comprehend the dynamics of polarisation and emergent union between materialism and spirituality. While I focus on what I see as the positive and creative aspects of this union, they are mirrored by a dark and destructive alternative that is also emerging out of apparent polarisation. Figure 5 shows this broad pattern, with a myriad of materialist and spiritual philosophies, applications and movements located along these pathways. Although my labelling and placement of these conceptual elements may be somewhat arbitrary and not necessarily understood, I find this graphical representation useful in locating permaculture in the conceptual storm of our times.

Care for the Earth

Spaceship Earth

People often associate Care for the Earth with some sort of planetary stewardship, reflecting the concept of Spaceship Earth first popularised in the late 1960s and early 1970s by Stewart Brand.[5] These ideas have been powerful in galvanising an understanding of the global environmental crisis and other ethical crises, but they often remain abstractions separate from us. Further, Spaceship Earth implies that we have the power and wisdom to manage the earth.

The Gaia hypothesis of James Lovelock and Lynn Margulis[6] has provided a brilliant example of whole-system science that makes it clear that the earth is a self-organised system. The evidence of 4000 million years of evolutionary history is that, if we get to a point of seriously affecting the fundamental life-support systems of the planet, we will be "neutralised" by one or more co-evolutionary mechanisms (such as climate change or disease). The Gaia hypothesis has also spawned a countercultural revival of the almost universal view of the earth among indigenous and peasant peoples, as our living, all-powerful mother. Care for the Earth in this global context is not only due to ethical restraint and respect but also to fear of motherly rejection and annihilation.

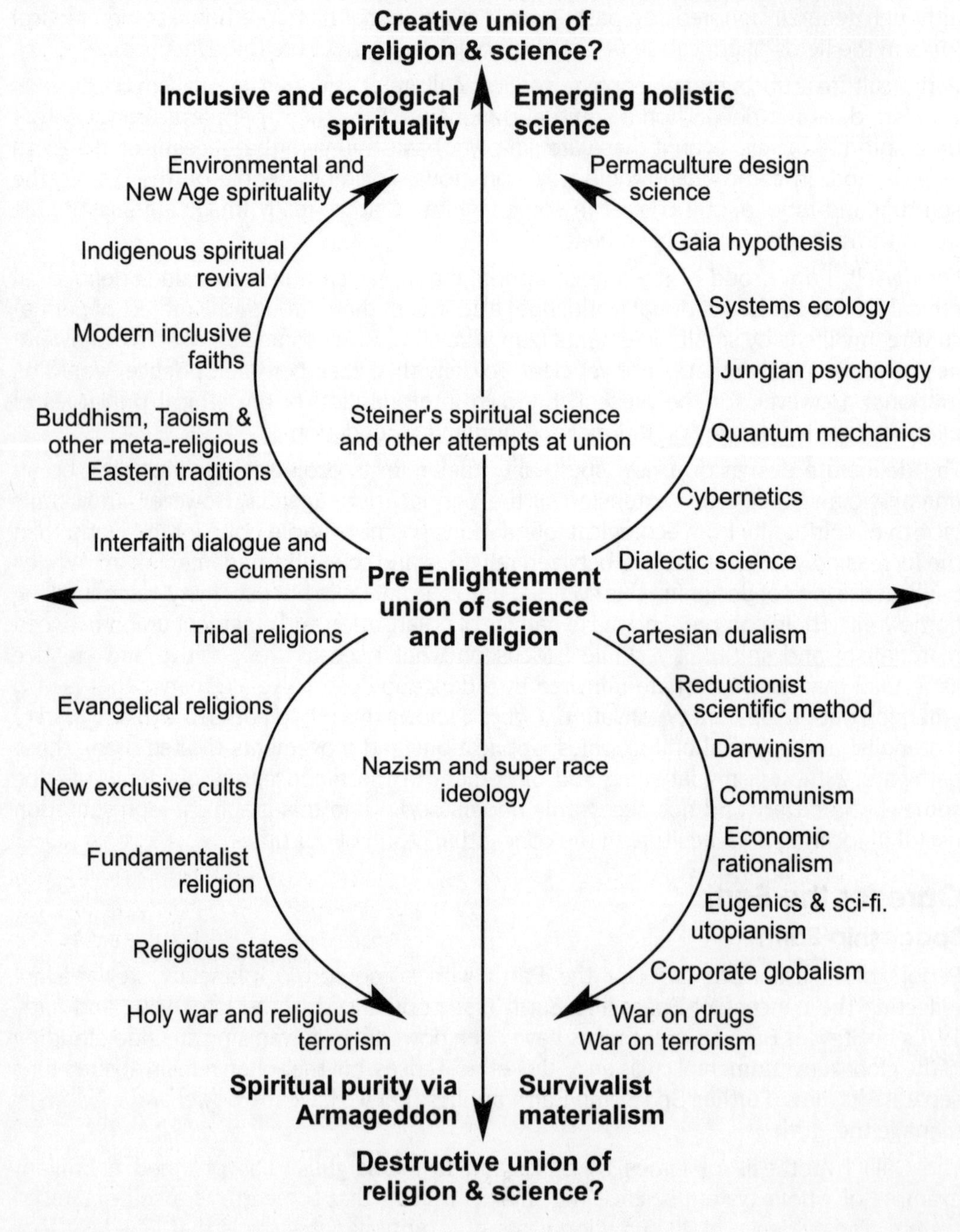

Figure 5: Emergent Union of Materialism and Spirituality via Creative and Destructive Pathways

Living soil

In the most grounded sense, Care for the Earth can be taken to mean caring for living soil as the source of (terrestrial[7]) life and for which we have the greatest responsibility. In this sense, Care for the Earth builds on both the scientific and ethical traditions of the larger and older organic (biological) agriculture movements. There are good scientific and historical reasons for regarding the state of our soils as the best measure of the future health and well-being of society, but reverence for the soil tends to be forgotten in the excitement about more easily understood, "sexy" issues.

How to care for the soil remains controversial. These technical questions are entwined with ethical ones because we do not know how far it is possible to improve the soil's capacity to support nature and provide for human needs. It is certain though, that careless and profligate use of the soil can result in a rapid loss of its capacity to support life.[8]

Stewardship

As we lift our focus from the earth beneath our feet to the ring of the horizon, Care for the Earth means looking after home, place, country or territory, as understood by indigenous cultures and, more recently, through the concept of bioregionalism. This understanding acknowledges our individual and collective responsibility in the care of particular natural resources about which we have some understanding and power.

Wendell Berry, the American writer, organic farmer and environmentalist, has eloquently criticised the notion of planetary stewardship as a product of our arrogant disconnection from nature and belief in our own power. I quoted him in my book about our own home in central Victoria:[9] "The question which must be addressed … is not how to care for the planet, but how to care for each of the planet's millions of human and natural neighbourhoods, each of its millions of small pieces and parcels of land, each one of which is in some precious and exciting way different from all the others."[10]

The stewardship concept demands that we constantly ask the question: Will the resource be in better shape after my stewardship? One cannot go far in this process without challenging the ethical validity of the ownership of land and natural resources that lies at the heart of our legal system. Control of land and natural resources has been central throughout history; in a low-energy future it will again become the primary focus for ethics, politics and culture. Indigenous land rights and agrarian land reform in poor countries are two issues that continue to challenge the prevailing ethics about land. The ethic of earth stewardship provides a moral imperative to continue to work out more creative ways for vesting control of land in collective structures, rather than taking as natural the individual ownership of land that goes with our Western industrial culture. Efforts to do this over the last hundred years show that it is not an easy task.

Biodiversity

Care for the Earth can also encompass the notion of caring for all the diverse lifeforms that inhabit the earth. This care is not dependent on the current usefulness to us of those lifeforms, but accepts them all as valid parts of the living earth with intrinsic value. There is much controversy among environmental ethicists about caring for other species, which is reflected in a general uncertainty in permaculture and the wider environmental movement about how we should deal with this ethical and practical issue.

Our ability to directly care for the diversity of lifeforms is as limited as our ability to care for the whole planet. To assume responsibility for the fate of all species is beyond our power or intelligence. If permaculture is a philosophy, then it is a pragmatic, down-to-earth one, which takes the ecological limits of our power and intelligence as foundations for anything we do. The traditional saying "live and let live" encapsulates a more modest notion of avoiding harm where possible. Permaculture principles and strategies provide ways to meet our needs while allowing other species to meet theirs (see **Principle 10**: *Use and Value Diversity*).

Living things

In meeting our needs, the killing of other life (individuals) is inevitable, even if we follow a vegan diet. In traditional Tibetan society, the killing of rodents and wolves by the peasants was a tolerated exception to the Buddhist teaching of the sanctity of all sentient beings. It is easy for most of us, living separated from nature, to agree that all life is sacred because we do not have to deal personally with the killing done consciously and unconsciously on our behalf.

Most indigenous peoples see individual killing as a natural and integrated part of life, but regard any attempt to exterminate a whole population or lifeform as unethical. Today, Australian scientists are trying to use genetic engineering to make the fox sterile, in the hope of saving native animals from extinction. To many, this is a benign and ethical approach, a view that stems from an extension of the Enlightenment idea of rights and values of the human individual to individual animals. On the other hand, traditional Aborigines in central Australia now see donkeys, rabbits and other introduced animals as mwerranye (Arrernte for "belonging to the land"),[11] even though they understand that these species were introduced by white people. While they are happy to use these animals, they see the mass extermination programs (Landcare) as immoral waste and disrespectful to individual and collective animal life.

This is a good example of traditional indigenous views providing an important perspective to inform permaculture ethics. I interpret the care of other lifeforms to mean:

- we accept all lifeforms or species as intrinsically valuable, no matter how inconvenient they are to us (or to other lifeforms that we value)
- we reduce our total environmental impact[12] as the best way to care for all living things, with no need to understand, have control over, or be responsible for the myriad of impacts of every individual action
- when we harm and kill other living things, we always do so in a conscious and respectful way; not to use what we kill is the greatest disrespect.

Care for People

The second ethic, Care for People, can also be interpreted at many levels. It firstly makes permaculture an unashamedly human-centred environmental philosophy which places human needs and aspirations as our central concern because we have power and intelligence to affect our own situation. At the most local level this means accepting personally responsibility for our situation as far as possible, rather than regarding external forces or influences as controlling our lives. The permaculture approach is to focus on the positives, the opportunities that exist even in the most desperate situation. The successful use of permaculture strategies in helping urban and rural poor in the Third World to become more self-reliant is partly a result of this focus on opportunities rather than obstacles.

Although it is naive to ignore the family, historical and political explanations for current conditions in which we find ourselves, these can easily become a source of bitterness and disempowerment. On the other hand, we should be open to understanding why those same external forces make it hard for others less fortunate than ourselves to take control of their lives.

Spiritual beliefs that encourage us to remain emotionally unattached to specific outcomes are helpful when they allow us to stand back from our own goals and desires to see that maybe life and nature is providing us with what we need. On the other hand, these beliefs can also lead to fatalism, in which there is no need to address practical or even ethical issues.

Care for self

Care for People starts with the self, but it expands in widening circles to include our families, neighbours, local and wider communities. In this sense it follows the pattern of almost all traditional (tribal) ethical systems. Figure 3 shows this as a meta-analysis of permaculture zone and sector. The greatest ethical concern is naturally focused close to the centre because that is where we have the greatest power and influence. To be able to contribute to a wider good, one must be healthy and secure.

At first glance this may appear to be a recipe for ignoring the gross disparities of wealth between rich and poor nations and people, especially when applied by the billion or so middle-class people across the planet who, more than the numerically few rich, consume the vast bulk of the planet's resources.

The fact is that our own comfort is based on the rape of planetary wealth, depriving other people (and future generations) of their own local resources. Our own "hard work" and the so-called "creativity" of our economy and "fairness" of our system of government are all secondary factors in creating our privilege. Once we understand the massive structural inequities between rich and poor nations, urban and rural communities, and human resources and natural resources, the emphasis on providing for one's own needs first is seen in a different light (see **Principle 4**: *Apply Self-regulation and Accept Feedback*).

As we reduce our dependence on the global economy and replace it with household and local economies, we reduce the demand that drives the current inequities. Thus "look after yourself first" is not an invitation to greed but a challenge to grow up through self-reliance and personal responsibility.

Non-material well-being

One of the best ways to apply this principle is to focus on non-material values and benefits. When we enjoy a sunset rather than watching a movie, when we look after our health by walking rather than consuming medicine, when we spend time playing with a child rather than buying them a toy, we are taking care of ourselves and others without producing or consuming material resources. There is increasing recognition that rising consumption is not improving well-being in rich countries. For the United States, alternative measures of well-being such as the Genuine Progress Indicator[13] have been declining since 1978, despite huge rises in per capita consumption. This is because much of the new wealth is being used to offset the adverse effects of over-consumption and over-development. This includes increasing expenditure on remedial health care, pollution control, crime control, litigation, and a myriad of other "bads" and "disservices" (as opposed to goods and services). Thus, the limits to material improvements are structural and internal to ourselves as much as they are external environmental and political ones.

In poor countries, most people know that it is the non-material aspects of well-being that are most valuable (even if they dream of their children having the opportunities to consume much more). This perspective reduces, or even eliminates, the apparent conflict between looking after nature and looking after us.

Set Limits to Consumption and Reproduction, and Redistribute Surplus

In *Permaculture: A Designers' Manual*, Mollison focuses on Setting Limits, but the principle is often taught with emphasis on the more positive Redistribution of Surplus. The apparently contradictory messages of abundance and limits encourage us to repeatedly ponder the meaning and expression of these two aspects of nature as a paradox which should continuously reshape our ethical response to life's opportunities and problems.

Abundance and limits of nature

A sense of abundance emerges when we experience the gifts of nature/god and human endeavour. This is often sharpest within a context of external or internal constraint. Something as simple as the taste of the first strawberry of the season is special when nature does not provide strawberries year round. Luxuries all serve this function, but they lose their power and value when they become everyday realities. The experience of abundance encourages us to distribute surplus beyond our circle of responsibility (to the earth and people) in the faith that our needs are provided for. The sense of abundance is lost through addictive indulgence to excess and waste. This excess and waste is only possible through power over nature and people.

A sense of limits comes from a mature understanding of the way the world works. We see that everything in nature, including ourselves, has a limited lifespan and a limited place. The view of our planet from space has given iconic power to the understanding of its material limits. The statistics on the growth in human consumption and numbers and the extinction of species make clear the impossibility of continuous growth in anything. Recognition of limits does not come from the experience of scarcity. Except in extreme famine and other natural disasters, scarcity is a culturally mediated reality; it is largely created by industrial economics and power, rather than actual physical limits to resources.[14] This manufactured scarcity encourages unrestrained consumption and reproduction in the hope they will deliver security.

Setting Limits to Consumption and Reproduction requires us to consider what is enough, and sometimes to make hard decisions. When we accept our own mortality and limited power, the setting of our own personal limits becomes a reasonable bargain with the world. We maintain our autonomy and self-control by exercising self-restraint, and so reduce the likelihood that some external force or power will force us to change. For example, some businesses have found that by developing zero pollution standards they have avoided the regulatory requirement for complex and costly environmental monitoring.[15]

In thinking about what is enough, we have to look at the needs and wants that drive material gain, and also at the capacity of earth and people to provide those needs and wants. The ecological footprint[16] is one relatively simple method to audit and reorganise our personal demand on natural resources. Such methods help us to set limits and monitor our performance. Often the processes of enlightened self-interest leading us to non-material values will provide a carrot (see Care for People). In other situations, we need to draw a line in the sand.

The issue of population growth is a vexed one with many different perspectives. The world is probably already overpopulated for the long-term good of humanity and other species. Some of the most detailed and authoritative nationwide studies of sustainability have been done in the tiny and relatively prosperous Caribbean country of Costa Rica. Modelling of the ecological footprint suggested Costa Rica could sustainably support 80% of its 1987 population at current levels of consumption, while EMERGY analysis suggested a figure of only 53%.[17] Although Costa Rica is more affluent than most countries, its per capita use of EMERGY is still one-quarter that of the United States.

From an ethical perspective, we need to focus on what is appropriate for us to do, rather than on what others should do. This is especially important in a world divided between rich and poor where the situations are often so different. There is much evidence from poor countries to show that birth rates drop rapidly when three factors apply: women are economically secure, they have control over their own reproduction, and infant mortality rates are low.[18] On the other hand, almost every child born in rich countries accelerates consumption.

For many people considering this question in rich countries, adoption and fostering provide opportunities to satisfy the natural instinct to have children as a part of our lives. Often these desires are seen as driven by women, but for men the idea of heirs and successors is often a powerful one. Learning to see all children, rather than just our own offspring, as our heirs is one of the great challenges for male culture around the world.

Redistribution of surplus

Redistribute Surplus requires us to share surplus resources to help the earth and people beyond our immediate circle of power and responsibility. It requires us to exercise that power to stimulate or fund the provision of needs beyond those of ourselves, family, community and region where there may be no mutual obligation or feedback mechanism to reward our benevolence. This provides the context for consideration of those apparent tensions and dilemmas between the personal and the collective, the local and the global, the current and the future, which I have raised in relation to the first two ethics.

What we choose to support with our surplus time, resources and wealth is an increasingly important life issue for many of the world's relatively affluent citizens. In the social domain, all cultures show a wide range of ways in which surplus is distributed for the benefit of others, present or future. In traditional pre-industrial and modern societies, the distribution of surplus is often codified in legal and religious institutions such as the tax system or the church.

Today the traditional institutions of church and state are losing their authority while corporations and other powerful economic institutions have gained enormous power with little if any ethical constraint. In this context, people are finding a wide variety of ways to help others, including overseas aid and development projects, social service clubs, philanthropic trusts, voluntary and community work. Even the arts and culture, which have no strong marketplace value, can be seen as current expressions of the redistribution of surplus. It is not surprising that redistribution of surplus should be a major feature of affluent modern societies. What is surprising is that many traditional and materially poor societies redistribute relatively more of their material wealth.

In most indigenous and peasant societies, caring for the land beyond the needs of family or descendants was embedded in nature-spirit traditions which required gifts, work or other expressions of devotion to maintain all of the living world. At a more practical level in many agricultural societies, the planting of long-lived and valuable trees and forests has been a traditional way of redistributing surplus time and resources for the benefit of future generations and the land itself.

Within the more recent tradition of the organic movement, land management to repair and improve long-term soil fertility, especially by increasing its humus content, has been regarded as a form of stewardship in which one leaves the land in better shape than it was when one received it. Many permaculture strategies, and even whole design principles (such as **Principle 2**: *Catch and Store Energy*), are predicated on the idea that increasing long-term soil fertility is possible. Although more modest soil conservation strategies imply an ethical responsibility not to further degrade soil, the more activist permaculture concepts demand that we make all reasonable efforts to increase, and even transform, the biological capacity of soil for the benefit of future generations.

Planting of trees and other perennial vegetation to restore the health of the land, without the need to gain an economic benefit, has been a central activity of the permaculture and broader Landcare movements. In its purest form, conservation of indigenous ecosystems in small reserves implies a commitment to manage, in perpetuity, systems otherwise unable to survive broad ecological changes.

In many cases, social and environmental good works can be seen as some sort of atonement for our collective sins, rather than necessarily providing what people or nature really need. Sometimes our giving is a disguised form of self-interest that creates dependency. In both the social and the natural domain, how to give in ways that show true altruism has always been problematic, but in the modern world it is an ethical question of great magnitude and complexity.

Permaculture as Tools to Assist in Ethical Decisions

In attempting to lead an ethical life we need conceptual tools that will allow us to find what is appropriate, is practical for the situation and context, and yet will have some enduring value in chaotically changing times. Permaculture, and especially permaculture design principles, are conceptual tools which many people are finding useful in this journey. Hopefully my own description of permaculture principles will deepen understanding of how these tools can assist in an ethical adaptation to ecological realities.

1 For an exploration of the evolutionary limitations of tribalism in the modern world see Article 26, "Tribal Conflict: Proven Pattern, Dysfunctional Inheritance" in *David Holmgren: Collected Writings* 1978-2000.

2 Dialectical materialism, although currently unfashionable because of the disasters of communism, still provides one of the most substantial critiques of philosophical and scientific understanding that is useful for a post-industrial, low-energy culture. See R. Levins and R. Lewontin, *The Dialectical Biologist* Harvard University Press 1985 for a dialectical perspective on many issues directly relevant to the permaculture agenda and John Bellamy Foster, *Marx's Ecology: Materialism and Nature* Monthly Review Press 2000, for an insightful re-evaluation of the ecological nature of Marx's thinking.

3 Levins and Lewontin, *The Dialectical Biologist* use the term alienated to describe this dis-integrated industrial ideology and reality.

4 "Reworking Success", ABC Radio National 1999.

5 In the 1968 *Whole Earth Catalogue*.

6 Another concept first published by Stewart Brand after rejection by *Scientific American* Magazine. For the story behind the Gaia hypothesis see Lynn Margulis, "Another Four Letter Word: Gaia" in P. Warshall, *30th Anniversary Celebration Whole Earth Catalogue* Point Foundation 1998. For explanation of the concept see J. Lovelock, *Gaia: A New Look At Life* Oxford University Press 1979, and other more recent books by Lovelock.

7 I use the word "terrestrial" here and elsewhere to mean belonging to the land rather than to the sea or the air. In **Principle 2**: *Catch and Store Energy* I discuss the fundamental forces constraining terrestrial life that do not necessarily apply to the very different oceanic ecologies. Despite humanity's vigorous exploitation of the open oceans, we can no more escape our dependence on land and soil than we can realistically escape the planet.

8 The permaculture idea that tree systems — less dependent on the "arable" soils required for annual crops, could help us escape the limitations of soils — may have contributed to later problems of achieving and maintaining soil fertility and balance by permaculturalists.

9 *Melliodora* (*Hepburn Permaculture Gardens*).

10 W.Berry, *Culture and Agriculture: The Unsettling of America* Sierra Club Books 1977 (written prior to the Gaia hypothesis).

11 R. Nugent, *Some Aboriginal Attitudes to Feral Animals and Land Degradation* Central Land Council, November 1988.

12 Using methodologies such as Ecological Footprint and EMERGY Analysis (see "Set Limits to Consumption and Reproduction").

13 See Redefining Progress website: http://www.rprogress.org

14 See Ivan Illich, "Beauty and the Junkyard", in *Whole Earth Review* no. 63, 1991, for explanation of this concept of scarcity as a creation of industrial economics.

15 See P. Hawkin, A. Lovins and H. Lovins, *Natural Capitalism: Creating the Next Industrial Revolution* 1999 for examples from the textile industry and other industries.

16 A spreadsheet for calculation of household ecological footprint is available from the Redefining Progress website, http://www.rprogress.org. The Ecological Footprint model of total environmental impact developed by Mathis Wackernagel is not as rigorous as EMERGY analysis but is much more easily applied and is gaining popular currency.

17 C. Hall (ed.), *Quantifying Sustainable Development: The Future of Tropical Economies* Academic Press 2000. EMERGY is all the available energy that was used in the work of making a product and expressed in units of one type of energy. Written in capitals (so it is not misread as energy) EMERGY is a contraction of "embodied energy", a more general term that covers several different methodologies. I sometimes use the more general term "embodied energy" where the differences between methods are not important in explaining general conclusions.

18 See *New Internationalist* (eg no. 235, September 1992) for the argument that consumption by the rich rather than reproduction by the poor is the key to the global unsustainability.

Observe and Interact

Beauty is in the eye of the beholder

Good design depends on a free and harmonious relationship to nature and people, in which careful observation and thoughtful interaction provide the design inspiration, repertoire and patterns. It is not something that is generated in isolation, but through continuous and reciprocal interaction with the subject.

Permaculture uses these conditions to consciously design our energy descent pathway.

In hunter-gatherer and low-density agricultural societies, the natural environment provided all material needs, with human effort mainly required for harvesting. In pre-industrial societies with high population densities, agricultural productivity depended on large and continuous input of human labour.[1] Industrial society depends on large and continuous inputs of fossil fuel energy to provide its food and other goods and services. Permaculture designers use careful observation and thoughtful interaction to reduce the need for both repetitive manual labour and for non-renewable energy and high technology. Thus, traditional agriculture was labour intensive, industrial agriculture is energy intensive, and permaculture-designed systems are information and design intensive.

In a world where the quantity of secondary (mediated) observation and interpretation threatens to drown us, the imperative to renew and expand our observation skills (in all forms) is at least as important as the need to sift and make sense of the flood of mediated information. Improved skills of observation and thoughtful interaction are also more likely sources of creative solutions than brave conquests in new fields of specialised knowledge by the armies of science and technology.

The icon for this principle is a person as a tree, emphasising ourselves in nature and transformed by it. It can also be envisaged as the keyhole in nature through which one sees the solution.

The proverb "beauty is in the eye of the beholder" reminds us that the process of observing influences reality and that we must always be circumspect about absolute truths and values.

Observe, Recognise Patterns and Appreciate Details

A process of continuous observation in order to recognise patterns and appreciate details is the foundation of all understanding. Those observed patterns and details are the source for art, science and design. The natural and especially the biological world, provides by far the greatest diversity of patterns and details observable without the aid of complex or expensive technology. Those patterns and details provide us with a great repertoire of models and possibilities for the design of low energy human support systems.

While good observation is the source of new insight and creativity, it is also the foundation for renewing the most basic abilities that we appear to be losing as fast as technology finds substitutes. For example, observation of a baby's pattern of bowel movements and early action to hold them over the potty at the right time can lead to easy and early toilet training,[2] saving endless work, water and energy.

Computerised Geographical Information Systems, while very useful, often substitute for, or cover up, a deficit in simple skills of reading the landscape.

Interact with Care, Creativity and Efficiency

There is little value in continuous observation and interpretation unless we interact with the subject of our observations. Interaction reveals new and dynamic aspects of our subject and draws attention to our own beliefs and behaviour as instrumental to understanding. The interplay between observer and subject can be thought of as the precursor to design. The accumulation of the experiences of observation and interaction build the skill and the wisdom needed both to intervene sensitively in existing systems and to creatively design new ones.

The Thinking and Design Revolution

Everyone knows about the breathtaking emergence of the information economy. The information and knowledge systems that direct and organise the physical economy of goods now have the greatest value and power. Computers are the most obvious feature of the information economy, but changes in the way we think, especially the emergence of design thinking, are more fundamental to the information economy than the hardware and software we use. Permaculture itself is part of this thinking revolution.[3]

A large part of the thinking revolution involves the emergence of design as a universal skill alongside those of literacy and numeracy. It is not so much that we are just beginning to design; rather, we are becoming more conscious of the power of our individual and collective design processes and how to improve them. Design is fundamental to humanity and nature, and yet it is so difficult to define.

Victor Papanek defines design as "the conscious and intuitive effort to impose meaningful order"[4]. This emphasises that design is not simply the result of rational, analytical and reductionist thinking, but also depends on our intuitive and integrative capabilities.

To design requires that we are familiar with models generated by nature and humanity (past and current solutions and options) as well as having an ability to visualise some new adaption, variation or possibility. The capacity to imagine other possibilities is another important aspect of design thinking. The most creative design involves the promiscuous hybridisation of possibilities from apparently disconnected, or even discordant sources to create a new harmony.

Papanek's "imposition of meaningful order" recognises the powerful nature of designing. The dangers of "playing god" inherent in this definition remind us of the potent nature of design. As Stewart Brand said in the original *Whole Earth Catalogue* (1968), "We are as gods so might as well get good at it."

From a systems ecology perspective, "design by nature" is not simply a metaphor but a result of the forces of self-organisation which can be observed everywhere in the living and wider universe. This imposition of meaningful order is a counter-flow to the prevailing entropic forces of disorder within nature and the wider universe. (For entropy, see **Principle 2**: *Catch and Store Energy*.) Self-organisation occurs wherever energy flows are sufficient to generate storages. *Designing is as natural as breathing and, like breathing, most of us can learn to do it better.*

Observation and interaction involve a two-way process between subject and object: the designer and the system. Perhaps because of the bias built into our culture by Cartesian dualism, we need constant reminding and examples of the true nature and consequences of this two-way process if we are to improve the quality of our design thinking and practical actions.

The maxim "everything works both ways" is a useful general reminder that finds expression in many diverse examples. The following more specific maxims provide the concrete guidelines and reminders which can help prevent us as designers from falling back into dualistic thinking.

- All observations are relative.
- Top-down thinking, bottom-up action.
- The landscape is the textbook.
- Failure is useful so long as we learn.
- Elegant solutions are simple, even invisible.
- Make the smallest intervention necessary.
- Avoid too much of a good thing.
- The problem is the solution.
- Recognise and break out of design cul-de-sacs.

Design thinking guidelines

All observations are relative

Observation can be a reflection of an internal state rather than objective fact. Even the concept of objective fact in science is now acknowledged as flawed; scientists know that observation, directly and indirectly, influences reality.

Given the limits to objectivity, it is better to be clear and articulate about our assumptions, preconceptions and values, and to acknowledge how these influence and structure how we see. Ethics and ideology act as filters that determine what and how we see. These filters are unavoidable — in fact, essential — but the rush to judgment of right and wrong frequently clouds our observation and prevents understanding. This commonly occurs in our attitude to pest plants and animals.

Top-down thinking, bottom-up action

In considering any subject it is always useful to step back and look for the connections and contexts which can reveal our subject as part of large-scale systems. This assists us to identify important inputs to the system that are outside system control or feedback effects and also to see outputs and losses that larger-scale systems are absorbing.

This "top-down"[5] systems thinking is a useful balance to "bottom-up" reductionist perspectives that seek to understand a subject by looking for its fundamental parts. On the other hand, bottom-up action focuses on the leverage points that are available for small-scale elements or individuals to influence large-scale systems in which they participate. This is especially important when we are trying to manage rangelands, forests and other wild landscapes where our "management" options are a small part of the larger system. Similarly, in trying to foster appropriate community change in a context of more powerful forces, recognising leverage points where we can make a difference is essential. Perhaps the current prevailing mode of action could be characterised as "top-down" (dominant) action typified by governmental and corporate management. Instead what is needed is more bottom-up (participatory) action at all levels in natural and human systems. The historical and political context for this maxim is further considered in **Principle** 4: *Apply Self-regulation and Accept Feedback*, especially through interpretation of the well-known environmental slogan "think globally, act locally".

The landscape is the textbook

The natural world provides such a vast diversity of subject material for the observer and designer that we can characterise the landscape as the textbook to follow. All the knowledge we need to create and manage low energy human support systems can come from working with nature.

By observation we usually mean using our eyes, but this just reflects how visually dominated modern people are, raised in a literate and now graphical world. All of our senses have great potential to provide valuable information. For example, smelling or tasting soil can reveal otherwise invisible aspects of its biological, physical and chemical balance. An experienced bird-watcher often learns more from songs and calls than from glimpses of birds that may be elusive. A good fitter and turner can feel the removal of a few "thou" (thousands of an inch) from a crankshaft turning in a lathe.

The development of good observation skills takes time and a quiet- centred condition. This in itself requires a change from a lifestyle that is indoor, semi-nocturnal and media-dominated to one that is outdoor, mainly daytime, and nature-focused. At Melliodora we try to balance indoor deskwork with observation and physical work in the garden that supplies most of our food. As well as feeding us, working with nature provides the inspiration for, and testing of, the more abstract ideas expressed here.

In consultancy work I have found that skills in "reading landscape"[6] to be the most important a permaculture designer can develop to be useful in advising others on the potentials, limitations, land use history and successional[7] processes of any particular parcel of land.

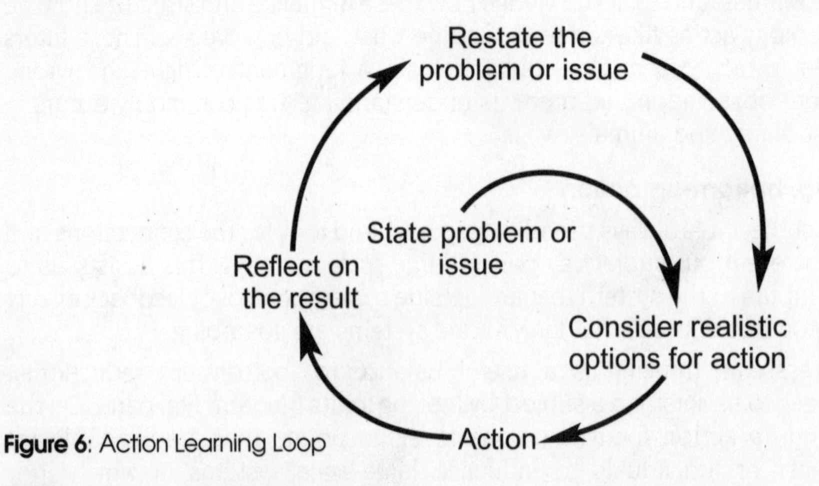

Figure 6: Action Learning Loop

Failure is useful so long as we learn

Planning and design processes (like life in general) often involve incremental adjustment in response to experience. Figure 6 shows a simple action learning loop that allows us to start wherever we are, no matter how little knowledge we have, and proceed from a very narrow perspective to a broader and more wholistic one by incremental adjustment.

Although this is an excellent and simple process, when working with complex natural systems we need to remember that we don't understand, let alone control, all the factors; and that cause and effect are often a loop or a web, rather than a linear chain. When you try some action, don't assume you were the reason for any success. Conduct small trials and think about other possible causes for success or failure.

Elegant solutions are simple, even invisible

In science the simplest answer that explains all the facts is regarded as having more validity than a complex answer. Similarly in design, enormous complexity often indicates poor design. A really effective design solution may be remarkably simple. This simplicity may be inherent, or it may arise because self-organising (living) complexity that works without us understanding or controlling it is doing most of the job (see **Principle 5**: *Use and Value Renewable Resources and Services*).

Really effective systems may work so well that we don't even notice them. This is common with free environmental services, such as purification of air and water or soil rehabilitation and good design solutions that go unnoticed until they fail from abuse. The saying "we don't know what we've got till it's gone" is relevant here. Careful observation and a respect for humble lifeforms and processes is a partial antidote for this perennial problem.

Make the smallest intervention necessary

In attempting to adjust systems to fix problems, we need to be careful that we don't damage or disrupt other processes that are working perfectly. Because much effective design is invisible, large-scale interventions are likely to do more harm than good, and they require large amounts of energy and resources to implement. Japanese natural farmer and philosopher Masanobu Fukuoka[8] has written eloquently about the value of doing nothing and the damage from large-scale intervention in nature. The Bradley method[9] of bush regeneration, based on the careful observation and minimal interventions of the Bradley sisters, has been acknowledged as highly effective in conserving bushland from spread of environmental weeds. Its focus on minimal disturbance to remove weeds from the most intact areas of native vegetation runs counter to the prevailing approach, which is to attack the weeds head-on with herbicide and other high-impact methods.

Avoid too much of a good thing

When we experience a positive result from an action, there is almost always a powerful temptation to repeat the action, working on the often misguided idea that if some is good, more must be better. In sustainable societies and nature, limited resources often acted as a natural constraint on such actions, but in the modern world these temptations are everywhere. One of the universal effects of affluence is the change in diet towards consuming more sugar, fat and protein. Foods with these concentrated energy sources have always been desirable, but in the past natural limits protected us from the adverse effects of overindulgence.

Any gardener or farmer who has a great harvest following use of a fertiliser knows the temptation to use more next time. The slowing in growth of world food yields, despite increases in fertiliser use, shows the problem is widespread. As in the case of over-indulgence in sugar, fat and protein, overuse of any fertilisers creates imbalances that reduce long-term yields and plant health.

Ethical constraints on greed are part of the antidote to the problems created by too much of a good thing. The recognition that these situations are characteristic of affluence, and of any natural system subject to surplus energy flows, helps us recognise new cases before we fall into the trap.

Many of these statements that caution us against bold and impulsive action reflect a general observation: that success or function in any system may represent failure in connected systems that are larger or smaller in scale.

Despite these cautions against bold and confident action that pushes ahead based on positive feedback, we need also to be open to the signs that radical change in our perspective and action is needed.

The problem is the solution

This slogan expresses the idea that things are not always as they seem; the things we view negatively may have a positive aspect that is more important than, or at least compensates for, the predominant negative perception.

The most common examples of this idea relate to weeds or other lifeforms which we regard as pests, which may be positive. For instance, weeds or pests may be:

- environmental indicators of need for management change
- agents repairing damaged soil, etc
- resources which, for economic or cultural reasons, we fail to value.

Pests can be viewed as the surpluses of nature, which need to be used rather than destroyed. (see **Principle 3**: *Obtain a Yield* and **Principle 12**: *Creatively Use and Respond to Change*.) This idea of pests as surplus natural abundance that we can use has been a major theme in my work over many years.[10]

An open, inquiring attitude to problems is almost always more fruitful than an urgent demand for solutions. The latter is often driven by fear and an unquestioned consensus about the nature of the problem.

Another aspect of this maxim is that the best solutions to problems can be found in places and cultures where the problem is extreme. In these situations co-evolution over time will develop the best responses. In places where the design problem is less severe, people often ignore it or come up with solutions that use more effort or resources to overcome the problem. For example, in my brief time on the precipitous Amalfi coast in southern Italy, I became aware of several creative solutions to the problems and limitations of very steep land; some traditional and others modern and imported. Conversely, while travelling in the Mediterranean I was surprised by the poor development of rainwater tanks and small earth dams compared with Australia. Despite the dry summers, the prevalence of permanent streams, springs and good-quality groundwater in the Mediterranean has reduced the stimulus to harvest rainwater that has operated in Australia.

Recognise and break out of design cul-de-sacs

It has been suggested that "solutions usually come from people who see the problem only as an interesting puzzle, and whose qualifications would never satisfy a selection committee"[11].

When we recognise the potential to turn problems into solutions, we often encounter the inertia of fundamental beliefs, system architecture and power structures that stand to lose by the innovation. Although modern culture appears to encourage innovation (within narrow limits), we need to be aware why conservatism — resistance to radical change — is an important characteristic of both natural and human systems.

In nature and in human behaviour, proven solutions tend to become entrenched, while recent innovations are easily swept away by unfavourable conditions. In the history of life, mutations in the basic chemistry of cellular life almost never manage to get past early embryo stage because evolution over hundreds of millions of years has perfected these processes and any variations are usually lethal. More recent evolutionary patterns have rarely been vested with fundamental functions, so occasional mutations survive, at least initially. For example, variation in the number of nipples, and even of digits, is common in mammals. Stress and competition tend to suppress such variations through low success in reproduction.

In families, most parents deliberately or unconsciously teach their children the lessons (good and bad) that they learnt in childhood from their parents, in preference to their own insights as adults. The more stressed parents are, the more this becomes true. The cycle of battered child to battering parent is now well recognised. Similarly in human culture, patterns of proven behaviour or knowledge become entrenched through tradition and institutions. Over long periods of relative stability, institutions rather than individuals are the critical keepers of culture.

This conservatism is functional when rates of change in the surrounding environment are slow. In these conditions, internal novelty is more often a malignant aberration than a useful innovation. Incremental small improvements are integrated where they can be implemented without altering the fundamental structure. These incremental changes can lead systems to optimum design for a particular set of conditions. However, rapid change in the environmental conditions can make the "optimum solution" a design cul-de-sac where there is no way forward and no going back.

The story about the frog that fails to leap out of the water as it slowly rises to a lethal temperature illustrates how incremental adjustment fails to get out of the trap created by a design dead-end.

Amory Lovins talks of "tunnelling through the cost barriers" to gain large improvements in energy and resource efficiency, rather than fiddling with small improvements. Although common sense suggests this will be very costly, many examples in industrial design and manufacturing show the costs can be less than those produced by small incremental adjustments.

What is needed in these situations is the capacity to think laterally, readiness to abandon the proven and take risks and the incentive to get to something better.

Broad experience and observation of the world outside the particular situation or system can allow recognition of the patterns of design cul-de-sacs and the general nature of transformative solutions. In most cases these solutions will challenge and undermine existing structures of power and wealth in society and are therefore strongly resisted.

Formal Education and Communications Media

Today it is common to hear the suggestion that appropriate education and training is the key to allowing people to contribute to a more sustainable future. Although I recognise the value of formal education and training and communications media, they are also part of the problem preventing us from a more direct connection with, and experience of, the natural world through observation and experience.

Much could be said about how we need to modify formal education and the media so they can contribute better to the thinking revolution, but we need to recognise several fundamental problems.

First, almost all education, and especially higher education, is based on knowledge from secondary sources interpreted through existing frameworks of understanding. Academic training that requires that every idea and concept be referenced to refereed published sources is the extremity of this approach to learning.

Second, information technology has accelerated the speed of circulation, retrieval and reproduction of information. The professional packaging of secondary experience in media promises faster and better ways to learn. For instance, in comparison to the wildlife program on TV, looking at the real thing seems to be a dull and fruitless activity. This focus on the representation of existing information has been to the general detriment of observational skills and original thinking.

Third, almost all formal scientific education is based on bottom-up reductionist thinking. While the cultural harvest from specialised reductionist thinking over the last few hundred years has been great, it is the integration and cross-fertilisation of concepts and ideas that is now providing the most fruitful results for dealing with the systemic problems of the environmental crisis. Much of this integrated thinking is happening outside educational institutions (see **Principle 11**: *Use Edges and Value the Marginal*). Ironically, information and communications technologies are increasing these possibilities, although this benefit is minor compared with the prevailing tendency of the technology to favour presentation over content, and replication over creation.

Fourth, and most problematic of all, formal education at the higher levels is focused on digesting massive amounts of information within the constructs of separate disciplines and cultures (such as science and art). This education lacks wholistic integration with living experience, and its insights and truths often make little impression on the underlying structures of understanding and belief created by family upbringing, mass media and popular culture. In many fields we find the solutions to problems are known but are not applied for a myriad of specific reasons. The core of these problems is that knowledge which people appear to hold has not been integrated with and reinforced by their personal experience.

Observation and the Limitations of Direct Experience

In pointing out the limitations of formal education, I do not ignore the limitations of observation and direct experience as a foundation for ecological design. The proverb about reinventing the wheel reminds us of the limitations of observation and design isolated from other sources of knowledge.

Observation, trial and error and direct experience are often seen as slow and inefficient ways to learn for two reasons. First, the consequences of failure can sometimes be so great that we dare not experiment. This was well understood in most traditional sustainable societies; they tended to be conservative and frown on innovation as involving too many risks. Second, it is common for people to repeat mistakes over and over again for a variety of reasons, from failure to observe or ask the right questions through to fear of oppressive authority.

If grassroots approaches to energy descent are to succeed, they must find more efficient ways for people to learn through observation and direct experience to create ecological support systems. To move from the relatively limited tool kit approach to permaculture design and towards a diverse and flexible pattern language will require an enormous reinvigoration of observational and innovation skills.

Modern Context for Experiential Learning

The expanding possibilities created by the Industrial Revolution and its continuing consequences have provided a myriad of opportunities for overcoming the limitations of direct experience. Firstly, market economics have for several generations rewarded materially advantageous innovations, creating a cultural climate which sees (at least some) innovation in a favourable light. Secondly, it has been possible for more people to experiment or engage in lifelong self-education, increasing and diversifying the pool of direct experience from which innovative creativity can arise. Thirdly, relative freedom and democratic pluralism have fostered the tolerance of difference and even of dissident behaviour. The seemingly invincible nature of capitalist materialism has contributed to the sense among the elites that they gain more from a free pluralist society than a rigidly controlled one.[12]

Finally and perhaps most importantly, affluence has provided a safety net or insurance against the adverse consequences of failures which are inevitably high with any experimental or innovative activities. Generational and family wealth, the welfare state and philanthropy are all aspects of this insurance system.

These factors have led to an explosion in the arts and sciences and creativity generally, most of which until recent decades was tolerated, if not actively nurtured, in educational institutions, government organisations and private corporations. The wealth and power of these organisations provided some latitude for individual experimentation and innovation (along with some simple laziness and waste), but during the 1980s and 1990s economic rationalism stripped away most of that latitude in the interests of so-called economic efficiency. At the same time "centres of excellence", "creative departments" and "innovation units" have been set up to capture these values for narrowly defined, short-term goals. Many commentators have argued that this approach is knocking the creative guts out of our society.

Today it is individuals who are pursuing a great diversity of interests, values and lifestyles, outside the constraints of the marketplace or educational institutions.

Included in the great diversity of personal, amateur and hobby activity are opportunities to become skilled observers of nature. This knowledge can be applied in gardening and other working relationships with nature that provide the foundations for an energy descent future. The contribution of amateur observers to scientific knowledge is dramatically illustrated by the work of the Australian bird observers. The *Atlas of Australian Birds*, first published by the Royal Australian Ornithological League in 1981 and recently updated,[13] was made possible by the observations of tens of thousands of amateurs across the continent.

Permaculture has provided a focus for many people motivated to apply the intellect to what have traditionally been regarded as humble and simple activities of necessity in peasant societies. The desire to solve problems, to experiment and to design is one of the defining characteristics of the permaculture gardener. A great example of the contribution of the amateur enthusiast is the work of Melbourne doctor Louis Glowinski. In his suburban backyard, Louis has created one of the most important horticultural trial sites in southern Australia for new fruit and nut crops in temperate climates that has inspired the Melbourne permaculture network. His book *The Complete Book of Fruit Growing in Australia*[14] is a delightful mix of original observation, personal opinion, and thorough research and documentation.

Amateur experimenters often suffer from a lack of resources, skills or communication with peers, but these apparent negatives can free the amateur to explore untrodden paths, not restricted by the fear of failure or the boundaries of a particular discipline.

Observations and Interaction in a Post-modern World

Permaculture could be described as a post-modern concept, in which all assumptions are open to question and elements from different systems and traditions are combined without regard for any fixed aesthetic or tradition. Indigenous and traditional cultures of place have provided much of the inspiration, elements and design solutions, both in the original conception and in the ongoing evolution of permaculture. However, the speed of change in recent decades has resulted in much of traditional knowledge and culture being either lost, or taken into the modern world through travel experience, documentation and migration. To some extent, modernity has already culturally appropriated almost everything in the natural and traditional worlds. (For further discussion of permacultural appropriation, see "Cultural Globalisation and Renewal of Cultures of Place" in **Principle** 10: *Use and Value Diversity*.)

Modern society, as well as being a great collector and appropriator of everything in the natural and traditional worlds, seems to be a "fast-breeder" system that generates new information, knowledge, innovation and culture. In this context, the educated, travelled and well-informed person is a new "resource" created by the high-energy industrial culture. (see **Principle** 4: *Apply Self-regulation and Feedback*.) Unfortunately most of this diversity will prove highly dysfunctional in the energy descent towards sustainable society.

But even this waste and inefficiency can be understood as reflecting natural processes and humanity's way of adaptive change to unknown future conditions. When any natural system (organism, ecosystem, individual or society) receives a sudden increase in net energy supply, the system is transformed by rapid growth. If the high-energy supply is maintained, diversification tends to take over from simple growth. Negative feedback controls are suspended, and any and every developmental possibility is allowed to persist. The history of the Roman Empire gives us a great example of cultural growth and diversity based on the resources of the whole Mediterranean. In the garden, extra water and fertility increase the vigour and profusion of plants that grow, including a myriad of weeds.

If energy supply is too great, the system usually dies from self-pollution (in the way a yeast culture dies from its own alcohol waste) or through cancerous growth that causes major system malfunction. We are familiar with the stories of people who win the lottery and go on a spending spree, finding hell rather than heaven.

Eventually energy supply again becomes constraining, either because the store of energy is exhausted, or because the system has grown to fully use the flow available from the new energy source. In these conditions, negative feedback controls again operate to constrain or eliminate system elements or organisation that do not contribute to system survival in an energy-efficient way.

While this basic picture may be a grim prognosis for much in human culture that we have come to cherish and depend on, there is an optimistic angle. The proliferation that occurs under free energy supply will, by various evolutionary mechanisms (see **Principle 12**: *Creatively Use and Respond to Change*), create possibilities that, although they could never emerge in a context of energy constraint and negative feedback controls, may nevertheless prove to be adaptive once the energy supply is again constrained. By this process, bursts of free energy act as fast breeders for evolutionary change. For example, permaculture gardeners experimenting with many new and climatically marginal crops can be seen as a hobby activity reflecting personal and societal affluence, but it will also lead to advances useful in a low energy future.

Of course, the changes can be so great that, from the perspective of the original system, they represent cancer and death rather than adaptive change. For example, voluntary frugality threatens to undermine the foundations of the consumer economy.

Scavenging on the Cultural Rubbish Tip of Modernity

In a low-energy future, most of today's technology and culture will end up in the dustbin of history; but some aspects of modernity may provide critical components of an energy descent culture. Permaculture principles provide one lens for helping to identify the useful pieces of modernity and combine them with those from nature and from pre-modern cultures in a new designed synthesis.

In this task we need to become detached observers of our own cultural context, looking for patterns that reflect principles rather than simply the familiarity of upbringing, habit and convention. One image that comes to mind is looking at what we do and why, in the way an anthropologist would study indigenous people. Another image is the scavenger at the rubbish tip looking for gems among the possibilities; while shocking to some, this image is one that many permaculturalists relate to. Perhaps these images come together in the archeologist who sifts through the deposited remains of past cultures to understand their life and their demise.

It is as if we are wandering about a landscape littered with the pieces of many different jigsaw puzzles. Our task is to pick up as many pieces as seem possibly useful, limited in the end by how many we can recognise, and carry them to a place we don't yet know, where we must construct a new jigsaw puzzle from what we have.[15] Good observation and design skills will be the key to our success.

The Value of Scepticism

Personal, political and religious freedom that we often take for granted is itself a foundation for the thinking and design revolution. A relatively free society allows us to remain sceptical about any of our current collective and individual beliefs. In *Voltaire's Bastards*[16] John Ralston-Saul makes an eloquent plea for scepticism and questioning as antidotes to the excesses of certainty and fundamentalism. He documents how extremist ways of thinking and exercise of power have grown like monsters from rationalism since the Enlightenment.

Scepticism and resistance to dogma are just as important in the development of a society adapted to energy descent as they are in the maintenance of a free one. Nature is always providing us with a myriad of indicators, warning signs and questions in response to our actions driven by certainty and belief. Adaptation to energy descent depends on a dynamic balance of values and contextual knowledge rather than holy crusades of good over evil or simplistic, universal solutions.

The Importance of Interaction

These and other insights, drawn both from the interpretation of traditional systems of knowledge and from the great thinkers of the modern world, provide a wealth of ideas that can help us make sense of observation and experience. But unless we get out there, and open our eyes and use our hands and our hearts, all the ideas in the world will not save us.

Thus the thinking and design revolution, of which permaculture is a part, only makes sense when it reconnects us to the wonder and mystery of life through practical interaction.

1 See F. H. King, *Farmers of Forty Centuries* 1911 facsimile edition Rodale Press for a description of Chinese agriculture at the turn of the 20th century as an example of a sustainable society dependent on maximum use of human labour.

2 Su Dennett, pers. comm.

3 See Edward De Bono's many books about "lateral thinking" and more broadly the "thinking and design revolution". For the connection with permaculture, see Article 23 "Permaculture: Thinking and Acting for Sustainability" in *David Holmgren: Collected Writings* 1978-2000. See also *Natural Capitalism* and other books by Amory Lovins et al on the design revolution in business and industry.

4 V. Papanek, *Design For The Real World* (2nd edn.) Thames & Hudson 1984.

5 The term "top-down" is used by Howard Odum to describe systemic thinking that gives an overview of major features of a system without being confused by detail. The saying "unable to see the wood for the trees" reminds us of the problems of being too close to something to see the larger systemic pattern. The term "top-down" conveys the meaning we associate with the view from a mountain, the air or even heaven.

6 See Article 3 "An Eclectic Approach to the Skills of Reading Landscape and Their Application to Permaculture Consultancy" in *David Holmgren Collected Writings* 1978-2000.

7 The progressive replacement of one plant community by another in development towards climax vegetation. For a full discussion of the succession and climax concepts and their application in permaculture, see **Principle 12:** *Use and Respond to Change Creatively*.

8 M. Fukuoka, *The One Straw Revolution* Rodale Press 1978.

9 See Joan Larking, Audrey Lenning and Jean Walker, (eds.), *The Bradley Method Of Bush Regeneration* Lansdowne Press Sydney 1988.

10 The following articles from *David Holmgren: Collected Writings* 1978-2000 are particularly relevant: Article 1 "Permaculture: Design for Cultivating Ecosystems" gives examples of replacing negative responses to weeds with more positive approaches; Article 19 "Permaculture and Revegetation: Conflict or Synthesis" and Article 20 "Inquiry Into Pest Plants In Victoria - Submission To Environment and Natural Resources Committee of Parliament of Victoria" expand on this positive view of weeds and pests.

11 From J. Gall, *General Systematics* Harper & Row 1977.

12 Many would argue that pluralism and democratic freedom are now contracting rapidly and that many of the elements are now in place for a new totalitarianism in the Western world (see J. Ralston-Saul, *The Unconscious Civilisation* Penguin 1997).

13 See web site http://www.birdsaustralia.com.au/atlas/index.html

14 Louis Glowinski, *The Complete Book of Fruit Growing in Australia* Lothian Books 1991.

15 This image came to me from reading the American historian William Irwin Thompson's accounts of Pythagorus in ancient Egypt and the monks of Lindisfarne in Celtic Britain, who as social radicals reinterpreted some of the essential truths from their respective decaying cultures and provided the seeds for a new culture.

16 John Ralston-Saul, *Voltaire's Bastards: The Dictatorship of Reason in the West* Penguin 1993.

Catch and Store Energy

Make hay while the sun shines

We live in a world of unprecedented wealth resulting from the harvesting of the enormous storages of fossil fuels created by the earth over billions of years. We have used some of this wealth to increase our harvest of the earth's renewable resources to an unsustainable degree. Most of the adverse impacts of this over-harvesting will show up as available fossil fuels decline. In financial language, we have been living by consuming global capital in a reckless manner that would send any business bankrupt.

We need to learn how to save and reinvest most of the wealth that we are currently consuming or wasting so that our children and descendants might have a reasonable life. The ethical foundation for this principle could hardly be clearer. Unfortunately, conventional notions of value, capital, investment and wealth are not useful in this task.

Inappropriate concepts of wealth have led us to ignore opportunities to capture local flows of both renewable and non-renewable forms of energy. Identifying and acting on these opportunities can provide the energy by which we can rebuild capital as well as provide us with an "income" for our immediate needs.

This principle deals with the capture and long-term storage of energy, that is, savings and investment to build natural and human capital. The generation of income (for immediate needs) is dealt with in **Principle** 3: *Obtain a Yield*.

The icon of sunshine captured in a bottle suggests the preserving of seasonal surplus and a myriad of other traditional and novel ways to catch and store energy. It also reflects the basic lesson of biological science: that all life is directly or indirectly dependent on the solar energy captured by green plants.

The proverb "make hay while the sun shines" reminds us that we have limited time to catch and store energy before seasonal or episodic abundance dissipates.

Energy laws

To move beyond the metaphors of capitalism and financial planning, we need a basic understanding of energy laws, the foundation for all that is possible in nature and human affairs. Understanding of these energy laws was fundamental to the development of the permaculture concept.[1]

We are used to thinking of sources of energy as fuels that are supplied to us through the economic system, but energy (in a diversity of forms) is the driving force behind all natural and human systems. Food, which we think of as body fuel, is the most important energy that people (like all animals) catch from their environment.

Throughout the universe, energy is always spreading from centres of concentration to vacant regions, where it tends to remain dispersed and diluted. In addition, energy of high quality degrades into lower-quality forms, thus reducing its power to affect change or do "work" in the sense that physicists and engineers use this word. This tendency to disorder and eventual death is called entropy; it affects every living and non-living system.

However, self-organising systems (primarily living ones), can capture and transform a limited proportion of energy they absorb. This energy is then held in storages of varying form and durability for use in self-maintenance, growth and capture of more energy. This stored energy is generally higher in quality than the source from which it was derived and is thus capable of driving a wider range of processes than the original source energy.

For living systems (from single cells to Homo sapiens to the whole living planet), available energy flows are mostly erratic, limited in quantity, and low in quality. Living systems that are "designed" to optimise the efficiency of energy transformation and storage tend to prevail through evolution.

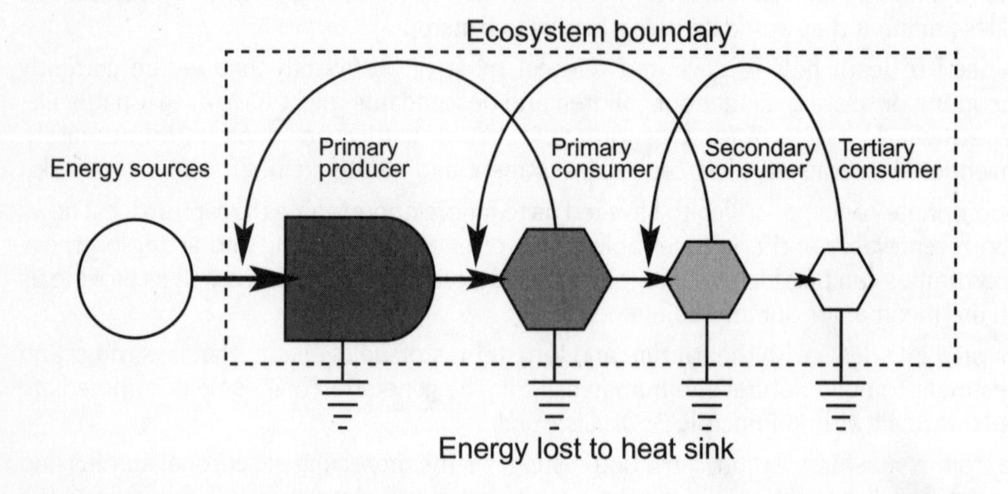

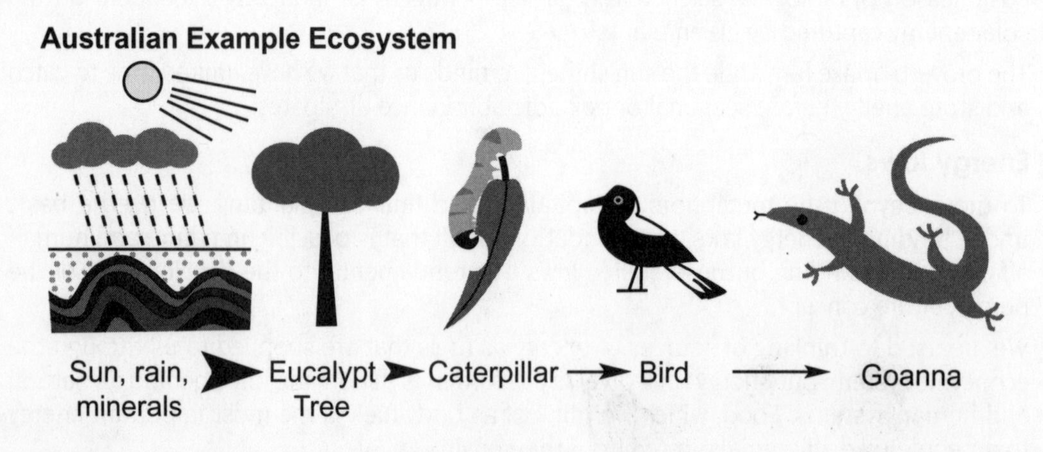

Figure 7: An ecosystem seen as an energy circuit.

Figure 7 uses the energy circuit language developed by Howard Odum to show the relationships between external energy sources, primary producers (plants) and consumers (animals) as an energy hierarchy or food chain. It shows energy flows as arrows with the following general pattern:

- reducing energy and mass contributed along the food chain
- increasing power and value per unit energy and mass along the food chain
- essential energy dissipation losses to the environmental heat sink (electrical earth symbol)
- feedbacks of high quality energy to stimulate inflows.

A simple Australian example of a food chain illustrates these relationships.

Gardeners and farmers have for centuries captured the food energy in seasonal agricultural crops which they have stored or preserved for later consumption. Saved seed was a particularly potent storage of energy, essential for next season's crop. By catching and storing the energy in seed, growers maintained a genetic and cultural lineage from ancestors to descendants.

All biological and mineral resources can be considered (and measured) as embodied energy. The tools, infrastructure and technology that are needed to support a human society, whether simple or complex, all derive from these primary energy sources collected from the natural environment. The more useful and durable forms into which we transform those sources of energy are stores of high quality energy — or, in ordinary language, "real wealth".

The fundamentals of this energetic view of the world are taken as given by scientists. But because of a major disconnection between the biophysical sciences and the social sciences, most notably economics, the energetic view has little impact on our normal understandings of value and wealth.

The modern pattern

In modern affluent societies, the flow of energy in forms useful to people (food, materials and services) has become so reliably available that energy capture — and even more so, energy storage — has ceased to be a major concern. So long as people have a flow of money to buy, the provision of basic needs has been left to farmers, mining engineers, and so on. In the extreme development of modern urban living, no food or fuel is stored in the house and purchasing power is dependent on credit, which itself is dependent on permanent employment.

At the same time, economic rationalism in business and government has led to the decline of large-scale stores of food, fuels, materials, essential spare parts and permanent employment in the interests of economic efficiency, which in turn increases the likelihood of disruption and even disaster. The vulnerability of modern systems to slight disturbance was highlighted in the late 1990s by the concern over the possible impacts of the Y2K computer bug and the enormous cost of reactive prevention strategies.

In an emerging low-energy world, we will rediscover opportunities to harvest and store immediately available (on-site) renewable energies and wasted resources across our rural and urban landscapes and in our households and local economies. This will be essential to avoid disasters from inevitable disruptions to energy and resources supply lines.

Sources of Energy

What constitutes a usable source of energy will vary greatly with the context, and it is natural for people (and all organisms) to focus on the sources that have been sustaining in the past. After a period of unprecedented energy flow, people are particularly blind to new or novel sources of energy, especially if they are modest and situation-specific.

Important sources of energy are currently poorly used but are commonly available for building greater personal and local self-reliance:

- solar energy: simple drying of wood and crops, passive solar design, and devices such as dryers, water heaters and photo-electric panels
- wind energy: pumping and power generation
- biomass: the sustainable management of trees and forests for fuel and construction materials
- run-off water: irrigation, aquaculture and power generation.

In addition, we are surrounded by wastes from agriculture, forestry and industry, particularly organic matter, that can be used for soil improvement, fuel and animal feed, and also as building materials, containers and so on. These manufactured resources are not necessarily renewable, but they can be tapped in the transition to low-energy systems (see **Principle 6**: *Produce no Waste*).

At Melliodora, we have made some use of all five of these energy sources to establish and maintain the system. The particular mix and ways in which we do so are unique to our site and situation. The slogan "the problem is the solution" helps us recognise opportunities to harvest sources of energy that are specific to our site and situation. For instance, occasional gully floods provide nutrients and sediment; hot drying winds are an opportunity to dry fruit and other produce; blackberries and other weeds serve as mulch, animal fodder and so on. Most fundamentally, every sunny winter day charges our mud-brick, passive solar house so we stay warm overnight. These examples reflect the way ecosystems develop in response to the diverse and irregular energy availability across landscapes.

Landscape Storages of Energy

Figure 7 shows that the distinction between sources and storages of energy is vague: what is a storage for one element or organism in a system is a source for another. Because humans are not confined in a fixed and simple ecological niche,[2] what is a source in one situation is a storage in another. However, for the purposes of explaining the dual actions of this principle, the distinction is useful. Permaculture strategies of landscape development and evolution range from the back doorstep to the horizon and deal with today's opportunities through to those of our descendants. Many, if not most of these can be grouped as rebuilding the natural capital of landscapes in four key energy storages: water, living soil, trees and seed. To understand why so much permaculture literature and action focuses on these strategies, we need a grasp of how the energy sources of sun and earth have driven the evolution of terrestrial[3] life on earth over millions of years and why these imperatives are critical to the continuity of human civilisation in a future of declining energy.

How Nature Catches and Stores Energy

For billions of years life was confined to shallow seas that fringed barren eroding land masses. About half a billion years ago this marine life began to colonise the land and create soil life capable of taming this wasteland. The evolution of terrestrial ecosystems and landscapes has maximised the power of nature to use both climatic and earth energies.

Solar energy (in the form of visible light) is used by plants to transform water and carbon dioxide from the atmosphere into carbohydrates by the photosynthesis process. These carbohydrates are the start of the chemical energy supply chain that provides for the needs of all other living things,[4] as well as (indirectly) creating the fossil fuels of coal, oil and gas.

PHOTOSYNTHESIS (in green plants)

Carbon Dioxide + Water + Sunlight —> Carbohydrates + Oxygen

RESPIRATION (in plants and animals)

Carbohydrates + Oxygen —> Carbon Dioxide + Water + Metabolic Energy

Solar energy also drives the weather and climate systems that deliver energy in the form of rain, wind, lightning and fire. The climatic energies influence not only the type of plants that grow but also the nature of soils and the shape of the land, from river catchments to drifting desert dune fields. Sunlight for photosynthesis is a limiting factor in very cloudy and high latitude climates during winter, but a much better indicator of biological productivity is the amount of rainfall an area receives. The pure water in rainfall is embodied solar energy because it is solar heat that evaporates the water to create the atmospheric moisture for precipitation.

Most people are aware that the sun is a thermo-nuclear furnace but few realise that a much closer, but slower, source of nuclear power inside the earth is equally important in sustaining life. This power drives the movement of the tectonic plates of the earth's crust. Uplift and vulcanism at the plate edges build and reshape mountains and deliver rock minerals critical to soil fertility and all living things. Subduction[5] of oceanic sediments and surplus organic materials exposes them to heat and pressure which re-forms rocks, creates fossil fuels and concentrates rare minerals into ore bodies.

Changes in either the climatic or the geophysical energies, or both, result in a radical reorganisation of terrestrial landscapes and ecosystems. These large-scale system changes can cause massive erosion, physical destruction, biodiversity loss, habitat destruction and fragmentation, as well as building and fertilising new soil and allowing rapid invasion and evolution of new life forms. For example, the continental ice sheets and montane glaciers expand and contract in pulsing rhythms that we call ice ages. The ice destroys whole living landscapes but it also grinds vast amounts of rock into glacial dust, a mixed mineral fertiliser of great potency which sustains new life. The time-scale for these changes in landscapes ranges from millions of years down to human lifespans.

Humans, like all other animals, have evolved to take advantage of the resources created by these large processes; but we ourselves have also become agents of change, now geological in scale, through our harvest and use of fossil fuels and minerals. Within a few human generations, the low-energy patterns observable in natural landscapes will again form the basis of human system design after the richest deposits of fossil fuels and minerals are exhausted.

Although it is tempting to think of these natural landscapes as reflecting a stability in climatic and geologic forces, long periods of climatic and geophysical stability actually result in a rundown of the energy available to ecosystems and people. Geologically young regions with recent[6] mountain building and vulcanism tend to be much more biologically productive and have supported large populations of people despite their vulnerability to natural disasters. Geologically old regions (like most of Australia) tend to have low biological productivity and supported fewer people.

The biological productivity and human support capacity of these geologically young landscapes derive from their ability to capture and hold water from the atmosphere, essential minerals from the earth, and the organic matter that their resultant ecosystems generate. These large stores of soil minerals, organic matter and water can support large human populations; they led to the development of agriculture and what we call civilisation.

The capacity of terrestrial systems to hold water, mineral nutrients and organic matter is limited and is always being eroded by the inexorable force of gravity. Eventually these storages of energy are lost to terrestrial life (in river flows to the ocean or deep earth storage). Atmospheric oxygen continuously oxidises (locks up) minerals and breaks down organic matter, slowly, or rapidly through fire. These forces have been so consistent in their effects since life emerged out of the sea that all terrestrial ecosystems and landscapes can be seen as design systems to overcome, or at least limit, the effects of these forces. Thus we can say that terrestrial ecosystems have co-evolved to catch and store the energy in water, mineral nutrients and organic carbon as effectively as possible.

Water storage in landscapes

Water is perhaps most easily understood as a limiting factor, especially in Australia, the driest inhabited continent. The erratic nature of the rainfall and the constant needs of microbes, plants and animals for moisture have caused landscapes to evolve as efficient rainwater storages. Vegetation holds substantial quantities of rainwater, both in its tissues and in the humid air and moisture trapped by forest canopies and understoreys. Forest litter and mulch act as an open sponge, absorbing and holding water. The balance of soil chemistry and the development of humus in topsoils provide a more stable moisture supply for plants, while deep subsoils, especially clay, provide very stable moisture, although it is harder to access. The capacity of the soil to store water is a major factor determining the productivity of ecosystems and the sustainable support base for people.

Rainwater that infiltrates into the subsoil beyond the reach of plants contributes to catchment productivity as it is released slowly through springs and soaks lower down the landscape, especially along gullies and watercourses. These springs and soaks sustain the base flows in streams and rivers between "run-off events" — precipitation heavy enough to cause run-off. Deep-rooted plants can recycle water from underground aquifers, but some water moves down into deep groundwater storage beyond the recycling mechanisms of plants or landscape.

The pattern of these storages is from more ephemeral to more permanent as water moves down under the influence of gravity.

These micro and local storages of vegetation, soil and sediments are set in catchment landscapes with topographical forms capable of storing water. Lakes, which are analogous to constructed reservoirs, are most common in geologically young landscapes, such as glaciated and earthquake-prone mountain landscapes. In drier climates, small deep pools in gorges as well as gravel and sand streambeds are important water storages. Within rivers and streams, the pool and riffle pattern acts to filter and oxygenate water.

Plumbing drainage model

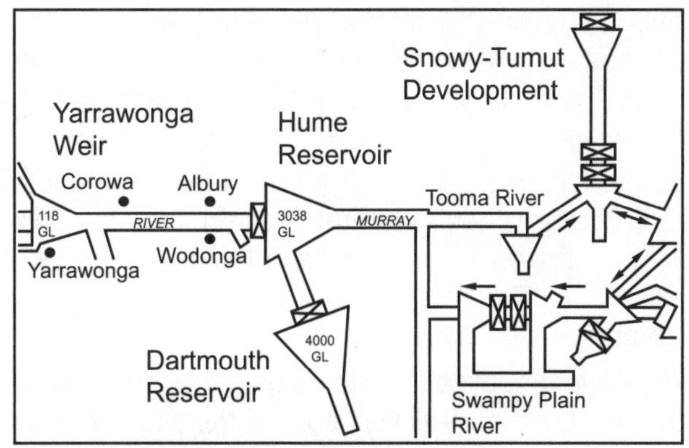

River Murray (Snowy - Yarrawonga)

Hydrographic Model

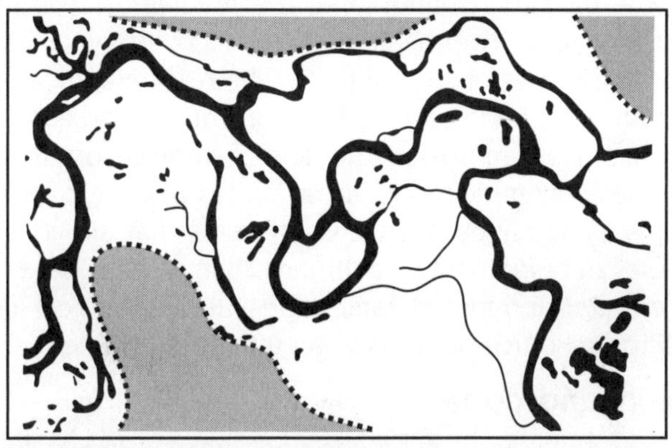

River Murray (Albury - Wodonga)

Ecographic Model

Riverine Floodplain (Koorinesia)

Figure 8: Plumbing drainage, hydrographic and ecographic models of river systems (after Tané 1996)

Swamps and wetlands are even more important as temporary storages and filters. They have been called the kidneys of catchment landscapes because of their role in filtering and purifying water. Streams and wetlands also ameliorate the destructive forces of large flood flows which recur at intervals ranging from one year to a thousand or more years. During the flood, some of the energy is captured as new alluvial soils deposited by slow-moving floodwater. In drier and flatter coastal landscapes, sand dunes dam the course of streams, creating fresh and brackish estuaries and coastal lakes. Peak floods break the sand barriers, releasing estuarine fish trapped for years or decades. Storms and winds then rebuild the sand bars to create new lakes.

Concepts about the management of catchments have shifted radically in recent times, from a prevailing engineering model of draining water away fast to a hydrological model of slowing and filtering the flow of water from the mountains to the sea. This shift in thinking is graphically illustrated by Tané's plumbing drainage, hydrographic and ecographic models of river systems.[7]

The old plumbing drainage model was depicted by the Murray Darling Basin Commission in its system description for the managed flow of the Murray River. It has been largely discredited as destructive of natural resources. The storages (public dams) are too large and low down in the catchment landscape; the regulated flows are too fast and too constant for the sustenance of river health and productivity, which has evolved to use the seasonal variations and pulses of natural flows. The hydrographic model depicted by the mapping of floodplain water bodies recognises the sinuous and complex nature of natural river floodplains, which slow and divert the flow of water as optimal for maintenance of natural resources. The ecographic model incorporates the latest understandings of rivers and their floodplains as highly productive ecosystems that are constantly renewing and rebuilding themselves in response to catchment change. Tané sees Aboriginal "abstract" paintings of these water-dominated landscapes as a graphical description of these physical and biological resource patterns as an integrated whole.

Nutrient storage in landscape

The evolution of ecosystems and catchment landscapes in response to the limited availability of mineral nutrients is more difficult to understand than water. Mineral nutrients are largely invisible, and they control the productivity of every ecosystem in subtle but important ways. The essential elements of carbon, oxygen, hydrogen and nitrogen are abundant in the atmosphere; they are supplied to living things through the energy-harvesting system of photosynthesis and other associated processes in plants. The mineral nutrients of calcium, magnesium, potassium, phosphorous, sulphur, along with trace elements essential to living things, occur in small and varying amounts in the diverse rock types which make up the earth's crust. Plants can easily absorb these nutrients in water-soluble forms, but solubility also results in leaching of the nutrients beyond the reach of plants. Consequently, soil ecosystems have evolved to catch and store plant nutrients in non-soluble but available forms.

Ecosystems develop against a geochemical background of nutrient imbalance and deficiency. They seek to overcome these through mechanisms that mine bedrock and other inert sources, and also catch nutrients leaking from adjacent systems and atmospheric transfer in dust, smoke, pollen and other lifeforms.[8]

Over long periods of geological time, there is a loss of mineral nutrients from all ecosystems through the forces of gravity and leaching, as well as periodic fires, droughts, floods and other natural disasters. In addition, the chemical bonding of nutrients (by natural processes) into highly unavailable forms makes it more difficult for plants to obtain balanced mineral nutrition. Unless a landscape can mine or catch more than it loses, there is a progressive decline in productivity; species requiring high levels of nutrients are replaced by those adapted both to low levels of mineral nutrients and to chronic imbalances.

In Australia much of our exquisite biodiversity results from endless adaptations to low and unbalanced mineral nutrients. Unfortunately humans, by our very nature, are completely dependent on very high and balanced mineral fertility in our foods. Without adequate, balanced mineral nutrition, human societies from hunter-gatherers to great civilisations have crumbled because their soils could not provide them with the quantity or quality of food required.

Soil is the most important storage for nutrients in temperate climates. Humus is perhaps the greatest "invention" of nature, because it increases the capacity of soils to store mineral nutrients (as well as water and carbon). There are good ecological reasons for the veneration of humus in the organic, biodynamic and permaculture movements. But in the wet tropics, the rates of oxidation and leaching are so high that natural systems rely more on storage of nutrients in living plants. Long-lived trees are the most abundant and stable living store of nutrients in tropical systems. Even in temperate ecosystems, plants and especially trees represent an important store of nutrients which can be recycled to the soil by annual leaf drop, browsing by insects and herbivores, or fire.

Carbon storage in landscape

The storage of mineral nutrients in both soil humus and living plants is co-dependent on the primary storages of organic carbon in plant biomass. This organic carbon is produced by photosynthesis in green plants[9] and provides the chemical building blocks of life.

Ecosystems with actively growing plants can accumulate several tonnes of carbon per hectare every year. Trees are especially significant as carbon stores because of their ability to keep accumulating carbon as wood for hundreds — even thousands — of years. This long-lived storage of carbon in woody biomass is one of the best measures of the ability of terrestrial ecosystems to continue to catch and store energy and to resist seasonal variability and other disturbances.

Concern about the greenhouse effect has combined with the understanding that trees store carbon to produce a huge increase in research into "carbon sequestration" by trees. This interest in using trees as a "sink" to get rid of unwanted atmospheric carbon dioxide has increased awareness, as well as scientific knowledge, of the role of trees in storing carbon. From a permaculture perspective, the debate and activity is back-to-front; it focuses on the problem: carbon dioxide pollution, rather than on carbon as a source of fuel for new life.[10]

For humankind, the storage of carbon in living plants and soil humus is much more than disposal of unwanted atmospheric carbon dioxide. Carbon in staple food crops provides the body fuel needs of people and is literally "the staff of life". But the small quantities of carbon in human food are a minor part of the carbon storage that humanity needs. Much larger quantities of carbon in fibrous fodder plants sustain grazing animals, which in turn provide us with a myriad of renewable products and services, from protein-rich foods and wool to "horsepower". Particular forms of cellulose and lignin from some plants provide us with the material for fabrics, paper and ropes, as well as the myriad diversity of timber for every imaginable use. Last, but perhaps most important to the post-fossil fuel age, plants (trees) provide a renewable fuel for cooking, heating, smelting and other tasks.

Timber and fuel forests — and, to a lesser extent, pastures, fodder trees and fibre crops — can be grown on marginal soils which lack the depth, structure or fertility to support human food crops. This is the single most important reason why the storage of carbon by perennial plants, and especially trees, is central to permaculture strategies for catching and storing energy.[11]

Soil humus as carbon storage

The value of carbon-rich plant materials in directly providing for future human needs cannot be overestimated. But an equally valuable storage of carbon is created when we simply allow plant materials to rot back into the soil. Organic matter, especially carbon-rich bulky plant materials, is the fuel for soil micro-organisms, which in turn are the key to the cycling and availability of plant nutrients. After processing by worms and other soil organisms, organic matter is converted into polysaccharides, proteins and other fast-turnover products that support soil micro-organisms and plant life. Consequently, much of the carbon cycles back to the atmosphere as carbon dioxide within a season or so from respiration by the teaming microbial life in healthy soil. Some of the carbon in organic material is distilled into complex organic compounds that are more stable, such as humic and fulvic acids, which increase the capacity of soils to hold nutrients, water and oxygen. Under favourable conditions these humus storages can be stable for hundreds and even thousands of years. Wes Jackson of the Land Institute in Kansas[12] has referred to the loss of this ancient humus from American prairie cropping soils as "the mining of young coal". The rebuilding of humus in the world's cropping soils should be seen as the other great task for humanity, alongside reforestation of our catchments and degraded rangelands.

Rebuilding Natural Capital in Landscapes

Using these understandings of how nature catches and stores energy, we can rebuild the natural capital in managed landscapes so that they can provide for future human needs as well as essential environmental services. Permaculture strategies for catching and storing energy in landscape can be grouped under four broad headings: water, living soil, trees and seed.

Water

The value of catching and storing water is well understood (especially in Australia). More recent concerns about the adverse environmental impacts of catching and storing too much water in the wrong landscapes emphasise the importance of focusing on the greatest overall benefit.

By building reservoirs, dams, swales, tanks, cisterns and other structures, people increase the biological potential of landscapes to store water and therefore to support other biological processes. If it is done in the right location and proportions, the environmental benefits of water storage are greater than the impacts.

For a given quantity of stored water, small catchment dams have more benefits and less adverse environmental effects than large reservoirs in major river valleys. Large reservoirs in suitable environments, such as steep mountain valleys, are less environmentally damaging and are analogous to lakes naturally created by glaciers and landslips. Large reservoirs and tanks can store and maintain water in a near pure state, which is valuable for its biochemical potential to support living (and manufacturing) processes.

High water (in tanks, dams and reservoirs) has potential energy to drive mechanical processes such as high-pressure irrigation, fire fighting, and generation of mechanical and electric power. Opportunities for micro-hydropower, especially in mountainous tropical countries, are probably collectively greater than the hydropower from damming large rivers, and the adverse environmental impacts of the micro solutions are much less. The higher the water source, the greater the pressure and its flexibility in use. For example, cheap hoses of 12 millimetres diameter are adequate for garden watering from a high-pressure source; more expensive (and heavier) hoses of 18 or 25 millimetres are needed for low-pressure hand watering.

Nutrient-rich, relatively shallow water (in dams, wetlands, ponds, swales and paddies) provides storage of chemical energy to support highly productive aquaculture systems. In fact, shallow aquaculture systems on fertile lowlands always produce more protein than grazing or even dairying. This ecological efficiency of aquaculture for protein production was the reason for promoting aquaculture in early permaculture literature. In the wet tropics, growing rice in paddies integrated with fish and duck production is one of the most highly productive and sustainable agricultures yet developed.[13]

Living soil

Living soil with a good structure and high humus content has a huge capacity to store water, mineral nutrients and carbon.

The differences between soils in their capacity to store water, nutrients and carbon is the greatest single factor in the productivity of terrestrial ecosystems and agriculture. As we have burnt nearly half the world's fossil carbon in oil (and somewhat less of the coal), we have "burnt" over half of the soil carbon in the world's farmlands.[14]

This invisible burning of "young coal" has been caused by ploughing that inverts the soil and the use of artificial (soluble) fertilisers. These accelerate the smaller losses created from export (without recycling) of organic material from farmland. Soil carbon is the fuel of the largely invisible soil ecosystem that in turn regulates plant nutrition.[15]

Increasing the humus content of agricultural soil has always been a principal objective of organic agriculture. Changing the management of farmland to use organic and permaculture strategies and techniques can rebuild this storage of soil carbon, fertility and water to close to those of natural grasslands and forests. It is arguably the greatest single contribution we could make to ensure the future survival of humanity.[16]

Building soil organic matter

Mainstream agricultural research now recognises the loss of organic matter from cropping soils as one of the greatest threats to agricultural sustainability. Strategies and techniques to increase soil organic matter are no longer regarded as the peculiar obsession of organic farmers, but the problem of how to describe, measure and value differing forms of soil organic matter leads to much confusion. Most soil-testing laboratories measure Total

Organic Matter, rather than attempting to distinguish differing forms and their relative age and turnover time.

Soils with a build-up of partially decomposed mulch and compost may indicate a mineral imbalance.[17] Soils that have no visible compost layer but are very dark and well structured may have a high humus content, reflecting past "digestion" of large amounts of organic matter.

Where supplies of organic materials are abundant (as in gardens supplied by surrounding low-density urban or rural landscapes), a favourable mineral balance and microbial population are the critical factors that enable soils to digest organic matter. Where supplies of organic matter are limited to what can be grown on site (as in broadacre farms), appropriate crop rotations, pastures and tree and shrub systems are as important as the mineral and microbial factors.

Robbing Peter to pay Paul

It is often stated that agricultural crop waste, especially grain straw, could provide a huge renewable source of carbon for fuel and fibreboard products in the future. Although such schemes might be preferable to the practice of burning grain stubble, they amount to "robbing Peter to pay Paul". To maintain, let alone increase, the level of soil humus, cropland everywhere needs full recycling of crop waste, either through grazing animals or directly by soil microbes, or both.

In north Germany grain straw is burnt in high-efficiency furnaces in order to heat buildings; a "renewable energy" that replaces fossil fuel. Traditionally this straw would have been used as feed and bedding for livestock contained in large barns over winter. The resulting compost was then returned to the fields in spring, thus maintaining the humus content and fertility of the predominantly sandy soils. Today the slurry washings from the barns are stored in large tanks during the winter and then sprayed onto the fields. Although this slurry provides some organic matter, it is not enough to maintain levels of soil humus or to prevent leaching of nutrients into the groundwater, which supplies the local towns and cities.

Concerns about nitrate pollution have led to slurry quotas, which in turn have led to trading in quotas and even to "slurry marriages" between farming families trying to maximise their production. The next step could be to follow the Dutch solution of exporting animal manure to Spain, which would use much more fossil fuel than is saved by burning straw for heating. This story illustrates the complex, interconnected nature of environmental problems and the need for a wholistic framework for moving towards real solutions. Understanding and applying this principle may help prevent such absurd circular problems.

Brown coal as new soil humus

Ironically brown coal, with appropriate processing, is emerging as one of the most valuable resources for building the long-term humus content of agricultural soils. High-sulphur brown coals[18] are especially valuable because sulphur is a valuable plant nutrient. Coal-based fertilisers are increasingly being used in the conversion of conventional agriculture to organic methods; they offer the hope of building the long-term, stable humus content of soils at a faster rate than is possible with traditional methods. We need to remain cautious about the enduring value of any novel fast track to fertile and balanced soil, given the history of our collective failures. However, using fossil fuel directly to rebuild the natural capital of our farmlands sounds a better bet than burning it for industrial and consumer electricity or, alternatively leaving it in the ground as a toxic addition which we need to reject.

Balance between woody and herbaceous biomass

In clearing forests for agriculture, humans have mobilised the nutrients in large woody biomass for uptake by crops. Annual crops provide no permanent nutrient store, but where perennial pastures or other perennial crops are grown, nutrient storage in plant biomass can be as great as that in native forest and woodland (see Plant and Animal Biomass as Indicators of Fertility in **Principle 7**: *Design from Patterns to Details*).

A balance between stored and available nutrients in all systems is an important measure of the balance and tension between long-term stability and short-term productivity. For example, grazing by livestock converts nutrients in perennial pasture into more concentrated and useful forms (urine and faeces), which feed micro-organisms and more palatable and nutritious grasses, which in turn support more livestock. But these nutrients are also more mobile and easily lost through leaching and gassing off. Letting grass mature and rot back into the ground and slowly grow over to woody vegetation is a safer but slower strategy; it is useful where overgrazing has been a problem but fire hazard is not. Managed rotational and cell grazing systems have been shown to be a productive balance between overgrazing and destocking.

Soil humus as carbon sink

The focus on the greenhouse effect has produced some research and policy discussion of agricultural soils as carbon sinks, although this has not received as much publicity as the role of vegetation clearing and tree plantations. This research is providing quantitative evidence to support the long-standing claim of the organic agricultural movement that rebuilding agricultural soil humus levels is the greatest contribution to the survival of humanity.

Alan Yeomans, who manufactures the famous Yeomans soil-conditioning plough originally developed by his father P. A.Yeomans, has argued[19] that loss of humus from agricultural soils is as large a contributor to greenhouse gas emissions as motor cars, and that achievable increases in humus across the world's farming soils could reabsorb the whole of the damaging imbalance of carbon dioxide in the atmosphere. Study of his remarkably simple back-of-the-envelope calculations suggests that we are at least talking about quantities in the same order of magnitude.

Research and debate about actual and potential carbon cycles and storages in agricultural soil will no doubt continue. The greenhouse issue simply gives us another good reason to get on with the job of rebuilding the natural capital of soil humus as essential for humanity's survival in the post-fossil fuel era.

We can do this in many ways, both directly and by supporting farmers and land managers who are doing so — largely, but not exclusively, organic and biodynamic farmers. The methods are:

- return all organic wastes to productive garden and agricultural soils
- eliminate all intensive forms of livestock husbandry or factory farming (which consume excessive fossil fuels and reduce soil humus by increasing the demand for field crops)

- provide for (reduced) meat consumption in rich countries from conservative management of natural rangelands (mostly native animals such as kangaroos) and extensive management of grazing animals to build soil humus through perennial pastures
- use rotations of leguminous pastures to build arable soil humus, rather than continuous cropping supported by herbicides
- replace soluble fertilisers aimed at crop feeding with rock mineral fertilisers and coal humus for soil feeding (see below)
- establish large-scale tree systems as an integral part of all farm landscapes, especially in high rainfall areas (with emphasis on soil-building fodder shrubs, tree crop species and long-lived timber trees, with a lesser role for soil-depleting and fire-encouraging species such as eucalypts and conifers).

When the earth beneath our feet is less like a dead concrete slab and more like a dark, moist living sponge, then we know we are on the right track.

Mineral balance

The American soil scientist William Albrecht[20] was one of the first to recognise that it is possible to create an ideal balanced soil in which all crops give high yields of good quality, and he did the pioneering scientific work to identify the mineral and biological characteristics of such a soil. Albrecht's ideal mineral balance also increases the capacity of a soil to store water and resist erosion by creating an open absorbent fabric. In addition, this ideal mineral balance optimises the conversion of soil organic matter and litter to humus.

My own observations suggest that it is reasonable to extend the concept of Albrecht's ideal soil for all crops to suggest that this represents a biological optimum soil in which all plants will thrive. Within the constraints of climate, this balanced soil will support the most productive biological system in terms of total energy capture and storage. Thus balanced and fertile soil is nature's integrated and self-reinforcing design solution for maximum power of terrestrial life.

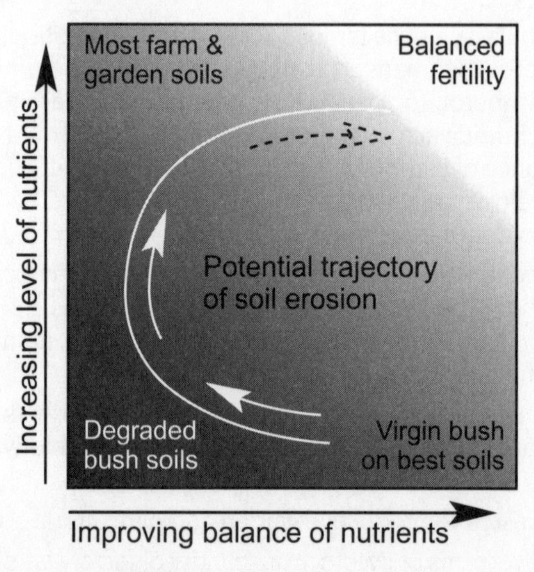

Figure 9: Soil fertility and balance matrix

Figure 9 shows soils on a matrix that combines level and balance of mineral fertility. Most agricultural (and garden) soils have evolved through two stages of development, with a rare few reaching the ideal of a high level and balance of fertility. Virgin fertility (on the better soils at least) tended to be reasonably balanced. Pioneer exploitation quickly or slowly depleted the level of nutrients and created imbalances. Modern fertilising has managed to increase the level of nutrients, and thus production. However, imbalances typically remain or new imbalances have been created that are reflected in the poor quality of food and increased rates of fertility loss.[21] Poor management and abandonment can see this cycle reversed as natural vegetation slowly rebuilds balance at a lower level. In Australia and other geologically old regions this rebuilding may be very slow and may never recover the balance of virgin soils. As a result of persistence and/or luck, some land managers achieve the Holy Grail of balanced but high fertility.

In future (perhaps within a hundred years), after the fossil fuel energy subsidy to agriculture has declined, the mineral fertility and balance of our farmlands and entire catchment landscapes will become one of the most important issues in resource management and economics, and yet the powerful means that are currently available to achieve this on a large scale will be very costly or simply unavailable.[22] In this situation we will once again be dependent on the slower, low-energy processes of building and balancing fertility.

Trees

Trees and other long-lived perennial plants are essential to sustainable agriculture partly because of their ability to efficiently absorb and store water and nutrients that might otherwise be lost by annual plants. This idea was central to initial articulation of permaculture and earlier ideas of tree cropping.[23] More recently it has been promoted through the Landcare movement as critical to dealing with salinity, acidification, eutrophication and other serious land degradation problems in Australia. Tree crops that provide human food tend to be more demanding of mineral fertility; they are often less vigorous and grow more slowly than hardy forest trees capable of generating the most woody biomass.[24] Although permaculture strategies have tended to focus on the first class of trees, it is the latter group that is most important in reclaiming vast areas of degraded land and providing the traditional forest yields of timber, fibre and fuel, as well as secondary products such as honey, fungi, herbs, meat and hides.

These timber forests are especially important in a world of declining fossil fuels for several reasons:

- they can grow on our poorest land unsuited to other food or even fibre crops[25]
- actively growing forests can accumulate biomass at rates of 5–35 tonnes per hectare per annum. This rate is similar to that of grasslands, but in contrast to grasslands the wood in trees is a long-term storage which can be stable for centuries
- wood in straight and tall trees continues to increase in value long after the rate of growth has slowed because large, slow-grown trees provide sawlogs with a great variety of durable product uses
- where a sophisticated market for forest wood products exists (as in Europe), trees capable of yielding sawlogs are worth ten times more than those yielding paper pulp or fuel wood.

In a low-energy future the value of mature forests capable of sustained yield of a diverse range of wood products will be very high. As in the past, the wealth of nations will be measured by the quantity and quality of their forests.

The dependence of European nations on their forests for the building of wooden warships is the great example from history, but the breadth and depth of dependence on forests applies far more widely. In the same way that steel replaced wood for ships and other uses as fossil fuel energy increased, wood will progressively replace steel, concrete, aluminium, plastics and other energy-intensive composite materials as fossil fuel energy declines. But this will only be possible if we grow those forests at least a generation in advance.

Without any particular focus on greenhouse gas amelioration, the principle of building natural capital suggests that we should grow long-rotation, mixed species forests for multiple values. The greenhouse outcomes of this approach would be more useful than the knee-jerk industrial monocultural mentality that drives much of the current design and investment in plantation forestry.

Such long-rotation mixed forests have various greenhouse advantages:

- the best way to establish most long-lived timber species is to grow them with fast-growing, soil-improving nurse species (such as acacias), which also have the effect of increasing uptake of carbon in the early years
- continuous careful thinning of long-rotation forests can maintain good growth rates in some of our most useful timber species for at least a century by which time fossil fuel use should have dramatically declined (see discussion under "Long Rotation Forestry" in **Principle** 9: *Use Small and Slow Solutions*)
- well-managed forests of long-lived, high-quality timber trees protected from fire can last for hundreds of years before their timber value and carbon storage begin to decline
- houses and other high-quality products made from this type of timber can last for centuries more
- some decomposition of leaves, bark and wood from older trees accumulates as soil humus capable of lasting for thousands of years
- such forests can be regenerated without the massive release of carbon dioxide associated with clear-felling and burning.

The reasons to grow long-rotation mixed forests are many; their function as a sink for greenhouse gases is simply one more reason to get on with the job. Permaculture has contributed to the now widespread notion that "reforesting the earth is one of the few tasks left to us to express our humanity".[26] While it is appropriate to focus on the often invisible environmental services (such as catchment protection and greenhouse gas sequestration) that forests provide, few people recognise that it will be the capacity of forests to store carbon as structural timber and fuel which may allow humanity to be sustained by renewable resources in a low-energy future.

Seed (especially of annual plants)

Even with increased use of perennial crops, annual and biennial[27] vegetables and field crops remain essential to sustenance and culture. Most of these crops produce large amounts of seed, a superabundance that astonishes novice gardeners used to a few seeds in a packet. Maintenance of a seed line by regularly growing and saving seed is one of the most important examples of catching and storing energy. Even though the total amount of energy in seed is small, its density and potential value are very high. For a limited range of locally hardy crops, simply letting some self-sow each year is all that is required. With others, isolated growing and careful selection of seed from many plants is necessary to prevent undesirable cross-pollination that could degrade the seed line. Seed of some species can be stored for years, even decades, while others last only a single season. Although we can think of saved seed as a household storage of energy (see below), it is the regular growing that maintains the seed line. Thus, the permaculture food garden can be thought of as a landscape storage of very special energy — genetic information. Stored seeds are the stable and durable form taken by that information at one stage in the cycle.

Within the permaculture movement, the saving of seed — especially of old, local and rare varieties abandoned by agribusiness — has inspired activists, connected them to networks of seed savers,[28] and reawakened a cultural practice fundamental to future survival. The rapid buy-out by the agribusiness multinationals of most of the established seed companies in the 1970s and 1980s and their promotion of junk hybrids jump-started the modern seed-saving movement.

One of the advantages of perennial plants highlighted by permaculture is that they do not require saving of seed and growing again every few years. Once established, most perennial plants have plenty of years to produce seed and other propogules[29] adequate to ensure reproduction. However, conservation of valued genetic diversity in trees and perennial plants is also an important example of catching and storing energy. In rural Australia, many Landcare groups have been involved in collecting seed from the last remnants of locally indigenous species in their locality. The plantations and shelterbelts established with that seed are living storages of that genetic material.

Plantings of useful species established in the past are storages of great value that we can draw on. Perhaps the best example is the street and other plantings of Canberra, which were an integral part of the visionary planning by Walter and Eliza Burley Griffin.[30] The many useful and uncommon species of oaks in Canberra streets, which have been a source of true-to-type seed for permaculturalists over the last two decades, are of particular value to future tree crop systems in southern Australia.

Scion and bud wood[31] from rare fruit trees in abandoned gardens and horticultural research stations[32] have been collected by many gardeners. New permaculture-inspired gardens and orchards are the living stores of this heritage.

At Melliodora all these storages of energy were central in the original development strategy for the property and continue to define how we measure the strength and resilience of the system.

Characteristics of Natural Capital

Water, living soil, trees and seed all have the following characteristics that are important in any low energy sustainable society:

A degree of self-maintenance

Living storages such as soil and trees are largely self-maintaining and continue to grow over time. The quality of water in dams and even tanks can be self-maintaining through the living systems they contain. Vegetable species which self-seed and stay true to type are a self-maintaining genetic resource.

Low depreciation rate

An energy storage with a low depreciation rate will last for a long time, only slowly degrading in quality and quantity, as well as requiring minimal additional energy to maintain. If we can establish such stores when energy, information and labour are abundant, then a future low-energy society should be able to maintain them. The leaching of nutrients from well-balanced soils protected by perennial vegetation is extremely slow. The timber in mature forests of long-lived healthy trees degrades very slowly, despite the propaganda of the forestry industry. Well-designed earthworks to store water require little maintenance. Seeds of some vegetables, such as tomatoes and beans, are very easy to save and stay true to type.

Easily tapped without specialised or expensive technology

If an energy store can be tapped with simple means, people in future will be able to make use of it almost irrespective of their level of technology and other sources of wealth. Throughout history, water, fertile soil, seed and good timber forests have been real wealth in any culture or language.

Resistant to monopolisation, theft and violence

Stores that are naturally dispersed and diffuse make centralised and inequitable control more difficult. This is especially true of seeds, despite the enormous effort by the agribusiness corporations.[33] Stores (such as fertile soil, water and forests) that are large and low-value (per unit of weight) are resistant to theft. Resistance to the violence of civil unrest, terrorism and war is more problematic, but these stores are less vulnerable to violence than the usual symbols of wealth in buildings and consumer goods.

Catchment and Regional Planning

The key landscape storages of natural capital and the above criteria are the basis for a new way of thinking about catchment and regional planning. They provide a wholistic lens through which we can more clearly consider both proposals for development (generation of real wealth) and concerns about conservation (of what we already have). In this way we could integrate and summarise a vast complexity of environmental regulation and provide a more fundamental expectation that development should generate real natural capital rather than the current hotchpotch of impact amelioration and cosmetic cover.

In considering new and existing land uses and management we should:

- identify the mechanisms and storages for water, nutrients and carbon in the system
- identify the likely leakages of water, nutrients and carbon from the system
- compare the relative efficiency of storage and risks of losses with natural and wild (weedscape) systems evolved under similar energy and resource regimes.

Even where no hard data is available, good skills of reading the landscape allow eyeball assessments that are useful at all stages of a project — strategic planning, design and implementation.

Perhaps most importantly, the incremental modification of existing land uses can allow ecological succession and co-evolution processes of wild nature which are already efficiently catching and storing water, nutrients and carbon. (see **Principle 12**: *Creatively Use and Respond to Change*. These can assist and reinforce existing natural and wild factors that are already achieving these functions.

Many strategies and techniques described in permaculture literature are excellent examples of ways to increase efficiency in the catching and storing of water, nutrients and carbon. They include perennial crops, keyline soil conditioning and water harvesting, swales, food forests, tree fodder, cell grazing and long-rotation forests.[34]

EMERGY evaluation

Howard Odum and his colleagues have developed and applied EMERGY evaluation for environmental impact assessment. This provides a more quantitative and rigorous example of the wholistic thinking that permaculture attempts to foster. (see "EMERGY Accounting" in **Principle** 3: *Obtain a Yield*). I see the two approaches as complementary, one providing a check on the other. The EMERGY accounting helps us refine our broad principles and strategies; the permaculture principles provide commonsense rules of thumb to identify missing aspects in any model used to quantify complex human environmental systems.

Rebuilding Household Storages of Energy

One permaculture strategy is to encourage household and local economic self-reliance (see **Principle** 3: *Obtain a Yield* and **Principle** 4: *Apply Self-regulation and Accept Feedback*). This rebuilds traditional forms of energy storage in food, fuel and other resources that complement the landscape storages of energy. Many images of traditional rural life come to mind.

At Melliodora, we see the seedbox, the cellar full of produce and the stack of firewood as traditional storages of modest wealth which we enjoy and celebrate. Order and diversity in the seedbox reflects efforts to save, exchange and purchase the potential for abundance in future seasons. A pantry and cellar full of preserves capture the essence and substance of a bountiful season and provide insurance against a bad one next year. A large wood stack, with green wood added at one end and two-year seasoned wood used from the other, represents both the abundance of nature and good honest work all slowly maturing through the agents of weather and sunshine.

These types of household energy storages are by their nature:

- diverse
- small
- dispersed
- easily used
- not rich or portable enough to attract much attention from thieves and those seeking to monopolise wealth.

A nation full of such wealth is much more secure and stable than one dependent on the high-throughput, centralised systems for supplying food and fuel of the supermarket and electricity grid (see "Self Reliance and Disaster Preparedness" in **Principle** 4: *Apply Self-regulation and Accept Feedback*).

The Built Environment as a Store of Energy

The energy stores in the seed, food, fuels and other resources considered so far are fundamental to any society, but they are only part of the picture. The transformation of energy into progressively more elaborate and valuable forms extends well beyond physical resources to the things we make with those resources — tools, buildings, and infrastructure such as roads, power and telecommunications.

In modern society, the huge harvest of fossil and natural energies over the last few hundred years has moved down the "industrial food chain" resulting in the explosive development of cities, technology, and infrastructure for power and communications. People in the affluent world now live primarily in a built, rather than natural, environment; they are constantly shedding "junk" to make way for new possessions that are made possible by technology, and "necessary" by lifestyle and culture.

In the systems sense, modern society has been very active in "catching and storing energy" in all of its "stuff". Unfortunately most of these stores are useless without the rest of the industrial complex, and they mostly encourage more energy consumption and waste. For example, energy-consuming buildings and freeways rely on a constant flow of new fossil energy. Even computers, heralded for allowing us to do more with less, may be dependent on continuous redundancy and replacement to achieve economies of scale and stave off accumulating software corruption and other problems.

The ways in which permaculturalists deliberately pursue the catching and storing of natural site energies can be extended to apply to the development of energy-efficient buildings, appropriate technology, and all things manufactured and constructed.

When considering the development of the tools, buildings and infrastructure, we should aim to emulate, where possible, the characteristics listed previously for natural storages of energy. The following design criteria are relevant:

- modest in scale
- well-designed for long life and/or made from easily renewable materials
- simple to maintain (not necessarily maintenance-free)
- multi-purpose and easy to adapt to other uses.

Along with the creative retrofitting of existing buildings and infrastructure, applying these design criteria in all new developments could help contain the overdevelopment problem in the Western world, where the costs of maintaining what we have are beginning to erode our economic wealth. Maintenance engineering has a bright future, but the solutions will not be easy to find because most of our built assets have been designed and constructed assuming cheap energy (see **Principle 6**: *Produce No Waste*).

At Melliodora we have made maximum use of earth and stoneworks to construct terraces, dams and paths, and resisted the tendency to build too many fences, outbuildings and structures which need continuous maintenance and renewal. Our permanent buildings are no more modest in size than is typical on small rural properties, but we pack more functions and hours of use into them with our working-from-home and self-reliant lifestyle. Each building is multi-purpose and constructed to be easily retrofitted internally.

In building construction we have made extensive use of earth and wood as locally available and renewable materials. Although we have used modern industrial building materials for long-life functions, we have minimised the use of pumps, active devices and technology in performing essential building functions.

Energy Storage in Culture

Despite the massive scale of technology and built assets today, the largest stores of high-quality embodied energy are in forms that people do not recognise as containing energy or even much material. For example, information and organisation in government, economy, community and culture are thought to be largely ethereal "goods" unrelated to energy and resources. But it is no accident that the proliferation of these "things" has been an essential element in our high-energy fossil-fuel-based society.

In nature, long food chains and webs involve the transformation of solar energy into more and more complex forms, such as the organisation of a beehive, the physical structure of an old-growth forest, the hunting skill of a long-lived predator and biodiversity in general. Howard Odum and other systems ecologists[35] have shown that the EMERGY (embodied energy) required to create these complex forms of organisation and structure is very large. Similarly, studies of human systems have shown how the creation and maintenance of government, economy, education and culture follow the same energy rules as natural ecosystems. The diversity and complexity of these ethereal assets reflect the past harvesting and storage of energy in useful forms.

The much-heralded "weightless economy" of information and services (concentrated in centres of wealth) that does not involve the processing or transport of much energy or even materials, is novel and unprecedented, but it does not free us from the laws of energy, as most economists seem to think.

Financial capital is, of course, the most obvious and potent form of non-material wealth in the modern world. In recent decades, the rapid movement of capital around the world and its focus in destructive and short-term investment has been widely recognised as the most dysfunctional element of modern capitalism. There is obviously a need to better regulate and direct this most volatile form of wealth for long-term benefits. But in fact, much of the value of financial capital may be illusory, because it has become so disconnected from the sources of real wealth. Financial crashes of many different types have been increasing since the 1980s; there may be many more to come before financial wealth comes to more realistically represent potential material wealth.

Ethical investment is one of the most rapidly growing areas of financial services. This growth is constrained more by a shortage of appropriate businesses and projects that meet social and environmental criteria and appear financially sound than by a shortage of investors. Nevertheless, returns on investment have generally been as good as, or better than, market averages.

The rapid evolution of more appropriate governance, economy and culture, including art and myth, are central to adapting to energy descent. To claim that it is possible to use permaculture principles to design a sustainable culture and society may be too big a leap, but I believe it is possible at least to use them for evaluating the diverse cultural phenomena we find ourselves participating in.

Attitudes and values adaptive to descent

Some of the attitudes and values that I think are contributing to the development of a new descent culture include:

- acknowledging and supporting useful contributions from outside any particular intellectual discipline, profession or field of knowledge we are involved in
- learning from, and valuing, knowledge systems and ways of understanding outside scientific rationalism (our dominant cultural paradigm and the one from which permaculture emerged)
- remaining sceptical of official authority and formal qualifications in any field unless they are supported by more localised, networked credentials and demonstrable performance

- acknowledging the validity and value in all pre-existing cultures of place (different from the international culture of no place) and freely adopting bits that seem useful in a local context
- contributing to the evolution of a culture of place by supporting and celebrating the local, instead of the international, in knowledge, food, art and culture
- using the immense power of media and information technology, with a sceptical caution that resists total commitment and loss of other ways of communication, memory and interpretation — I think of information technology as the icing on the cake, rather than the cake itself.

These attitudes and values are relevant to the ways we organise our lives, earn a livelihood, raise children, deal with sickness and crisis, contribute to community life, and redistribute wealth and power.

Future sustainable culture

As the rate of change during energy descent slows, more enduring (sustainable) and diverse (bioregional) cultures will emerge. I expect those diverse cultures to have the following characteristics:

- bioregional political and economic structures, giving a renewed geographic diversity
- biogenetically, racially, culturally and intellectually cross-fertilised, giving natural hybrid vigour
- accessibility and low dependence on expensive and centralised technology
- capable of being developed by incremental steps with feedback and refinement

See **Principle 8**: *Integrate Rather Than Segregate* for examples of the application of these characteristics to rebuilding community in the transition to and acceleration of descent.

Appropriate Use of Non-Renewable Resources

I have discussed various strategies for investing existing wealth to rebuild natural and human capital. They all imply some, if not considerable, use of fossil fuels and other non-renewable resources. The transition to declining energy availability provides a unique strategic opportunity to make the best use of existing wealth and non-renewable resources to rebuild natural and human capital.

In general, the best use for non-renewable resources and technology should be to establish a system, rather than to maintain or harvest it, even if the "establishment" process is a gradual one that takes place as a transition over a lifetime (or even generations). For example, a passive solar house uses the high-embodied energy in glass to capture solar energy. If it does this while providing the other functions of glass in buildings, such as natural lighting and views, this is good use of embodied non-renewable energy.

Bulldozers and other large earth-moving equipment are perhaps the most dramatic example of appropriate use of fossil fuels and technology promoted in permaculture. They are used to make well-designed systems to harvest, store and distribute water, as well as creating vehicle access and house sites that increase the productivity of urban and rural landscapes. These earthworks are capable of being maintained indefinitely, by human labour if necessary. Another example is the soil rehabilitation by deep ripping of hard pans, followed by well-designed and managed plantings to maintain soil structure in the long term.

Mineral fertilisers for building soil

Mineral fertilisers are a special case of using fossil fuels (in crushing and transport) and often limited non-renewable sources (such as phosphate rock) to improve the long-term productivity of soils. The history of modern agriculture shows that most attempts at improved fertility are generally very short-term (crop feeding, not soil feeding); in addition, they often have other detrimental effects such as soil imbalances or pollution. However, there is evidence that, if the principles of balancing soil minerals can be better understood and applied, soil re-mineralisation can create a permanent improvement in biological productivity that does not need to be repeated, except to replace minerals exported in produce.

Permaculturalists have tended to ignore the powerful positive changes to long-term soil productivity and health that can be effected by carefully selected and timed mineral applications. This is understandable, given the history of detrimental effects of incorrect use of (mostly soluble) minerals and the emphasis in permaculture on biological solutions. Rock minerals are powerful medicine that can easily be overused or misused. Nevertheless, permaculturists who ignore the potential benefits they offer may design systems that do not provide the mineral balance essential for livestock and human health.

Better understandings of soil improvement based on the use of chemistry (rock minerals) and biology (plants, animals and microbes) are emerging, which can hopefully provide more reliable ways to achieve the "Holy Grail" of organic farming, high productivity and healthy balance. At Melliodora we have found that many of the emerging limiting factors to health and productivity of our place have had their origin in mineral imbalances that we had not adequately dealt with.

In recent years, we have been using soil testing (based on Albrecht methods), refractometer testing of plant sap and broadly based observation skills to guide remedial applications of minerals. Refining and extending our use of biological approaches (including biodynamic preparations) is our current focus, given that we think the basic mineral balance is reasonable. Only time will tell whether our current understandings lead to the Holy Grail.

The following points summarise my current understandings about the role of mineral balances in optimising fertility.[36]

- The level and balance of mineral nutrients are two important but different measures. Both are necessary to understand and maintain fertility.
- Bioregional and soil-type patterns of mineral imbalance are important, but intensive use of land, especially gardening, can create quite different imbalances.
- In the ideal balanced soil, the full range of crops that can be grown in the climate will be productive, healthy and produce good-quality food.
- Although wild and indigenous plants are adapted to particular imbalances, all will do well in a balanced soil.
- The most important soil mineral balance to get right is that between the alkaline mineral nutrients of calcium, magnesium, potassium and sodium.
- Different ways of testing measure balance. Measuring the acidity or pH balance is of some use but can be very misleading. A balanced soil has a pH of approximately 6.5 but a pH of 6.5 does not necessarily indicate a balanced soil.
- The ideal soil has the following balance by percentage base saturation:[37] calcium 68%, magnesium 12%, potassium 2–5% and sodium <1%.

- In clay soils, the balance between calcium and magnesium is the most powerful factor in determining how friable and open the soil is to air and water, as well as the ability of the soil to store water, carbon and nutrients. These factors in turn control biological productivity, ease of management, and resistance to erosion and degradation.
- The balance between calcium and potassium is the strongest soil factor in determining how lush or woody vegetation will be.
- When calcium is relatively high, herbaceous vegetation will tend to be soft, lush, palatable to animals, and broken down rapidly to humus by bacteria. Fruits tend to be sweet and long-keeping.
- When potassium is relatively high, herbaceous plants are fibrous and less palatable to animals, resist breakdown, and fungal decay dominates over bacteria. Fruits tend to be more acid and do not keep as well; woody plants do well and wood is more durable; forest litter tends to accumulate as dry fuel rather than rot.
- Organic matter and composts made in different ways and with different ingredients vary greatly in their quality as sources of soil fertility. Organic materials produced from a balanced system will maintain that system in balance if carefully recycled. Recycling within an unbalanced system will be successful to varying degrees, dependent on the nature and severity of the underlying imbalances.
- Fertilising programs to maintain balance are quite different from those necessary to establish balance. Just because a soil fertiliser (organic, rock mineral or artificial) produces a good result does not mean that more will produce a better result.

Idealism versus Pragmatism?

In practice, it is hard and maybe unwise to completely avoid using the generous energy subsidy that comes with the power and convenience of fossil fuels for day-to-day living. Nevertheless, in making use of coal-generated electricity, petrol in the car or lucerne hay (produced with fossil fuel subsidy) for garden mulch, we should never take these underpriced resources for granted. Instead, we should design our systems as if these resources were much more highly priced.

At Melliodora, for example, we are connected to mains electricity. Because of the design of the house and our lifestyle, our use, at less than 3 kilowatt hours a day, is less than one-fifth of typical household use. The energy-demanding functions of space heating, water heating and cooking are provided with renewable energy (passive solar and wood). With an evaporative cool cupboard, a seasonal diet and low-energy food preservation methods, electricity for refrigeration is more of an extra than an essential service. By buying the slightly more expensive renewable energy tariff, we contribute to stimulating development of renewable sources of energy, but this is less important than our low consumption of electricity. Our power needs could be provided by photovoltaic cells and grid feedback electronics, but for the present we can make better use of the money that would be required in developing other aspects of our property. (See **Principle 5**: *Use and Value Renewable Resources and Services* for discussion of the merits of solar electric power.)

At one level, this might seem to be simply balancing principle with pragmatism, but I also mention it to counter the view that use of fossil fuels is bad, inefficient and immoral. In fact, fossil fuels are very useful, but they are massively overused and most of the uses to which they are put are destructive or, at best, banal.

An early realisation of the banality of our fossil fuel use occurred to me on a sunny Sunday in Hobart back in 1974. A friend who was an abalone fisherman had invited me out for a jaunt in his powerboat. As we sped down the Derwent estuary chasing seagulls, powered by twin 80-horsepower outboards, I remember thinking that our consumption of energy was far greater than that of the ancient kings. While the actions of the ancient kings had immediate consequences for people and nature, ours appeared to be of no consequence beyond momentary exhilaration (and incremental resource depletion).

Although it is hard to assess the long-term effects of any particular behaviour, in many cases, it is clear that we are wasting our lives and the earth. I remember a discussion with a fellow passenger on a packed business flight from Sydney to Melbourne in 1990 that illustrates this well. I was musing on the balance of values in my air travel from Victoria to Orange NSW, for a two-day workshop of writers of the first Australian postgraduate course in sustainable agriculture.[38] The woman next to me was returning from a day trip to Sydney selling desktop computers to small businesses. Her candid admission that the equipment she was selling was little different from competing brands made from the same components underlined her perception of the trip as pointless (beyond earning her daily bread). Her master's degree in mathematics only emphasised the waste of human as well as natural resources. The comparison with her trip diverted my attention from my own ethical questions to wondering about the possible value from a whole planeload of Melbourne–Sydney business day-trippers.

Conclusion

Long-term asset building for the benefit of future generations has been a focus for ethical behaviour down the ages. In a time of rapid change and short-term thinking we need to rebuild the aspect of our culture that emphasises caring for the future, as well as deciding what is worth investing in for the benefit of our grandchildren and descendants. This principle provides a framework for considering what may have value in an uncertain world.

On a Permaculture Design Course I was demonstrating form-pruning of box[39] eucalypt forest regrowth when I was asked whether it would not be better to let the trees grow crooked and branched so that future generations would not cut them down for timber. My reply was that we needed to consider what future generations would think of us, who, having lived high on the hog of fossil fuel affluence, decided to leave them nothing in the way of high-quality renewable resources because we didn't trust that they would use them wisely.

On another course, a sceptical participant queried this focus on rebuilding stores of energy in forests and other biological resources. He referred to planting of the oak forests by the British to provide timber for wooden ships, which they never needed because new energy sources and technology allowed steel to replace wood in shipbuilding. My answer was that, in the unlikely event of a high-technology future where we didn't need natural resources, we would have all these beautiful forests which, like Britain's "ancient" oak forests of today, would be a home for wild nature and an inspiration to the soul. Not a bad result for getting it so wrong.

1 See Article 10 "Development of the Permaculture Concept" in *David Holmgren: Collected Writings* 1978-2000.

2 In ecology, "niche" means the role of an organism within its natural environment that determines its relations with other organisms and ensures its survival. The more common usages in relation to business and personality derive from this ecological concept.

3 The argument that the oceans could provide vast new resources for humanity is overstated, being based on a simple numerical equation about the proportion of the earth's surface covered by the oceans. It ignores the facts that most of the deep oceans are ecological deserts and that the energy cost of exploiting mineral resources in the ocean rises exponentially with the depth of water. Almost all of the exploitable biological and mineral resources of the sea are in the shallow waters of the continental shelves, where the water is up to hundreds of metres deep rather than thousands.

4 A few other minor biochemical processes allow some microbes to collect chemical energy in other ways.

5 Subduction describes the dragging down of one tectonic plate beneath another.

6 Generally the last 100,000 years.

7 H. Tané, "The Case For Integrated River Catchment Management" Keynote Address, *Proceedings of the International Conference on Multiple Land Use and Integrated Catchment Management* Macaulay Land Use Research Institute Aberdeen, UK, 1996.

8 Including migrating birds and fish. Eels, salmon and other species of fish that return from the sea to ancestral spawning grounds are a special case of transfer of valuable minerals from the sea to the top of catchment landscapes.

9 Atmospheric carbon dioxide is the raw material that plants use to create organic carbon compounds, starting with simple sugars.

10 Although this might seem like simply a matter of semantics, the dominant corporate strategies for carbon sequestration show all the signs of creating more, rather than fewer, problems; they include proposals for massive monocultural, short-rotation plantations and for pumping carbon dioxide from power stations underground.

11 Some activists within the indigenous ecological restoration movement have dubbed permaculturalists as "biomass junkies" for placing so much importance on woody plant growth as an ecological good. (For a discussion of a more balanced view of plant biomass, see "Plant and Animal Biomass as Indicators of Fertility" in **Principle** 7: *Design from Patterns to Details*.)

12. W. Jackson, *New Roots for Agriculture* Lincoln University of Nebraska Press 1980. For more information about the Land Institute and research in perennial grain crops see website http://www.landinstitute.org

13 For excellent documentation of "the state of the art" see T. Furuno, *The Power of Duck: Integrated Rice and Duck Farming* Tagari 2001.

14 Fertile arable soils in the temperate zones contain over 10% humus, while the majority now contain less than 5%

15 Relatively recent advances in soil science are confirming and clarifying the organic principles of plant nutrition and exposing the conventional soluble absorption theories as simplistic or even wrong.

16 See Article 25 "Why Natural Landscapes Catch and Store Water, Nutrients and Carbon" in *David Holmgren: Collected Writings* 1978-2000.

17 Acidity is well recognised as slowing the breakdown of organic matter, but this is more the symptom than the cause. A low ratio of calcium to potassium is the more fundamental cause.

18 High-sulphur brown coal is highly polluting when burnt in power stations and furnaces.

19 See book downloadable from Yeomans web site www.yeomansplow.com.au/

20 Charles Walters Jr, (ed.), *The Albrecht Papers* Acres USA, 1975.

21 Leaching (typically of nitrogen, potassium and calcium) can be accelerated by poor soil structure and loss of nutrient-holding humus, to which mineral imbalances are contributing factors.

22 Rock phosphate is one of the most important, and severely depleted, mineral resources. In countries where phosphate use has been widespread there are often abundant reserves locked up in agricultural soils, which can potentially be released by soil microbes; but in many poorer countries where fertilisers have not been widely used these soil reserves do not exist.

23 See Article 10 "Development of the Permaculture Concept" in *David Holmgren: Collected Writings* 1978-2000.

24 Eucalypts, casuarinas, acacias and conifers are typical of these timber trees.

25 For discussion of the relative merits of hemp and trees as fibre crops and the importance of sustainable forestry to the permaculture agenda, see Article 17 "Hemp as a Wood Paper Pulp Substitute: Environmental Solution or Diversion from Sustainable Forestry?" in *David Holmgren: Collected Writings* 1978-2000.

26 Bill Mollison in the video *Bill Mollison The Permaculture Concept: In Grave Danger of Falling Food*, Julian Russell and Tony Gailey, ABC Video 1989.

27 Biennials take two seasons to produce seed and complete their life cycle but may be harvested at the vegetative stage after one season (eg carrots and cabbages).

28 In Austalia Jude and Michel Fanton of the Seed Savers Network and Clive Blazey of Diggers Seeds have been some of the activist entrepreneurs at the forefront of seed-saving movement. For a valuable and practical guide to seed-saving see M. and J. Fanton, *The Seed Savers Handbook* The Seed Savers Network 1993. See The Diggers Club website www.diggers.com.au and *Seed Annual* for Clive Blazey's unique blend of environmental campaigning and integrated marketing. Kent and Diane Whealy, who started the Seed Savers Exchange in Iowa USA, in 1973 were early and continuing activists in this field. See website www.seedsavers.org

29 Propogules include seed, bulbs, runners, suckers, cuttings, and other vegetative parts of plants capable of reproduction. Plants produced from seed are often not true to type due to cross-pollination. Vegetative (clonal) reproduction is true to type.

30 See L.D Pryor and J. Banks, *Trees and Shrubs in Canberra* Little Hills Press 1991. An excellent reference book that includes the species planted in every Canberra street.

31 Cuttings used for grafting and budding to reproduce varieties true to type.

32 Cutbacks in state Agriculture Department budgets and collapse of public interest in favour of corporatised goals in the 1980s led to the abandonment of many collections and arboreta of fruit and nut varieties. Without action by individual departmental officers and private growers, these valuable collections would have been lost. The Heritage Fruit Group of Permaculture Melbourne and permaculture entrepreneurs Jason Alexandra and Marg McNeil manage and maintain one of the few remaining public collections of heritage apples in Australia at Petty's Orchard in Melbourne. See report on "Earthbeat" ABC Radio National http://www.abc.net.au/rn/science/earth/stories/s495362.htm

33 The agribusiness multinationals, backed by government legislation, have made enormous and innovative efforts to control seed supplies, including plant patenting and genetic engineering. The battle continues.

34 See B. Mollison, *Permaculture: A Designers' Manual* for an overview of strategies and techniques. See D. Holmgren, *Trees on the Treeless Plains: Revegetation Manual for the Volcanic Landscapes of Central Victoria* Holmgren Design Services 1994, for more detailed designs for tree systems to achieve these ends within existing farming systems.

35 See Article 22 "Energy and Emergy: Revaluing Our World" in *David Holmgren: Collected Writings* 1978-2000.

36 For further exploration of this subject, see Article 25 "Why Natural Landscapes Catch and Store Water, Nutrients and Carbon" in *David Holmgren: Collected Writings* 1978-2000.

37 As calibrated by the Brookside Laboratory in the USA, using percentage base saturation of Total Cation Exchange capacity allows for minor bases 5% and hydrogen 12%.

38 Postgraduate Diploma of Sustainable Agriculture at Faculty of Rural Management, University of Sydney, Orange NSW.

39 Box eucalypts are slow-growing but yield very hard and durable timber eminently suited to bridge construction and other heavy engineering uses.

Obtain a Yield

You can't work on an empty stomach

The previous principle, *Catch and Store Energy*, focused our attention on the need to use existing wealth to make long-term investments in natural capital. But there is no point in attempting to plant a forest for the grandchildren if we haven't got enough to eat today.

This principle reminds us that we should design any system to provide for self-reliance at all levels (including ourselves) by using captured and stored energy effectively to maintain the system and capture more energy. More broadly, flexibility and creativity in finding new ways to obtain a yield will be critical in the transition from growth to descent.

Without immediate and truly useful yields, whatever we design and develop will tend to wither while elements that do generate immediate yield will proliferate. Whether we attribute it to nature, market forces or human greed, systems that most effectively obtain a yield and use it most effectively to meet the needs of survival, tend to prevail over alternatives.[1] A yield, profit or income functions as a reward that encourages, maintains and/or replicates the system that generated the yield. In this way, successful systems spread. In systems language, these rewards are called positive feedback loops, which amplify the original process or signal. If we are serious about sustainable design solutions, then we must be aiming for rewards that encourage success, growth and replication of those solutions.

The original permaculture vision promoted by Bill Mollison of growing gardens of food and useful plants rather than useless ornamentals is still an important example of the application of this principle. The icon of the vegetable with a bite taken shows the production of something that gives us an immediate yield but also reminds us of the other creatures who are attempting to obtain a yield from our efforts.

Models from Nature

All organisms and species obtain a yield from their environment adequate to sustain them. Those that fail in this task quickly disappear. There could hardly be a more fundamental lesson from nature, one that reinforces our basic survival instincts.

Darwin's emphasis on competition and predation as the driving forces in natural selection was based on observation of natural systems, but it was also drawn from Darwin's personal experience of the competitive ravages of early industrial England which predisposed him to look for similar models in nature. In turn, the Victorian industrial elite used Darwinian ideas to support their social and political views. One hundred years ago the Russian geographer and anarchist, Peter Kropotkin,[2] provided a refutation of the social Darwinists with very diverse examples of co-operative and symbiotic relationships both in nature and in human history. (see **Principle 8**: *Integrate Rather than Segregate* for exploration of this important aspect of permaculture design.)

In the last two decades of the 20th century, unbridled economic competition again became the sacred cow of the political mainstream. The misguided and highly selective use of the capitalist ideology of economic competition should not blind us to a balanced appreciation of the need to obtain a yield and the role of competition in testing alternative design solutions, processes and systems.

Benefits of Competition

Competition in nature helps test the vigour and fitness of individual organisms, or a species, for particular conditions. Predation, where it removes the weaker individuals, also contributes to "survival of the fittest". For example, direct seeding to produce dense stands of plants (be it radishes or oak trees) encourages the fastest-growing and most vigorous individuals to prevail. We can help that process along by thinning, as the more vigorous individuals become obvious. By doing so, we are acting as selective predators. Australian graziers who let mismothered lambs die may be seen by some as callous or lazy, but they are also allowing a positive selection pressure for adequate mothering in their flocks. (See **Principle 8**: *Integrate Rather than Segregate*, **Principle 10**: *Use and Value Diversity* and **Principle 12**: *Use and Respond to Change* for more elaborate ways we can make use of competition and predation.)

In human systems, we understand that comfort and excessive protection from challenges and competition can lead to self-satisfied, lazy and eventually dysfunctional behaviour. We can see this in the raising of children, the evolution of organisations and the history of civilisations.

Maximum Power Law

The entropic loss of energy as low-grade, waste heat unable to drive any further processes was explained in **Principle 2**: *Catch and Store Energy* as an inevitable outcome of energy conversion in all physical processes. This inevitable loss of energy reduces the efficiency of conversion to useful work. The rate of input energy and the efficiency of conversion determine the rate of useful work, or power, produced by any process.

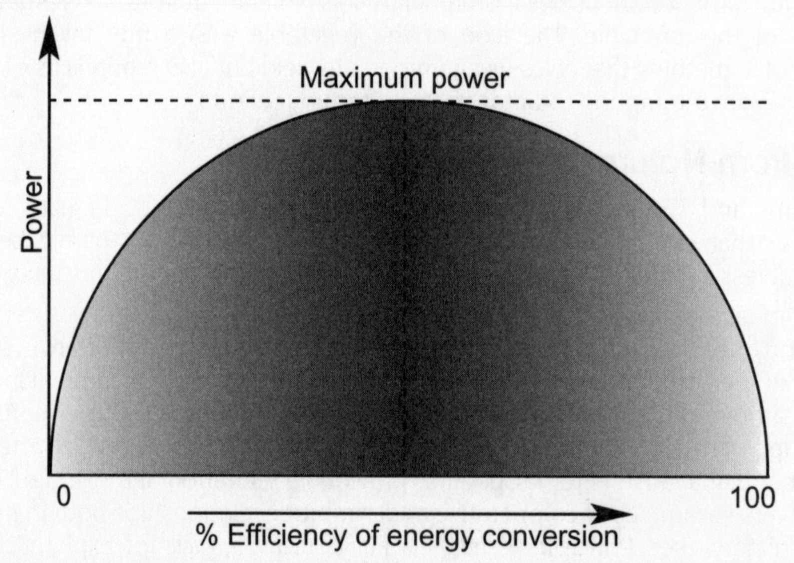

Figure 10: Maximum power for a heat engine

The classic thermodynamic example of efficiency and power of energy conversion is a simple heat engine converting fuel energy to mechanical energy. Figure 10 shows power output against efficiency of conversion for a heat engine using high-quality fuel. If the engine is running free with no load, energy is being consumed but there is no useful work, thus power output and efficiency of conversion are both zero. If the engine is loaded towards the point of stalling, efficiency of conversion approaches 100%, but power output falls to zero when the engine stalls. Maximum power is achieved when maybe half of the energy is being converted into mechanical power and half is being dissipated as waste heat and noise.

While the optimal efficiency for maximum power will vary greatly with the quality of energy used, all biological and self-organising systems show this same basic pattern in the conversion of energy to useful work. Everywhere in nature there is a balance between simply sucking in and dissipating energy for no real purpose, and attempting to get maximum efficiency of energy convesion by squeezing every ounce of possible value out of an energy source. Evolution in nature and innovation in society tend toward this balance between energy dissipation and efficiency of conversion that gives maximum power from any particular process or situation.

Consumer versus conserver values

The modern consumer economy and culture encourage consumption for its own sake, despite all the talk about efficiency and productivity. This is analogous to the engine running with no load, consuming energy but doing little of value.

The conservation ethic focuses on attempting to get maximum efficiency out of every use of natural resources, which is necessary to correct the dysfunctional nature of consumption (see **Principle** 6: *Produce No Waste*). However, at the extreme, this approach is analogous to the engine loaded to the point of almost stalling — efficient, but not so useful either.

The application of the Maximum Power Law to complex ecological, let alone human, systems is controversial. Many reasonably argue that the environmental crisis results from the excesses of humanity "obtaining a yield", and that permaculture and other sustainability concepts are about moderating these excesses in a more long-term self-interest.

The Maximum Power Law may appear to support rampant market capitalism as the most natural and functional evolutionary path for humanity. Although some systems energeticists make this assumption, I see the law as useful in framing a more down-to-earth but wholistic permaculture design principle. By this principle we recognise that we should design systems and organise our lives so that we obtain a yield in ways that maximise the power of useful work from whatever we do.

The concept of useful work or yield is central to the explanation of maximum power and permaculture. It is easy to see that mechanical energy is the useful output of an engine and that heat, exhaust gases and noise are useless; but separating the useful from the useless in complex biological and human systems is more difficult.

Howard Odum gives a list of ways in which all successful and persisting self-organised systems maximise power to meet the needs of survival. They

- develop storages of high-quality energy
- feed back work from the storages to increase inflows

- recycle materials as needed
- organise control mechanisms that keep the system adapted and stable
- set up exchanges with other systems to supply special energy needs
- contribute useful work to the surrounding environmental systems that helps maintain favourable conditions.

Rather than confirming some sort of dog-eat-dog world, this list provides abundant scope for including ethical and co-operative behaviour that meets the needs of survival.

It is also encouraging that quantitative analysis of complex biological and human systems to predict states that represent maximum power is tending to support many of the values embodied in the environment movement and especially permaculture. (See discussion below about EMERGY accounting.) On the other hand, sometimes these evaluations challenge us to question whether some apparently beneficial processes and designs may exemplify the excessive pursuit of efficiency but not achieve an outcome that is truly useful in a wider sense. (See discussion of biomass fuels below and solar cells in **Principle 5**: *Use and Value Renewable Resources and Services*.)

Positive Feedback

Positive feedback is a system mechanism that amplifies a process or effect, particularly the harvesting and use of energy. For example:

- in a bushfire, the heat from burning preheats green vegetation, making it more readily combustible; in extreme cases this creates massive updrafts, which suck in more oxygen to feed the conflagration
- by the use of energy captured from sunlight, plants grow more leaves, which in turn capture more energy.

Landscape change is often driven by positive feedback processes, as occurs when the growth of reeds in a billabong[3] slows the water flow and increases sedimentation, leading to more plant growth and a change from creek bed to swamp.

In human society, law, religion and the marketplace work (more or less) to distribute rewards to those who provide the most valued outcomes, thus encouraging them to provide more. For example, when we spend money we are providing a strong signal to encourage the production of more apples, cars, massages, or whatever we are buying.

In systems well adapted to their energy base, positive feedback can be thought of as an accelerator to push the system towards freely available energy, while negative feedback is the brake that prevents the system falling into holes of scarcity from overuse. (See **Principle 4**: *Apply Self-regulation and Accept Feedback*.)

Staple foods and net energy supply

Even in pre-industrial times there was a natural tendency for people to focus their energy and effort on what provided them with their most important yield. Most traditional peasant societies had one or more "staples", high-yielding carbohydrate-rich crops[4] which provided that important yield. In energy terms, these crops provided a clear net energy yield. Because the yield was relatively high, investing more work in selection, cultivation, protection and storage of these crops was worthwhile, and thus these areas were a focus of labour and technology. The value of these staple crops was often reflected in cultural and religious practices, such as the worship of maize as a god in pre-Columbian America and the John Barleycorn rituals of pre-industrial Britain.

The net energy gain from staples provided people with the sustenance and thus allowed them to use other important resources that could not provide a net energy return. For example, detailed energy analysis of traditional New Guinea swidden agriculture[5] showed very high net energy gains from food gardens, but raising pigs was a net loss in energy terms. However, pigs provided needed protein and important social functions, as well as being linked to localised warfare in an ecologically functional way. This cross-subsidising of important functions from the net energy supply can be seen at all levels in human and natural systems.

Many other yields and passive functions in traditional pre-industrial societies could be obtained with relatively little work or effort from self-regulating semi-wild systems. For example, medicinal herbs, animal fodder and fuels were available from plants on common land that did not require planting, fertilising, watering or much tending except harvesting. These benefits did not require energy subsidies, but neither did they provide a high energy yield. These modest self-maintaining systems were good examples of obtaining a yield in ways that are neither complex nor even obvious. Good design solutions can work so well as to be almost invisible.

Hardy (self-reliant) species

Hardy and self-reliant species are important in any low-energy sustainable system. By selecting hardy, locally adapted and self-reproducing plants wherever possible, the designer can minimise the resources required to maintain gardens, farms and forests.

These species can be thought of as "self-reliant" or "competent" in "obtaining a yield". In fact it is their minimal demands, relative efficiency, or associated symbiotic microbes that fix nitrogen or tap locked-up minerals that give them a competitive advantage on poor soils over fertility-dependent species.

Although problems of rampancy of hardy species need to be considered, the obsession with maintaining control in gardens, farms and forests limits our ability to develop systems less dependent on endless inputs of non-renewable energy and resources. By using only species that do not grow too much or reproduce too freely, we ignore some of the most useful species. Thus the first priority in healthy broadacre farm landscapes, rangelands and forests must be vigorous and self-reproducing plants.

Farm landscapes where there is no need to remove some tree and shrub growth are generally degrading and dying landscapes. It is easier to remove excess tree and shrub regeneration with fire, grazing animals or machines than it is to plant more trees and natural regeneration generally produces stronger, more adapted stock than planting does. The hardy nature of local indigenous and other Australian native plant species makes them ideal for providing shelter and other functions in unirrigated and unfertilised soils.

The same principles apply to animal husbandry. In Europe, intensive management systems have resulted in breeds of cattle that can no longer give birth without assistance. In Australia, turkeys are bred with breasts so big that the males cannot serve the females, which have to be artificially inseminated. During energy descent, plant and animal species and breeds that require large external input will have been displaced by those more adapted to prevailing conditions.

Increasing fertility

While adapting to prevailing conditions by using hardy species is a classic permaculture strategy, increasing and maintaining fertility in order to grow a wider range of more useful and higher-yielding species is equally important. It is relatively easy in climates with adequate moisture (and where grazing animals are excluded) to grow bulk biomass in grass and trees without increasing fertility. But if our aim is to grow human food, especially on limited areas of land, then high and balanced fertility is essential. Food crops, especially annual field crops and vegetables, are much more demanding of fertility, especially if the yield potential of highly selected varieties is to be realised. Novice gardeners are often surprised that yields from vegetables can vary as much as two orders of magnitude, depending on fertility.[6]

These apparently opposite strategies for obtaining a yield are always in dynamic tension in permaculture design, but the conflicts between them are less than they seem at first. Firstly, the area we need to make most fertile to provide a substantial proportion of our food needs is small, so by concentrating our soil-improving activities in gardens and intensive farms we can complement large-scale areas where we adapt to prevailing conditions. (See Zoning in **Principle 7**: *Design from Patterns to Details*.)

Secondly, plants that are adapted to low fertility generally grow better with higher fertility, so long as that fertility is balanced. The popular myth that native and other hardy plants dislike fertility can be attributed to cases of:

- unbalanced fertiliser applications (in some cases resulting in death of mature trees)
- insect predation attracted by more palatable growth
- direct competition by fertility-demanding species.

So increasing and balancing soil fertility allows the widest range of species to thrive, although more fertility-demanding (and generally more useful) species will tend to dominate.

Utility versus cosmetics in horticulture

The focus in *Permaculture One* on the cornucopia of useful plants from which we can obtain food and other important yields is an example of the great opportunities in affluent societies to refocus on what is useful. Despite their emotional association with nature, ornamental horticulture and traditional landscape design contribute to resource depletion and simply provide a cosmetic cover over the disharmony and unsustainability of industrially determined environments.

Over the years I have often found visitors to Melliodora are surprised that we have some roses and even lawn, those great icons of ornamental horticulture, and some local indigenous species that do not produce apparently useful yields. Rather than some begrudging concession to cosmetics and indigenous purity, I see these design elements as reflecting a more complex and contextual notion of utility than some people have interpreted from permaculture.

In the article "Lawns, Mowing and Mulch in Permaculture"[7] I give my interpretation of a balanced use of lawns and pasture in permaculture design. In "The Role of Native Vegetation in Backyard Permaculture",[8] I respond to the use of native and indigenous vegetation in gardens as a perhaps more benign example of ornamental horticulture.

While it is important to recognise the value of aesthetic delight and other, hard-to-quantify yields and functions in any system, the culture of affluent consumerism can lead us away from functional and effective design. For example, it may be argued that the noise of a hotted-up motorbike or car driven by a young male is an output that performs the social function of attracting attention and releasing aggression. While this argument may have some merit, a society where people meet simple emotional needs in ways like this will not survive long in an era of declining energy.

Permaculture designers can still give priority to fundamental and resource-hungry needs such as food, clean reliable water supply and shelter, while providing complex but passive environmental services (eg. wildlife habitat) and social functions (eg recreation) as by-products of an integrated design. This contextual and balanced understanding of utility reflects the Maximum Power Law, while acknowledging that we do not live by bread alone. The multi-functional aspect of permaculture design is explained further in **Principle 8**: *Integrate Rather than Segregate*, while my perspective on the role of aesthetics in design is in **Principle 7**: *Design from Patterns to Details*.

The food production strategy

While the permaculture strategy of growing our own food may be a step towards appropriate application of this principle, it is how we gain a reward or yield from the strategy that is important. Growing our food can give us aesthetic delight and relaxation, a better understanding of how nature works, a greater sense of security and well-being, and an appreciation of farmers who earn their living growing food, but it is consuming the harvest that provides us with the visceral reward. If we fail to harvest anything that gives us sustenance and enjoyment and buy all our food from the shop, then our permaculture-designed garden won't last long as a display of ideals. On the other hand, if we enjoy abundant, high-quality harvests, this will sustain us once the novelty has worn off and through the inevitable seasonal ups and downs. *If we expose very young children to the delight of foraging food in a garden, they are more likely to grow up with a deep and intuitive understanding of our dependence on nature and its abundance.* Despite the distractions of youth, it is also true that those early connections lead to later interest and ease in growing food as adults.

Social relationships

If our personal and community relationships are only based on powerful but shifting emotional benefits and we lack the experience of more practical and concrete "yields", then it is difficult to sustain and strengthen those relationships over the long term. If, on the other hand, we actually depend on our family, friends and relations to maintain the house, fix the car, supply our food and so on, we are more likely to resolve the difficulties that arise in these relationships. This truth is more obvious in rural communities where everyone understands the realities of interdependence.

This principle forces us to become more aware of the real sources of our own sustenance and well-being. For peasant people connected to the cycles of life and death, this may be self-evident. In the modern world, complexity, scale and affluence have obscured these sources, making it hard to know if we are better or worse off, and at what cost or benefit to anyone else. In this context, permaculture can be thought of as "remedial wholistics" that are necessary because of generations of industrial affluence. As a design system, it leads by progressive steps to regaining control over our own sustenance and a realistic rather than romantic understanding of what it means to live with and from nature.

Timing and Flexibility

In obtaining a yield from any system, timing is critical.

Most natural systems go through phases of growth and accumulation, leading to abundance. Outside the wet tropics, seasonal cycles of hot and cold, wet and dry determine the patterns of seasonal abundance. We need to relearn to match our harvesting activities to these phases of abundance. For the cool-climate food gardener, the autumn surplus and spring scarcity are fundamental realities that determine the organisation of the year's food supply.

It can come as a shock to find that the enormous yield fruit trees can produce in good conditions can all be ripe for eating over one or two weeks and that birds may take the lot the week before you do. In less affluent societies the competition from other people is often more of an issue.

Drip-feed culture

At another level, we can see that the drip-feed culture of weekly wages and weekly shopping is increasingly out of step with economic and social realities. Flexibility, retraining and contract work are progressively replacing the full-time job for life. These changes may be forced on people by institutions dishonestly seeking to escape their previous social responsibilities to provide secure jobs, but this should not blind us to the opportunities to break our dependence on the drip-feed culture that has little chance of surviving the transition to declining energy availability.

This flexibility and openness to opportunity was one of the skills of poverty, which has been lost. In the 1970s I remember a friend, who did voluntary work at Hobart's first women's shelter, describing the residents' shock at her extravagance in buying a sack of potatoes for the shelter from a farmer. The residents' notion of frugality was to go the corner store and buy a packet of instant mashed potato for that night's meal. I am sure welfare agencies today are the source of many more bizarre stories about the absence of basic household skills. No-one is as disadvantaged as the poor without the skills of poverty.

For many people, permaculture has provided a framework for a personal transition from conventional, often city-based, employment to a more self-reliant, self-employed and often rural life, where the ups and downs of opportunities and income are the norm. The fluctuations of seasons, work opportunities and other sources of "income" demand that we design for a high degree of flexibility.

The ability to substitute and adapt recipes to what is seasonally available is a common feature of peasant cuisine around the world and is essential if we are to really enjoy the benefits of home-grown food. The jokes about 101 ways to use surplus zucchini show that, although attempts to moderate the up and downs of yields from natural systems are important, surpluses and gluts can be an incentive to find new and creative ways to obtain a yield.

House-building

Owner-builders who change and adapt their house designs as they build in order to take advantage of bargains in second-hand windows, doors and other materials are a good example of the use of flexible design to obtain a yield.

The professional builder buys house-lots of materials for a standard design that has been driven by efficiency in use of labour and (to a lesser extent) materials, and can turn out finished project homes at a surprisingly low cost. However, these business efficiencies depend on a constant supply of precisely specified components, available at a guaranteed price and delivered on time.

Just-in-time madness

The progressive elimination of materials inventories because of the "just-in-time" manufacturing strategy shows the extreme pursuit of high efficiency at the cost to self-reliance and flexibility.

In systems theory, the adage that "loose systems last longer and work better" suggests that flexibility can be more important than efficiency. Systems ecology recognises that stable conditions give advantage to highly specialised species, but that changing conditions favour species known as generalists that can adapt to different food, habitat or other factors. Specialisation comes at a cost to flexibility; generalisation comes at a cost to efficiency. In the example of the builders, the owner-builder is the generalist; the professional is the specialist.

Many permaculture strategies and techniques are generalist in nature, allowing a high degree of flexibility with less emphasis on efficiency.

Efficient Use of Resources as a Trap

Self-reliance is a generalist strategy, but in self-reliant lifestyle, motivated by environmental ideals, it is still easy to fall into that trap of chasing efficiency in the use of resources to an extent well beyond maximum power. I have personally fallen into this trap many times and see others doing the same.

In commercial production of fruit and vegetables, large quantities of undersized produce are left unharvested because of the fickle nature of the market and the need to maximise efficiency in the use of labour and machines, rather than the use of produce. The scale of waste on farms can shock even the most hardened economic rationalists when they see it first hand.

By contrast, in the home garden we can make use of undersized produce. But there are limits beyond which more efficient use of produce does not make sense. Harvesting and washing marble-sized potatoes might be something we do once but not again.

Owner-builders visiting a local sawmill often take advantage of off-cuts for all sorts of creative constructions. However, most find there are also limits to how highly we should value the efficient use of a natural resource compared with the investment of skill and time. If our creative constructions are neither functional nor durable, then this apparently efficient use of wood may be misguided.

The balance of efficiency is different in the country from the city and very different in rich and poor countries. At Melliodora we know that millions of people around the world walk miles to collect fuel no better than the great piles of sticks that our goats leave after eating the leaves and bark of the tree fodder we cut for them. We use some of this material as kindling, but the rest we burn in autumn equinox and winter solstice bonfires.

Permaculturalists who are used to constantly looking for creative ways to use the wasted resources of a throw-away society can easily become obsessed with collecting more stuff than they can use. We need to remember that degradation from weather and termites, or simply forgetting what we have and where it is stored, can prevent us obtaining a yield. "Scoring" some amazing cast-off can be exhilarating, but it is only when we make effective use of our find that we have obtained a yield.

Numeracy

The farmer or business person who keeps complete financial records and uses them for managing the business might not be considered a typical permaculturalist, but numeracy and accounting skills are important and complementary to the core observation and design skills (see **Principle 1**: *Observe and Interact*). Numeracy and accounting give us measures of yield which allow us to deal with complexity and quickly respond to novel situations and systems. They are critical skills if we are to build systems adapted to energy descent.

Money may not be an adequate measure of value in accounting, but this should not detract from the value of accounting itself. An accountant friend once suggested that accountants were not really the enemies of sustainability, they just needed to be given appropriate numbers to add up. The plague of economic rationality of the late 20th century led to reactions calling for the "triple bottom line" (financial, social and environmental) in corporate accounting. There is a great scramble to find appropriate methods for environmental and social accounting which is likely to accelerate as limited energy supply becomes entrenched.

The failure of financial accounting to effectively consider environmental and social costs has contributed to an undermining of faith in the value of numeracy. This is part of a deeper distrust of science and measuring things as processes to ascertain their value. This more fundamental criticism of numeracy is valid[9] but, most of the problems relate to what is being measured and what is being ignored.

Records of the measures of inputs and yields by weight, volume and/or financial cost are useful in managing any garden, farm or household because they complement memory and qualitative evaluations and provide indicators of success and failure in complex systems. These simple measures are also the base data for more complex forms of environmental accounting discussed below. Unfortunately measuring and record-keeping take time and energy, but learning is always a costly process.

With the complexity of the modern world where a myriad of cross-subsidies of energy and other factors operate at many levels, it can be hard to know whether we have obtained a yield, and if so, how much. Industrial recycling, pollution control and many other environmental sacred cows may or may not be good ideas, depending on a complete environmental and social accounting.

Ecological footprint

One interesting method of environmental accounting that is accessible and useful at the household scale is the Ecological Footprint.

This method converts all consumed resources to a figure representing the area of land required to generate those resources and dispose of the wastes. Like all environmental accounting methods, it depends on calculations using regional or national data and relies on assumptions that simplify complex relationships. It has gone through several cycles of refinement and is being widely applied to measuring the total environmental impact of nations and households.

Comparative figures for all countries are now available. They show a global average of 2.9 hectares of productive land being used to support each person, although only 2.2 hectares are available. In other words, we are eating into natural capital to support humanity. Close to the top in consumption are the United States, at 12.2 hectares per person, and Australia, at 8.5 hectares per person.

Using a spreadsheet for calculating household footprint,[10] at Melliodora we use 3.1 hectares per person. Besides confirming that it is possible to live in an affluent country and not feel deprived while consuming little more than one-third the Australian average, the interactive nature of the spreadsheet allowed us to identify the sensitivity of the index to various aspects of our lifestyle and consumption patterns.

EMERGY accounting

The EMERGY methodology referred to above is an example of a powerful accounting system that has been continuously developed by Howard Odum and colleagues around the world since its beginnings in the late 1960s. It is based on universal energy laws and uses an energy symbol language to describe natural systems. Within the field of energy accounting methodologies, it is the most wholistic but complex to understand. Unfortunately, few in the scientific community, let alone the public policy arena, are familiar with it. The measures generated by EMERGY studies of natural and human systems around the world have provided a major quantitative check in my continuing development and application of permaculture principles, but I have never developed the skills and resources to do EMERGY evaluations of our own small-scale systems.

The chapter entitled Ethical Principles compared the two systems of accounting, EMERGY and Ecological Footprint, for Costa Rica. (Ecological Footprint analysis suggested the country could sustainably support 80% of its 1987 population at current levels of consumption while EMERGY analysis suggested only 53%). This example illustrates that EMERGY accounting generally provides a deeper level of challenge to current conceptions of sustainability and the importance of the debate about how to best measure the environment.

Although EMERGY accounting may provide a more severe message about environmental cost accounting than the Ecological Footprint, it also has a positive aspect in that it reorganises our understanding of benefits, redefining what we mean by wealth and work. By redesigning both sides of the ledger, it allows us to better distinguish the differences between productive use of natural resources and wasteful ones, and to identify within any system where the gains and losses are accumulating. My interpretation of EMERGY accounting is that it reinforces the proactive developmental perspective of permaculture rather than the impact minimisation that lies behind Ecological Footprint accounting and the environmental mainstream.

EMERGY yield ratio and replacement time

One application of EMERGY accounting is the calculation of EMERGY yield ratio. This compares the EMERGY (inherent value) of a resource with the feedback of EMERGY from the economy required to produce that resource. A value greater than 1 indicates a net gain to the economy in EMERGY. A value over 4 is a high-value source that is comparable to many current resources, both non-renewable and renewable, that are fuelling the economy.

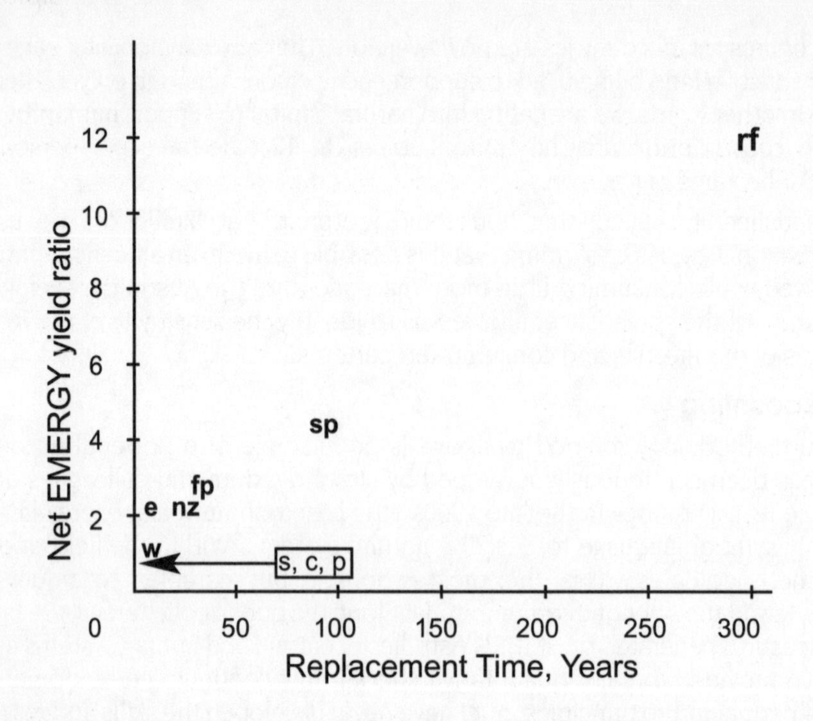

	Biomass fuel	Replacement time (yrs)	EMERGY yield ratio
rf	Rainforest timber, Brazil	300	12.00
sp	Spruce	90	4.10
fp	Slash pine, Florida	25	2.40
e	Eucalypts, Brazil	7	2.20
nz	Radiata pine, New Zealand	24	2.10
w	Willow, Sweden	6	1.34
s	Sugar Cane Alcohol	1	1.10
c	Corn	1	1.10
p	Palm Oil	1	1.06

Figure 11: EMERGY yield ratios of biomass fuels as a function of frequency of harvest (after Odum 1996)

Figure 11 compares the results from EMERGY studies of a range of existing biomass energy sources in different parts of the world with the replacement time (that is, the time taken for the crop to grow). Annual crops have yield ratios little better than 1, while wood plantations yield 1.5 to 4, and 300-year-old rainforest yields 12.

Perhaps it is not surprising that the more the work is left to nature, the greater the net yield but the longer the time required. It is unfortunate that people down the ages, from Neolithic grain farmers to forestry economists, like the proverbial dog chasing its own tail, have failed to recognise this basic pattern as they look for higher yield from shorter rotations of crops. Thus sometimes the most apparently productive and high-yielding sources of energy involve a lot of activity for little return, while long-term investments, especially in naturally grown forests, provide the greatest value for future generations. While catching and storing energy in these forests, it may be better to rely on (modest) use of fossil fuels rather than allocating fertile farmland to grow intensively managed annual and short-rotation biomass fuels.[11]

Simple embodied energy evaluations generally show more favourable net energy evaluations for renewable biomass as sources of fuel energy. But that dispute over methods is less important than the patterns revealed by these studies, which all show greater value in slow-growing, naturally produced biomass energy sources.

Environmental impact evaluation

An early study (1977) of the environmental impacts of a power station on local everglade swamps in Florida illustrates the challenges to conventional environmental protection.[12] A power company was using the swamp as a source of cooling water. The environmentalists wanted it to build the concrete and steel cooling towers that typify modern power stations. Odum, a world-renowned expert on the everglade ecology, was asked to consider the impact of the heated water on the lagoon. He estimated that the total biological productivity of 150 acres of swamp had been reduced by 50% (quantified as solar energy not captured). However, he also estimated that the embodied energy cost of the cooling tower alternative was 100 times greater. This study delayed the construction of the cooling tower by a decade until it was overridden by national policy. The finding in this case may have been one of the reasons that environmentalists have not championed this approach.

Odum also noted that, if the flow of warm water from the power station were not so erratic, the swamp ecology would adapt to this extra energy input and actually increase in biological productivity. Criticism could be made of this approach from both a biodiversity perspective (what if a species went extinct) and a political one (who gains and who loses). More recent development of EMERGY evaluation has began to quantify these factors and they do show that species extinctions, especially of what is known as keystone species, are very costly in EMERGY terms. On the political question, studies of development projects in poor countries typically show a gain to the donor country, a loss to the recipient country, and a huge loss to the local area and people who are the supposed beneficiaries. These may be one reason why developers, like environmentalists, have not championed EMERGY evaluations.

Voluntary frugality

Although an EMERGY yield ratio tells us nothing directly about the sustainability of a particular process, it certainly provides a useful quantification of whether we have obtained a yield. In a slowly evolving low-energy society, time eventually shows what works and common sense provides an intuitive grasp of probabilities. While EMERGY evaluations are very complex and contentious in their implications, I find the results tend to confirm my own evolving permaculture common sense which comes from decades of observing nature and operating from a basis of voluntary frugality. Unlike the real poverty that comes from a sense of no choices and constant self-comparison with the consumption standards of society, voluntary frugality allows us to learn to optimise allocation of resources. When we become permaculture jacks-of-all-trades, building houses, gardening and self-reliant as well as being self-employed and involved in community affairs, we have more chance of developing an intuitive sense of the patterns of resource allocation that work. When we work in a personal and business economy that grows organically from its own productivity rather than reliance on substantial debt, we can see these patterns become clarified. This constant learning mode works as an accelerated research and development process that accumulates integrated knowledge.

After three decades of personal commitment to voluntary frugality I am willing to concede that there are disadvantages in what some describe as a poverty mentality. My personal experience is that the biggest contradiction in this process is that the accumulated knowledge which one develops must always be given a modest value (monetary or otherwise) in small-scale self-reliant systems. Such systems are simply not wealthy enough (in money or EMERGY) to support a top-heavy design and management structure.

On the other hand, those same skills can overview complex systems and grasp the driving forces and pitfalls that characterise the current transition from growth to descent. These capabilities have immense value in steering the huge capital and other resources in large-scale economic systems towards better outcomes.

Having seen a number of self-educated permaculture colleagues move from small-scale self-reliance to relatively highly paid work informing and directing larger organisational systems, I am convinced that permaculture thinking and practice are one of the most efficient ways to develop the skills to direct large-scale systems to more appropriate and humane ends.

The Problems of Success

There is little doubt that obtaining a yield represents a structural dilemma for a transition to energy descent. If maximum power is an energy law governing all self-organising systems, then systems that yield the most (perhaps best measured as EMERGY) will continue to prevail. But the pattern of successful systems must shift from dependence on high and opportunistic yields from non-renewable resources to using low and sustained yields from renewable resources. The challenge to reconfigure success around declining energy is fundamental. It is tricky to determine how we (individually and collectively) can obtain a yield and reproduce our successes without creating more demand than is sustainable.

In systems modelling of sustainable transition, this phenomenon is described as "rebound". Design and behaviour changes that result in energy and resource savings tend to be immediately spent or invested in other sectors of the economy which further stimulates demand for energy and resources. This phenomenon was first pointed out to me by a colleague who had spent decades helping friends reduce their energy consumption with the purchase, modification and construction of energy-efficient technology. He had observed that his friends now drove their energy-efficient cars further and his sister's savings on electricity and gas for her new passive solar house gave her the money for an airfare to Europe.

The transition from permaculture self-reliance and small business to large organisational roles mentioned previously presents similar structural dilemmas and pitfalls. Creative, ethically driven people in large organisations tend to be co-opted and corrupted by the large-system forces within which they must work. The roles allocated to even the most powerful allow far less wholistic and integrated action than is possible at the personal level. Quality of change is traded-in for quantity of influence. When thinking and activity are highly focused around very specific, limited and conventionally measured outcomes, the degradation in quality is greatest.

Solutions to the Problems of Success

It is easy to see why systems that are under pressure to shift from high to low energy sources tend to do so by crashing and rebuilding. In order to bypass this path to the low-energy future, we must integrate into our thinking the apparent paradox between the abundance and the limits of nature referred to in the third ethical principle. We must transform ourselves so that we are happy, healthy and comfortable without the need to consume the planet and the future. Howard and Elizabeth Odum speak of the need to conceptualise the transition to an economy and society based on declining energy as "a prosperous road down".[13]

To some extent these problems can be dealt with by the progressively more wholistic approach embodied in the wholistic learning loop (see **Principle** 1: *Observe and Interact*). Where success in meeting one environmental objective leads to stimulation of other networked and progressively more integrated low-energy alternatives, the rebound from those successes can reconfigure the economy and society while decreasing the total demand for resources. For example, when the money saved from the more energy-efficient car or house is put into an ethical investment or spent buying local organic produce or employing a neighbour, then the adverse rebound effects will be reduced.

Further, the large-scale systems of intensive energy demand need only be starved of a small proportion of sales before they are forced to reduce production. When purchasing technology produced by large-scale systems such as cars and computers, second-hand goods are often better value and dilute the feedback of demand for new goods (also see **Principle** 4: *Apply Self-regulation and Accept Feedback* and **Principle** 8: *Integrate Rather than Segregate*).

Conclusion

The early vision of permaculture reclaiming the delusional ornamental landscapes of suburbia to create an abundant support base for its inhabitants has not eventuated to any great extent. However, that vision can also be taken as a meta-model for a larger and deeper process of change: from dependent and demanding consumers to interdependent and responsible producers. A global consensus about the reality of energy transition and descent necessary for constructive top-down change could emerge remarkably quickly in an electronically networked world. Permaculture is for those who already understand or sense the reality of transition and descent and want to give practical and integrated expression to that reality, whether the rest of society is ready or not to do so.

By paying constant attention to our successes and failures in design to obtain a yield and judging how close those designs are to maximum power for ourselves, our communities and the earth, we can resist both subsidised delusions of efficiency and rampant disregard for what we are consuming.

1 This is a rephrasing of Lotka's Maximum Power Principle. Howard Odum has suggested the Maximum Power Principle (or at least his EMERGY-based version of it) should be recognised as another Energy Law.

2 P. Kropotkin, *Mutual Aid* Heinemann 1902.

3 Creek anabranch.

4 Typically grains or tubers, but in some cases tree crops such as chestnuts or acorns. For a perspective on tree crops as staples see J. Russell-Smith, *Tree Crops: A Permanent Agriculture* Devin-Adair 1953. For my hypothesis on the reason for the decline of tree crop staples in favour of grains and other annuals, see Article 6 "Historical Precedents for Permaculture" in *David Holmgren: Collected Writings* 1978-2000.

5 R.A. Rappaport, "The Flow of Energy in an Agricultural Society" in *Biology and Culture in Modern Perspective: Readings from Scientific American* W. H. Freeman & Co. 1972.

6 For example, a tomato plant grown in poor central Victorian bush soil with adequate water and nothing to improve the fertility may yield less than 100 grams of fruit, while high-yielding plants in perfect conditions can bear 100 times this (10 kilograms).

7 Article 11 in *David Holmgren: Collected Writings* 1978-2000.

8 Article 18 in *David Holmgren: Collected Writings* 1978-2000.

9 See J. Ralston-Saul, *Voltaire's Bastards: The Dictatorship of Reason in the West* for an excellent historical overview of the problems of rational management in modern society.

10 Downloaded from Redefining Progress website http://www.rprogress.org/

11 I put this perspective in a submission to a 1983 draft report on biomass fuels by the Victorian Solar Energy Council but received no response or acknowledgement. The report suggested that Victoria could be producing 10% of its liquid fuel needs by 2000 from root crops grown in northern Victorian irrigation districts. In my submission I referred to a New Zealand study showing low EMERGY yield ratios for fuel alcohol crops and suggested a set of environmental impacts of the proposed production systems that appeared to have been ignored.

12 See H.T. Odum, *Environmental Accounting: EMERGY and Environmental Decision Making* Wiley 1996, p. 160.

13 The most recent book by Howard and Elizabeth Odum, *A Prosperous Way Down: Principles and Policies* Wiley 2001, is a readable and timely explanation for the lay reader of the EMERGY concepts and implications of energy transition for the economy, society and culture. It updates their much earlier, easily accessible text *Energy Basis for Man and Nature* McGraw-Hill 1979. Although I have never had any correspondence with the Odums, and the manuscript of this book was largely complete before the publication of *A Prosperous Way Down*, the common understandings informing the two are clear. The strategic difference in our responses to the reality of transition is the Odums' emphasis on top-down cultural and public policy change directed at a mainstream audience. Permaculture has historically focused on pushing the boundaries of innovative change at the cultural fringe and putting in place real but modest models of living from nature's abundance.

Apply Self-regulation and Accept Feedback

The sins of the fathers are visited on the children unto the seventh generation

This principle deals with self-regulatory aspects of permaculture design that limit or discourage inappropriate growth or behaviour. With better understanding of how positive and negative feedbacks work in nature, we can design systems that are more self-regulating, thus reducing the work involved in repeated and harsh corrective management.

Feedback is a systems concept that came into common use through electronic engineering.[1] **Principle 3**: *Obtain a Yield* described the feedback of energy from storages to help get more energy, an example of positive feedback. This can be thought of as an accelerator to push the system towards freely available energy. Similarly, negative feedback is the brake that prevents the system falling into holes of scarcity and instability from overuse or misuse of energy. Organisms and individuals adapt to the negative feedback from large-scale systems of nature and community by developing self-regulation to pre-empt and avoid the harsher consequence of external negative feedback.

Self-maintaining and regulating systems might be said to be the Holy Grail of permaculture: an ideal that we strive for but might never fully achieve.

Traditional societies recognised that external negative feedback effects are often slow to emerge. People needed explanations and warnings, such as "the sins of the fathers are visited on the children unto the seventh generation" and "laws of karma which operate in a world of reincarnated souls".

In modern society, we take for granted an enormous degree of dependence on large-scale, often remote, systems for provision of our needs, while expecting a huge degree of freedom in what we do without external control. In a sense, our whole society is like a teenager who wants to have it all, have it now, without consequences.

Much of the ecologically dysfunctional aspects of our systems result from this denial of the need for self-regulation and feedback systems that control inappropriate behaviour by simply delivering the consequences of that behaviour back to us. John Lennon's song "Instant Karma" suggests that we will reap what we sow much faster than we think. The speed of change and increasing connectivity of globalisation may be the realisation of this vision.

The Gaia hypothesis of the earth as a self-regulating system, analogous to a living organism, makes the whole earth a suitable image to represent this principle. Scientific evidence of the Earth's remarkable homeostasis over hundreds of millions of years highlights the earth as the archetypical self-regulating whole system, which stimulated the evolution, and nurtures the continuity, of its constituent lifeforms and subsystems.[2]

Nurture and Control in Nature

Systems theory, systems ecology and earth system science in the late 20th century provided evidence that higher-order "intelligence", in some form, is a universal characteristic of self-organising systems, even if those forms are more loose, contingent and free-flowing than those in an organism constrained by genetic inheritance.

Ecosystems provide for the survival and health of their member species by maintaining an environment that is, overall, beneficial and nurturing. Specific positive feedback from species high in the food chain encourages the life of species lower in the chain. For example, birds that eat berry fruits spread the seeds, which pass through their digestive systems unaffected (in a nice package of fertiliser). Grazing by animals helps maintain many grassland plants against ecological succession to forest.

Thinking of a nursing mother feeding her baby can bring home the image of nature as a nurturing mother. The baby sucking at the breast can be seen as a classic example of behaviour to Obtain a Yield: the sucking stimulates milk flow, while the mother provides both a nurturing and protective environment, as well as the essential sustenance which is a direct reward and positive feedback to the baby. In essence, this is the relationship of all organisms to earth and, to a lesser degree, to the ecosystems on which they depend. While the baby's dependence on the mother is total, over time a degree of self-reliance and eventually self-regulation emerges as the child develops.

As well as the positive nurturing feedbacks, negative feedback mechanisms act to constrain or control the parts of a system. In ecosystems, predators, pests and diseases meet their own needs of survival by controlling and regulating particular species or populations, but they also provide a service to the ecosystem as a whole by contributing to a healthy and functional balance. These negative feedback control mechanisms regulate, rather than destroy, the species predated or parasitised. They are everywhere we look in the natural world. In the human world, parents exercise negative feedback controls over the behaviour of child (hopefully in the child's long-term interest).

There is an inherent design tension between autonomy and higher-order system control. Each cell, organism and population is as self-reliant as possible. This self-reliance at the smallest practical scale provides benefits to the large-scale system. For instance:

- internal selection for maximum power in the cellular parts contributes to the general fitness of the larger system
- the failure of some of the cellular parts due to external stress does not adversely affect the resilience of the larger system. (For further discussion of cellular design see **Principle 7**: Design from Patterns to Details.)

On the other hand, cells within an organism that grow and reproduce without control can be fatal for the organism. We call this cancer. Similarly, at all levels of nature, including ecosystems and the living planet, the larger system controls its constituent parts for the good of that system.

We can think of a hierarchy of large-scale system controls that flow back down the energy hierarchy to constrain the exuberant and prolific life of smaller-scale systems. The large-scale system controls are relatively harsh, and at times destructive. For example, natural disasters, predators and parasites control plant and animal populations.

Most dramatically, the Gaia hypothesis has given us a scientific renewal of the idea of the earth as a nurturing mother who maintains favourable conditions for the diversity and renewal of life, but is ruthlessly harsh to individual species, and even whole ecosystems, in maintaining that balance. Many biologists are reluctant to accept the idea of any higher-order control beyond the organism level. I see this reluctance as reflecting the Cartesian mechanistic worldview, which most physicists have long since abandoned. Perhaps this reluctance is partly driven by a fear of the re-emergence of a spiritual wholism to explain nature.[3]

It is ironical that the life sciences are the last to discard this view and accept that higher-order system control operates at all levels in nature. In general, I have found systems thinkers in the fields of organisational and business management, computer science and engineering regularly use biological metaphors and models to explain systems concepts and are incredulous at the idea that biologists might not accept the reality of higher-order system control.

Self-regulation

One of the most important evolutionary responses of organisms to higher-order control is to develop internal self-regulation mechanisms, which control excessive growth or inappropriate behaviour before the harsher higher-order controls come into effect. For example, in a dam or pond the growth of fish and crustaceans is suppressed by their own wastes. This decreases the likelihood of disease or starvation killing all the fish. Kangaroos and other marsupial herbivores can respond to seasonal conditions by slowing the development of embryos, and so (at least partially) regulate their numbers before poor feed favours diseases and predator control.

Traditional societies had social and ethical constraints on population growth and resource use, which allowed communities and culture to persist over long periods without destroying the environment. This is a good example of self-regulation in human systems. I believe *these self-controlling aspects of human culture, rather than the expansion of technology for resource exploitation and growth, represent the highest evolutionary development achieved by* Homo sapiens*. The ways in which we apply these abilities to controlling the excesses of growth and expansion over the next century will be the greatest test of our evolutionary sophistication*.

Tripartite Altruism

Another way of thinking about this issue is to look at how any organism or population divides and allocates its available energy. Howard Odum has described a "tripartite altruism" in nature: approximately one-third of captured energy is required for metabolic self-maintenance (of an individual or population); one-third is fed back to maintain lower-order system providers; and one-third is contributed upward to higher-order system controllers. Figure 12 illustrates this using Odum's energy circuit language.

The behaviour of rabbits provides a simple example. Rabbits eat grass to live, grow and reproduce. Their manure fertilises the grass that feeds them, and the "sacrifice" of weak rabbits to predators helps keep the population fit and in balance. If these feedback mechanisms fail, the rabbit population will suffer. For example, if the rabbits deposit most of their manure under adjacent brambles in the process of avoiding predators, they may be progressively eliminated, both by overpopulation leading to overgrazing of a declining pasture and by the growth of fertilised brambles shading out their pasture food.

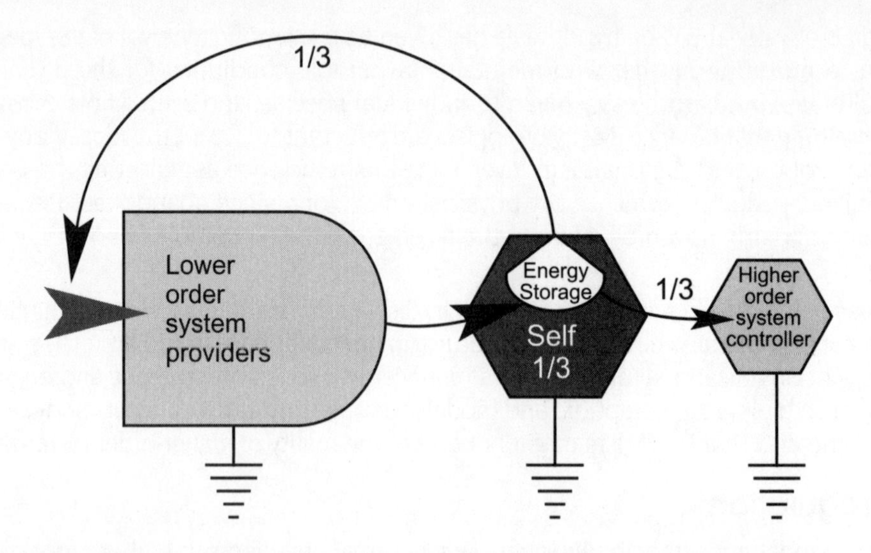

Figure 12: Tripartite altruism for distribution of available energy (after Odum 1983)

Nurturing, Negative Feedback and Self-Regulation in Managed Systems

As designers and developers of gardens and other productive systems, we can see our role as analogous to the nursing mother in a number of ways.

- In the early establishment phase, gardens are totally dependent on our care and attention. For example, newly planted trees must be protected from browsing animals and may require weed control, water and fertiliser to get started.
- The designer/manager who appears to be all-powerful, like the nursing mother, acts without being able to control or even understand all the factors that might impact on the garden.
- If the design is effective, the garden becomes progressively more self-reliant and less dependent on our care, although like a young child it may require intervention at times to save it from external and internal danger.
- If our design and care are truly inspired, the garden, as well as becoming more robust, will develop a degree of self-regulation and balance analogous to children growing through adolescence to be responsible adults.

As in the development of children, a myriad of circumstances, as well as the inherent potential of the site, will create a garden that is unique and differs from our plans and specific hopes.

In any managed natural system, especially on a large scale, much of the work of management is not making sure that plants and animals grow and reproduce but preventing excessive and inappropriate growth and reproduction. For example, using goats to control rangeland weeds or human work to thin dense stands of forest trees provides negative feedback to keep the system balanced. Even in the garden, much of the work we do is not planting, but removing or cutting back unwanted growth.

As indicated in **Principle 3**: *Obtain a Yield*, by making use of wild, hardy and self-reproducing species, we can avoid the need for high levels of inputs to maintain growth and reproduction, and so focus our intelligence and efforts on harvesting and controlling excess growth (preferably as another yield).

Failures in our efforts to obtain a yield are useful if we pay attention and learn in the ways described in **Principle 1**: *Observe and Interact*. If we believe our system or strategy is sound in principle, then we may need to invest more effort or resources to ensure it does eventually yield results. However, more work (such as weeding) or more resources (such as water) are not always the best response. Failure can alert us to a flaw in what we are doing. For example, if we harvest a little from our garden in the growing season and then go to the shop for the rest of the year, we might never notice that our produce doesn't keep in storage. Seeing all your pumpkins or garlic rot can be very discouraging, but it can help you to recognise the link to imbalances in soil minerals or other subtle factors and provide the spur to get it right.

The process of attempting to obtain a yield can also create long-term problems. If we have ways of responding to these problems, we can correct or adjust the system to make it more efficient or stable. For instance, if we grow a woodlot to supply our firewood, wasteful use will have an obvious negative impact. If, on the other hand, we buy firewood, we might never consider how much forest or woodlot is required to provide for our heating. If we get our heating at the flick of a switch and don't pay the real cost of power, then the impacts of our choices (such as global warming) will seem remote and abstract.

If we live a typical modern urban lifestyle, where electricity, water, salary, and so on arrive in a remote way, it is often hard to know when we are making unacceptable use of or exploiting common resources, whether water supply, peace and quiet, or the honesty of our neighbours. The market and the law are the main mechanisms for providing negative feedback, but these are often crude or ineffective for many social and environmental issues. In smaller communities, particularly rural ones, more intimate social mechanisms exist for giving us the signals for an appropriate level or time for the "harvesting" of common resources, be they physical or conceptual. For example construction of dams on small catchment gullies can lead to conflicts between neighbouring farmers but casual social interaction in the pub, volunteer fire brigade or sporting club at least allows the neighbours to understand each other's view. The more we come to depend on our own and local resources, the more likely we are to recognise problems, and institute corrective behaviours to deal with them.

The tripartite distribution of energy can be used to understand how relationships will naturally develop as a system goes through its own co-evolution, as well as to guide conscious design of systems. Perhaps most importantly, it can be used to guide our own allocation of personal, family or organisational resources.

Thus *the first priority is to survive (obtain a yield from captured energy), while the second is to pay for what we get in some way that helps maintain the future flow of energy. The third, is to contribute in some other way and direction, to the wider system, rather than seeing our own survival as an end in itself.* For example, food gardeners provide for themselves and their families through harvesting, while they "pay" for the harvested food with additional work to plant, weed, water, fertilise and otherwise maintain the plants to ensure future harvests. Some of the surplus food and seed might be given away to promote goodwill, to support those in need, and to maintain a seed-saving network.

In designing gardens and other food-producing systems, this principle suggests that we should, to the greatest extent possible, choose and organise elements, both plants and animals, that provide for their own needs, rather than being dependent on constant inputs. Deep-rooted plants that can draw on deep soil moisture, rather than shallow-rooted ones, are an example. If these plants help break open the subsoil layers, they contribute to maintaining or improving soil moisture by allowing better infiltration of rainfall. If they also provide food or some other yield to domestic animals, or us then we have both self-regulation and methods of control and balance through harvesting. For instance, dock grown in wet clay soils and eaten by goats or daikon radish for human consumption both help to break up subsoils and could be said to be feeding back energy to lower-order system providers (the soil). In providing a yield, they are contributing to the maintenance of higher-order system controllers (animals or people), which in turn maintain an environment favourable for the plant.

In modern economies people "earn a living" in diverse ways — salaried work, welfare, business, speculation. They use this money to pay for goods and services, since they have nothing directly useful to give the producers or suppliers; and in the process, they pay some of that money in taxation to ensure the wider society that they depend on continues to function. All these examples broadly reflect Odum's tripartite altruism. The modern economy, despite its complexity, fails to provide critically important feedback signals to ensure appropriate behaviour to decide important questions. For instance:

- how much is enough to consume? (When should the rabbit stop eating? How big a house do we really need?)
- how much work is enough? (How many vegetables is it sensible for one gardener to produce? How many hours do we really need to work?)
- how should we pay and how do we measure it? (What should the real price be? Do markets fail to provide correct information about values?)
- what should we contribute to the greater good? (Is paying tax, working for community groups etc., useful or sufficient? If not, what else should we do?)

With the progressive breakdown of many traditional and institutional functions, each person needs to consider how to answer these questions, since the current social and economic structures are clearly failing to provide appropriate and credible feedback and guidance. One example of the application of these concepts is to allocate one-third of our time to providing for our material needs, one-third to self-development and reflection, and one-third to wider societal benefit.

Energy Hierarchy

In nature and in low-energy pre-industrial societies, tripartite altruism is set into hierarchical structures illustrated in Figure 7. Those at the start of the energy chain (producers) are numerous but have little individual power, while those at the end of the chain (consumers) are less numerous but more powerful.

The trophic (food) pyramid in Figure 13 is another way to illustrate these relationships, where each level of the pyramid is founded and dependent on the level below. All levels are functional and complementary; none is more important than another, but individual organisms and people (as opposed to the whole group) at the lower levels have less influence and power than individuals at the higher levels.

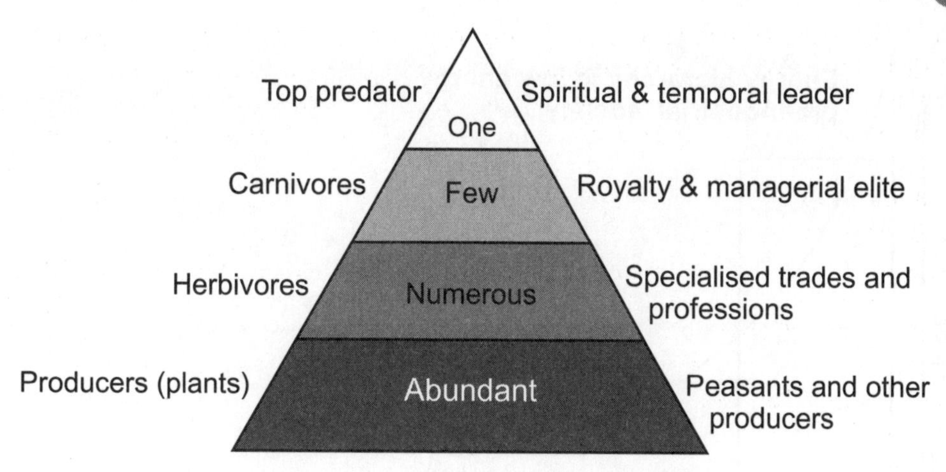

Figure 13: Trophic pyramid model of ecosystems and traditional societies

The saying "too many chiefs and not enough Indians" refers to an imbalance in this traditional organisational structure. Industrial energy and affluence have made this imbalance possible on a grand scale because energy slaves[4] (plus lots of almost-slaves in poor countries) support everyone. Transition from growth to descent requires that we rebuild the trophic pyramid to suit local energy realities and social justice.

Figure 14 shows the energy hierarchy as a hyperbolic curve with the abundant producers on the left and the less common consumers on the right. In contrast, industrialisation has created a great bulge of middle-class consumers with manual labourers becoming almost as few as the elite. This "normal distribution curve" has come to dominate our thinking about what we believe is a natural social structure. The evidence of nature and history reveal its novelty. We can think of the middle-class as a bulge in the low-energy curve created by the pulse of fossil fuel energy moving up the hierarchy.

While statistics of the modern nation state may reflect law, culture and history, economic and ecological reality is now global. When we consider all people today being part of a global society and economy, the middle-class bulge is very modest: rural and industrial workers in poor countries far outnumber the billion or so middle-class consumers mostly in rich countries.

Failure of Self-Regulation by Elites

The acceleration in difference between the rich and the poor on a global scale suggests we have a more extreme distribution of wealth and power than that observable in traditional low-energy societies.[5] This excessive accumulation of wealth and power by the rich coincides with a situation where the elites[6] are failing to perform their traditional role of providing control and guidance for the long-term good of society.

The historically recurring problem with elites is that they start to believe that the whole social ecosystem is simply for their benefit. This corrupts the process of acknowledging higher power (god or nature) and also fails to provide functional feedback control to keep the system healthy and adapted. This breakdown of self-regulation and tripartite altruism at the highest levels tends to cascade back through the social hierarchy until a corruption at all levels leads to some sort of breakdown and reform.

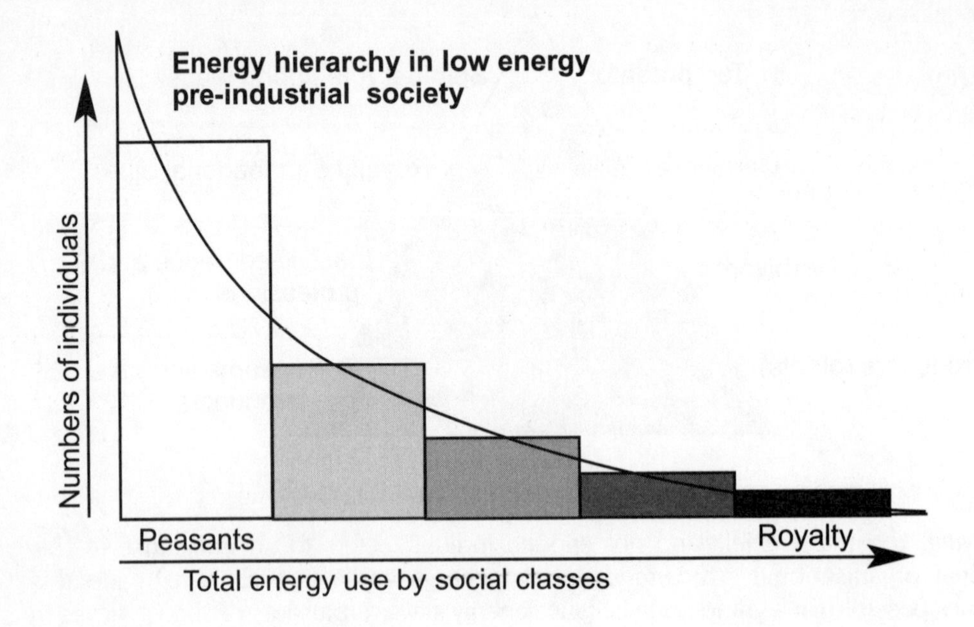

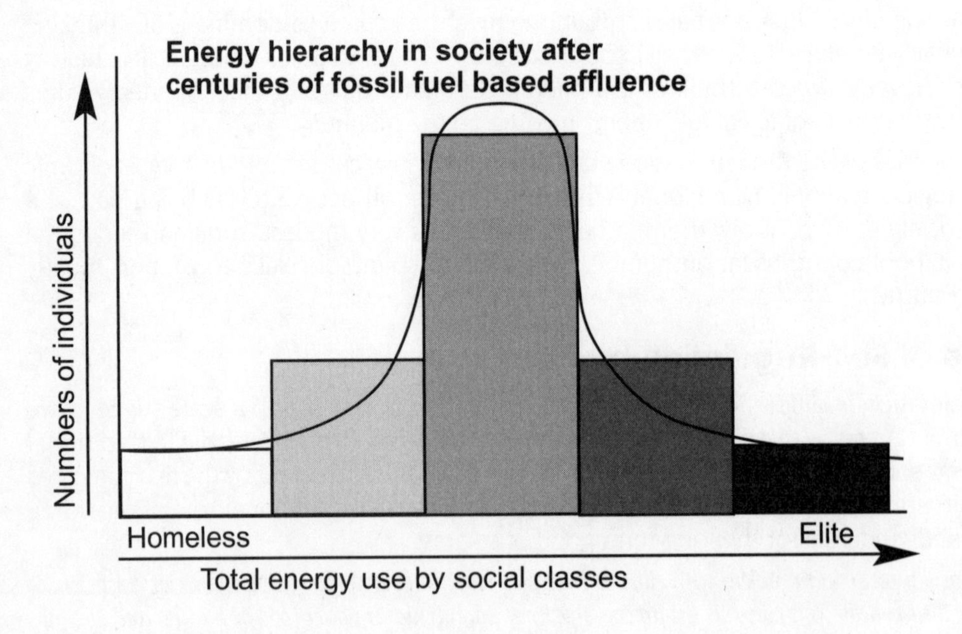

Figure 14: Energy hierarchy in pre-industrial and modern societies

Since the European Enlightenment, social democratic, Marxist and other ideologies have sought to overcome this problem with elites but the historical record of achieving effective guidance and wisdom without corruption have not been great.

Perhaps the energetic foundation of these failures is that the social hierarchy has developed so many levels in response to abundant supplies of high-quality energy. All of these levels are occupied by simple fallible humans. We simply do not have the wisdom to occupy the higher levels. The saying "power corrupts, absolute power corrupts absolutely" is relevant here.

The problem of the upper hierarchical levels is not that any one person at those levels exercises absolute power because they are constrained by a social network of others at that level. All the people at that level function as a self-reinforcing community. These networks are like (global) villages of parochial citizens unable to step outside their community's social norms; and yet they are stratospherically removed from the effects of their actions on people at much lower hierarchical levels of humanity.

We need to recognise that a high-energy society will inevitably develop a large number of hierarchical levels and then return to a low-energy base. This recognition may, in the long term, allow us to redevelop a flatter structure where elite functions and roles are not so inherently corrupting.

The trophic pyramid or energy hierarchy challenges some of our cherished beliefs about social equity but, we have a much better chance of building a humane and just transition to declining energy if we recognise the broader ecological "rules of the game". Without this recognition in rich countries we are likely to fall into more inhumane ways of restoring the trophic pyramid. Today's economic and social pressures are already squashing the middle-class bulge and increasing the numbers (for the time being at least) of both the rich and the poor. The prevailing focus on economics, and more specifically economic rationalism, can be seen at one level as a partly conscious recognition that we can no longer live by consumption and need to be more "productive". Unfortunately, most economic measures of "productivity" are hopelessly inadequate or downright wrong and they are irrational in giving reward to ecologically destructive processes. Much deeper and broader notions of "productivity" are needed.

Rather than a big economic stick to force the majority off the "gravy train" of affluence, we need a change in values and rewards to reinvest in creating livelihoods and lifestyles of humble husbanding of nature. In other words, we need a deep cultural revolution that recognises society's increasing dependence on nature. The mass of people will need to return to working with nature to provide the food and other resources necessary to support all the other levels of a future low-energy sustainable society.

Permaculture could be seen as ecological rationalism: it recognises the design rules and measures of value in energy descent, and it provides ethical and positive pathways for embracing those design rules.

Top-down and Bottom-up Strategies for Social Change

Strategies for social change are often broadly grouped as top-down or bottom-up. The top-down strategies range from lobbying decision-makers and influencing media through to revolution. What constitutes bottom-up and top-down depends, in part, on your perspective. For many people, the broad environment movement is a bottom-up grassroots movement, but the majority of the movement's activities are focused on changing the behaviour of government, bureaucracy, corporations and the media. More fundamental bottom-up strategies start with the self and develop by example and replication, moving towards mass change.

Permaculture, although complementary to many top-down approaches within the broad environment movement, is not primarily about lobbying government to change policies. Instead, it is concerned with facilitating individuals, households and local communities in increasing self-reliance and self-regulation. I see this process as the most potent way of reducing total environmental impact and transforming society by slowing and reorganising the production-consumption cycle. This approach is based on the recognition that a certain proportion of society is ready, willing and (most importantly) able to substantially change their own behaviour if they think it is possible and significant. This socially and environmentally motivated minority represents a leverage point for large-scale change.

Top-down Thinking, Bottom-up Action

In **Principle 1**: *Observe and Interact*, I proposed that bottom-up collaborative behaviour is a fundamental element of permaculture action, whether in the garden or the community, and that this behaviour is informed by wholistic systemic understanding which can be characterised as top-down. Thus, "top-down thinking, bottom-up action" is a restatement in systems terms of the environmental slogan "think globally, act locally". A historical rationale for this behaviour mode in permaculture can be made. This, in turn, depends on a top-down systemic view of pre-industrial, industrial and post-industrial societies which is illustrated in Figure 15 using the trophic pyramid previously described as a base model.

In traditional cultures before scientific reductionism, the behaviour of educated elites (royalty, priests, scholars) could be characterised as predominantly top-down, in the sense that they used wholistic understandings of human systems to inform their exercise of power. Others lower down the hierarchy, such as patriarchal heads of households, may have used wholistic understanding somewhat to inform their actions in controlling their limited territory of soil and family. However, in general, ordinary people had a much more fragmented understanding of nature and society; their power to influence things was small and it worked collectively through the aggregation of participatory bottom-up forces.

This pattern of powerful top-down thinking and action by elites, balanced by fragmented understanding and bottom-up action by the masses, is the basis of a common understanding of how traditional societies and nature work.

With the increasing power of scientific reductionism following the Enlightenment, the elites became more and more influenced by its methods, which analyse the smallest components of systems to predict and control larger-scale behaviour. This was especially true of the new ways of understanding markets[7] as the aggregated outcomes of many small, individually focused (selfish) actions, which collectively have understandable dynamics. These reductionist modes of understanding can be thought of as bottom-up, in that they build a world-view based on aggregating the properties of fundamental particles — atoms, cells, individuals — in order to describe and predict the behaviour of large systems — substances, organisms, societies.

In the 19th and early 20th centuries this way of understanding had become so powerful that the behaviour of elites could be characterised as "bottom-up thinking and top-down

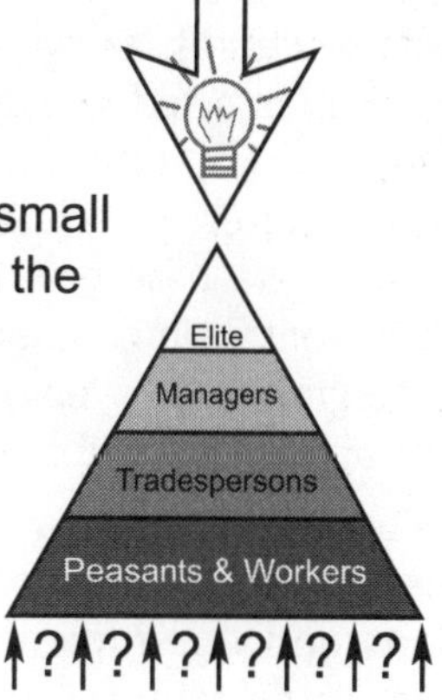

STEADY STATE

Traditional society with stable and small number of highly stratified levels in the energy hierarchy

Integrated Understanding, Top-down Action

Fragmented Understanding, Bottom-up Action

EXPANSION

Industrial society with expanding scale and number of loosely stratified levels in the energy hierarchy

Fragmented Understanding, Top-down Action

CONTRACTION

Post industrial society with contracting scale and number of consolidating levels in the energy hierarchy

Integrated Understanding, Bottom-up Action

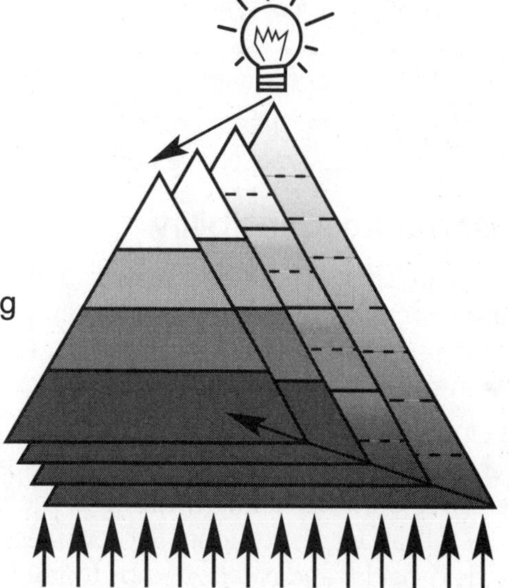

Figure 15: Trophic pyramid model of understanding and action in traditional, industrial and post-industrial societies

action". The elites still pulled the levers of power but they did so informed by the specialised reductionist knowledge of scientists and technologists. Progressively, the forces of democratisation acted as a moderating control using the same methods of thinking and action. This mode of understanding and behaviour has since become the norm for the middle-class masses. In school, at work, and in the home and garden, most people's understandings are based on a great amount of factual knowledge; but this knowledge remains fragmented into different intellectual disciplines and aspects of life with little systemic integration or overview. At the same time their actions tend to be managerial (top-down) wherever possible and they avoid collaborative and participatory processes.

But the elites themselves have learnt that reductionist logic and understandings are of limited use in comprehending and predicting complex systems the size of modern corporations and governments, let alone cities and national and global economies. Increasing power is returning to the generalists who can think wholistically rather than the technocratic specialists. Having gained an insight into the opportunities to use the power of top-down systemic thinking, it is natural to seek to apply those insights through managerial top-down action. The results in the corporate, public and community sectors of this top-down thinking and top-down action are limited and often counter-productive.

In the post-modern world of global energy peak, the most effective and powerful behaviour could be best characterised as "top-down thinking and bottom-up action". I believe this applies whether we are managing a corporation or a garden, and whether we are participating as a bioregional community in a globalised world or as an individual within a local community.

Although I am articulating "top-down thinking and bottom-up action" as a restatement of the environmentalist slogan "thinking globally acting locally", there is plenty of evidence that global power elites have learnt this new mode of understanding and action. A shift from evident, formal and managerial exercise of power to invisible, informal and collaborative modes allows elites to bypass the democratic and bureaucratic controls on their power demanded by the public. More fundamentally, when the powerful overcome their own egos to see they are just small players in these vast human systems of global energy peak, then they will recognise that collaborative and subtle exercise of power is more effective than threats and brute force. The ways in which this revolution in understanding and action must have permeated the highest levels are not yet clearly visible.

Personal Responsibility

The slogan "change the world by changing yourself" is recognised as a spiritual or inward-focused approach to working for a better world. There is compelling evidence though, that this idea is a principle for an externally focused concept such as permaculture with its roots in scientific objectivity.

The emphasis on personal responsibility in permaculture arises from the situation where much of the need to design for energy descent is informed by ethical principles that have leverage and impact primarily through the behaviour and actions of individuals. Despite all the talk about institutional and business ethics, it is only individual humans who can directly consider and be affected by moral concerns.

In taking personal responsibility for our needs and accepting the consequences of our own actions, we aim to change from dependent consumers of unsustainable products and services to responsible producers of appropriate wealth and value. Personal responsibility implies full awareness of the structure of our individual dependence on, and effect on, the local and the global environment, and local and global communities. Further, we need to change ourselves as our most substantial contribution to a better world.

Although many environmental activists regard this approach as politically naive and unrealistic, or simply too slow, there are sound political, historical and ecological reasons for this emphasis on personal responsibility in permaculture.

Individual wealth and power

Because of the high-energy nature of the modern world, each person, especially in rich countries, has much greater power and impact on nature and society than any previous — or probably, future — generation. Estimates of the resources that a person in the rich world commands are in the order of 100 energy slaves. Even the majority of citizens of poor countries[8] today may have the equivalent of several energy slaves. During the lifetime of an American baby-boomer born in 1950 and dying in 2025, over half the conventional oil reserves of the United States (the second-richest oil nation after Saudi Arabia) will have been consumed. The relative power between individuals and groups within generations is naturally the focus of economics and politics. Thus, we are constantly comparing ourselves to our peers, but we have many more abstract and vague concerns about future generations. Further, the culture of growth assumes that future generations will be more powerful than we are. Rapid energy descent means that our individual behaviour today may be more potent in determining the future than the behaviour of whole communities in that future.

Much of this power is expressed through our purchasing decisions in the monetary economy. In this way, it is the billion or so middle-class people around the world who are the engine of global destruction, rather than the numerically small elite, or the relatively self-reliant but increasingly destitute majority.

The rise of individualism in the modern world makes possible personal expression and action through lifestyle choice, even if few choose to do so in any more than superficial ways. This empowerment of the individual provides a unique opportunity for bottom-up change.

Institutional decay and organisational turnover

The second reason for the importance of personal responsibility is the relatively weak and decadent nature of many of our cultural institutions, including the church, parliamentary democracy, and even the legal system. Institutions fail due to cultural decay and the challenge of fundamental change that is too strong and fast for these normally durable but conservative social structures to adapt to. When they fail, individuals and small groups come to the fore to adapt and carry culture into the new age, where new institutions again emerge. The American historian William Irwin Thompson[9] has placed the current opportunity for social change by individuals and small groups of individuals in historical perspective by referring to previous periods in history.

For example, he describes Pythagoras as a radical initiate of the ancient Egyptian "mystery schools", who took sacred arcane knowledge from these decaying institutions and set up the world's first university in southern Italy, teaching mathematics and science. His followers, fleeing local political strife, settled in Greece. This was the culturally fertile ground for the germination of what we call classical Greek culture and the origin of Western civilisation. Thompson's own Lindisfarne Institute, which was influential in the philosophical foundation of the counterculture, was named after the libertarian monks of Lindisfarne who converted the whole of the British Isles to christianity well before the institutional church gained its hold.

The great power and global reach of recently developed organisations, especially the multinational corporation, tends to delude us into believing these structures are durable, almost permanent features of human society. In fact, the average life expectancy of the great corporations is less than a human lifetime. People alive today will outlive many of the institutions and organisations that currently dominate the world.

Lack of direct feedback mechanisms

The third reason for personal responsibility is that it is the fastest way to rebuild needed negative (and positive) feedback mechanisms. Globalisation causes economic and other benefits to accumulate at centres of affluence (rich countries and cities) and adverse social and environmental impacts to accumulate in the natural and human hinterlands and in poor countries. Because we are frequently not faced with the consequences of our decisions and actions, normal negative feedback and self-regulation mechanisms which act to prevent or ameliorate inappropriate behaviour in traditional societies, fail to function in modern society.

Despite the acclaimed power of modern media and information systems, people still live to a remarkable degree in the world that they can see, hear and touch, while almost all our food and other needs are provided by sources beyond the reach of our senses. This means we need more conscious and explicit awareness of our dependence and impact, followed by the reorganisation of our lives to close the circle of cause and effect so that our needs are provided closer to home.

As I have explained, the well-known environmental slogan "think globally, act locally" means much more than opposing inappropriate local development as a contribution to saving the planet. It requires awareness of the whole and of our relationship to it; it informs our reorganisation of the way we live so our impacts are closer to home. In practice, this may require acceptance of greater local environmental impact in exchange for more self-reliance and control.

For example harvesting trees from our own properties is generally a more environmentally sound solution than buying wood in the marketplace because we can control the process (and get the negative feedback of stuff-ups). In buying timber harvested from local forests, we have a better chance of understanding the complex issues and of influencing timber providers and forest managers than when we buy timber from overseas via systems that we cannot influence.

In describing feedback as one of the essential principles of the industrial design revolution, Amory Lovins used the example of the factory taking its water supply from downstream from its wastewater outlet as an incentive to reduce pollution to zero. Perhaps his most dramatic example of the power of feedback is a car with the exhaust fed back into the passenger compartment, rather than being aimed at the pedestrians.

Feedback fosters personal responsibility and personal responsibility builds in feedback.

The individual as a whole system

The fourth reason for personal responsibility is the need to develop the whole-system thinking that is central to design for energy descent. *Learning to think wholistically requires an overriding, or reversal, of much of the cultural heritage of the last few hundred years. With little experience of whole-system thinking and such cultural impediments, we need to focus our efforts on simple and accessible whole systems before we try to amend large and complex ones. The self is the most accessible and potentially comprehensible whole system.* This logic reflects the universal pattern of developmental learning where we need to crawl before we walk. But of course, the recognition of that development pattern is itself an example of systems thinking. Thus much of the reasoning behind the permaculture strategies that I am presenting will be seen as either self-evident common sense, or alternatively, as incomprehensible nonsense.

Self-audit

To an amazing degree, we are driven by processes that we are barely aware of or think very little about. In teaching Permaculture Design Courses, we have found that basic personal habits and behaviours are some of the most difficult and sensitive to deal with, and yet they are at the heart of the ecologically dysfunctional nature of modern society. For example, daily bathing or showering is now regarded by many people in affluent countries as an unquestioned, non-negotiable part of personal hygiene. The consequent escalation in demand for energy, water and access to bathrooms is a significant factor in driving activity in the wider economy.

Taking responsibility for the consequences of our personal habits and behaviours is facilitated by a self-auditing process, where we consider all our inputs and outputs, both material and otherwise, just as we consider the needs and yields of a species as part of the garden design process. Thus the self-audit is not a confession about our sins, but an ecological observation exercise designed to help us better understand the real structure of our existence. This is the top-down thinking implicit in "thinking globally, acting locally".

The self-audit process

- List or brainstorm all your needs, wants, addictions, abilities, liabilities, and responsibilities.
- Consider all the influences and connections.
- Map the material and energy flows and your personal movement patterns.
- Take responsibility without guilt or blame on others.
- Look for the easiest opportunities for reducing dependence, minimising harm and improving quality of life.
- Make small changes and review the audit regularly.

Another insight from considering these issues in Permaculture Design Courses is the degree to which fundamental material needs may be glossed over or taken for granted, while emotional, intellectual, and spiritual needs are highlighted. Many of these non-material needs are gained through, or in association with, material consumption; but they can be met in much simpler ways. This emphasises the degree to which family upbringing, peer pressure, ceaseless advertising and propaganda have linked emotional, intellectual, and even spiritual needs to consumerism. Every cultural and spiritual tradition in the world gives us the opposite message.

Addiction

Perhaps more fundamental than how we provide for our needs, is how we identify them as distinct from "wants" and the degree to which those wants have, in turn, become dysfunctional needs (addictions). Central to this process is the recognition of the role of addiction in the modern world and in ourselves. I have described permaculture (only partly tongue-in-cheek) as "aid for addicts",[10] and I use the growing awareness of the structure and processes of pharmacological addiction to understand the widespread addictions in modern life to shopping, television and other media, driving cars and other forces of unsustainability.

The term "hedonic treadmill"[11] refers to the phenomenon in affluent societies where continuous increases in material wealth are now needed to maintain societal happiness measures, but overall improvements in happiness do not seem achievable. I think we should call a spade a spade and recognise that this is addiction by another name. Suggestions that the majority of the population may be addicted to using their cars (for example) are often dismissed on the basis that there is no comparison to the emotional and physical dependence or level of harm suffered by a drug addict.

The answer to this is at two levels. First, a complete social and environmental audit of car transport can show the damage is deep and long-term, even if people suffer in ways not directly connected to their driving decisions. Second, the only way to test the emotional and physical dependence of commuters on driving is to withdraw the daily "hit". The irrational anger of people unable to get petrol during strikes is an indication of that addiction. I suspect many commuters unable to get petrol would be as dangerous as heroin addicts denied their hit; the same could be said for many television viewers and shoppers if they were denied their respective hits.

Beyond personal addictions, we can identify systemic addiction within organisations and institutions. A well-recognised example is the dependence of governments on revenue from gambling as well as alcohol and tobacco taxes.

Any discussion about transitions to organic and low input methods of broadacre cropping are pointless unless we acknowledge that current high-yield agriculture is addicted to inputs which maintain those yields, even if they are at the same time destroying the future productive base. Simply stopping the input can result in a cold turkey of collapsing yield.

Understanding the structure of these addictive relationships doesn't provide instant answers — but it can help us to see that irrationality is a greater driver in the design of our unsustainable society than scientific rationality — and allow us to design better adaptive strategies.

In dealing with our personal addictions or designing solutions for more complex large-scale ones, the lessons from dealing with drug addiction are useful:

- acknowledge our own addictive behaviours and admit that they are barriers to a better life and a more sustainable world
- recognise the emotional and other benefits or yields we get from the addiction
- avoid guilt and blame of others, including our parents
- disconnect from reinforcing relationships with addicts who are unwilling to acknowledge or deal with the particular addiction
- connect with ex-addicts who understand, and form self-help groups of addicts willing to change.

Self-Reliance as Political Action

Taking personal responsibility naturally moves us to be more self-reliant and less dependent on centralised sources of needs and responsibility. In the process, we discover that governments and corporations, while preaching self-reliance, actually need our dependence. This need at the centres of power has become so great that a slackening in the frenzy of consumption is called a "consumer strike". Environmental groups have found that focused selective boycotts of corporations such as McDonald's and Nike can have dramatic impact and force some beneficial and visible changes. *Self-reliance tends to work as a more generalised and invisible consumer boycott, undermining the market share and psychosocial dominance of the centralised and large-scale economies that support and maintain addictive and dysfunctional behaviour. At the same time, it tends to foster and stimulate new local forms of economic activity.* For example, I have argued that home food production has tended to foster, rather than compete with, small commercial organic growers serving local markets.[12]

The amorphous and informal nature of self-reliance makes it hard to identify how effective it is in undermining the power of dysfunctional and dangerous economic systems. This invisibility can slow the recognition of the power in people's hands, but it also has the effect of confounding the efforts of the media, corporations and government to track, control and subvert processes that might undermine their interests.

This resistance to subversion is one of the great strengths of apparently disorganised movements which might be described as anarchist in strategy if not conscious philosophy. The so-called anti-globalisation movement is derided in the mainstream media as directionless and without substance because the elites cannot identify a simple set of demands or any real leaders on whom to focus their denigration and subversion (and if necessary, negotiations).

Self-Reliance and Disaster Preparedness

While the value of self-reliance might be hotly disputed by many social and political activists, its importance in reducing vulnerability to precipitous change, whether from natural or human origin, should be more readily accepted. As I have said, the design of natural systems makes the elements of the system as autonomous as practical in order to ensure overall resilience in the larger system to disturbance. Failure in one element or group of elements doesn't lead to a domino-like collapse of the system. As well as providing resilience, this relative autonomy allows the elements to become starting points for the rapid evolution of new systems when the old systems become dysfunctional and unstable for the prevailing conditions. Ecosynthesis, the evolution of new ecosystems, is discussed in this context in **Principle 12**: *Creatively Use and Respond to Change*.

The organisation of modern society and the economy has made individuals, families and communities dependent on outside resources and services to an historically unprecedented degree. I remember being shocked at hearing on the news reports following the devastation of Darwin by Cyclone Tracy in 1976 that baby formula was one of the most urgently needed supplies because very few mothers breast-fed their babies. Perhaps this is one example where self-reliance may have actually increased due to the current advice in favour of breast-feeding, but in general, people are now far more dependent on centralised resources, information and authority than they were 25 years ago.

This dependence is in part due to the consumer economy that is always replacing non-monetary self-reliance with economically mediated goods and services. It is also a natural outcome of the society's capacity to ameliorate the impacts of natural and human disasters through high-cost, professionally designed and engineered, infrastructure and information systems.

Bushfire-resistant design

Permaculture One[13] included a section on design against bushfire as an integral part of all rural design in Australia, especially for the southeast section of the continent where most people live, which is acknowledged as the worst bushfire region in the world. This design-based approach was part of wider recognition that design against, and self-reliance in, the face of natural and human disasters is one of the essential elements of a society based on declining energy.

Since then, design against bushfire has been incorporated in mainstream information about bushfire preparedness in Australia but a more fundamental shift has also taken place. The experience of fighting large bushfires in the semi-rural fringes of cities in the 1980s and 1990s made fire services such as the Victorian Country Fire Authority realise the limits of top-down management[14] if the population being assisted has a culture of dependence. The CFA now facilitates local self-reliance through Community Fireguard groups and encourages prepared householders to stay at home in the face of the most severe fires.

Professional fire-fighting resources and personnel cannot substitute for personal, household and community self-reliance. Professionally managed volunteer rural fire-fighting resources in Australia have only been effective in co-operation with rural households and communities with a high level of self-reliance.

Most people believe the 1983 Ash Wednesday fires in Victoria were an historical benchmark of catastrophic fire but in fact they pale in comparison with 1939 and many other great fires of earlier times. In 1939 people lived and worked throughout the rural and forest areas devastated by the fires without the support of modern communications or professional fire services. All they had was their self-reliance. In 1983, the professional resources were some of the best in the world but, without the self-reliant households and communities, the top-down management of the fires came close to causing deaths of citizens by use of the strategy of evacuation. It is now broadly acknowledged that managed evacuation in large-scale, fast-moving bushfires creates more problems than it solves.

Systems under stress

This dynamic is not unique to bushfires, but it illustrates a more general principle about disaster management. *Very powerful systems can insulate themselves against the adverse effects of small disasters, but this can lead to a lack of fitness to deal with large disasters. Eventually a disaster arrives which overwhelms the top-down management systems, hastening catastrophe. In nature, small stresses generally function to keep the system fit and adapted, and therefore better able to cope with severe stress.*

The behaviour of organisms and ecosystems under extreme stress that threaten the integrity (life) of the system provide plenty of models of how to consciously design resilient systems. For example:

- bodies of all warm-blooded animals will conserve heat for the essential organs in extreme cold, allowing limbs to die if necessary
- in the evolution of Gaia (the living planet), it is now clear[15] that most of the critical planetary functions are performed by microbes capable of surviving the greatest

catastrophes. All the complexity of advanced plants and animals is the icing on the cake of life, which can proliferate and be trashed by ice ages, magnetic reversals, meteor impacts, etc. without the core functions of the living earth being affected

- in social unrest and famine, most social and economic functions break down while the family and the household manage to persist as the fundamental units of society. Experience in African famines suggests that, at the extreme, family relationships break down, but the last to fail is that between mother and child
- young parents going through the stress of caring for infants frequently abandon many of their own values and ideals and fall back on behaviour (good and bad) that they learnt by osmosis from their own parents at a young age.

In all these situations the most elaborate and most recently evolved structures and processes are discarded and there is a fallback to lower-order, small-scale and older, more proven structures and processes.

Modern society has lost or actively dispensed with most of its informal fallback strategies and back-up protection of essential functions. The defence forces are a costly and formal remnant of what were once a deep and broad structure of fall-back strategies and back-up systems built into every level of society.[16]

Today, the peasants in the world's poorest countries remain amazingly resilient in the face of natural disasters.[17] Despite the massive scale of international disaster relief, self-reliant response remains the predominant factor in restoration of normality in most parts of the world. In the relatively minor San Francisco earthquake (1989), community response in poor black neighbourhoods was more effective than in affluent white suburbs.[18]

The vulnerability to natural and man-made disasters in modern society is so great that it is conceivable that a series of natural disasters could lead to global depression or even the collapse of modern civilisation. As ridiculous as this seems, imagine the effect if the 1923 Tokyo earthquake had been repeated at the peak of the Japanese economic boom, when eight of the ten biggest banks in the world were Japanese and all were massively exposed to the breathtaking real estate values in Tokyo.[19]

The global insurance industry is already having a tough time due to the run of "natural" disasters in rich countries that have been attributed to the effects of global warming[20]. It seems that, as material assets proliferate and dependence on centralised systems intensifies, large-scale vulnerability to disaster compounds. Modern industrial society is setting itself up to play out the story of Atlantis.

My point here is not to get hysterical about any particular risk or its impact, but to emphasise some of the fundamental systemic values in self-reliance. Any society that fails to invest in self-reliance will eventually suffer for its short-sightedness.

By applying permaculture principles to our lives, households and community relations, we rebuild the deep structures that provide collective security and a degree of individual security against disasters. We accept that the endless pursuit of security, with shrill demands for paternalistic protection or macho survivalist plans, is unrealistic and counterproductive.

There are always limits to the efficiency and degree of insurance provided by any particular self-reliant behaviour, but on balance self-reliance remains one of the most important strategies for political and economic freedom, as well as for dodging the vicissitudes which nature delivers to us.

Conclusion

The ideals of self-reliant autonomy are compatible with open awareness and receptivity to the wider world. This principle helps us to balance and integrate those human imperatives in ways that fit a declining energy base. Whether we are selecting locally hardy tree species for a shelterbelt, using a compost toilet, choosing to have a home birth or, reducing the status of the television as a privileged member of the household, we are applying this principle of self-regulation and acceptance of feedback. In the process of empowering ourselves, we contribute to a more balanced and harmonious world capable of continuing to support life and humanity.

1 The return of part of an output of a circuit to the input in a way that affects its performance.

2 See J. Lovelock, *Gaia: A New Look At Life.*

3 This view has been expressed by eminent Australian biologist and theologian Professor Charles Birch (interview, ABC Radio National, 24 February 2000).

4 A human being requires about 2500 calories (10467 joules) per day of food energy. The embodied energy in natural resources indirectly consumed to support a modern person can be calculated and measured in joules. Dividing the total number of joules by 10,467 gives a sense of the human equivalent in energy expenditure. This type of measure does not attempt to evaluate the relative value of human effort based on age, skill, training, etc. The EMERGY methodology does allow some tentative measure of these values, thus providing a check on our economic and social valuation of different human contributions.

5 In 1991 the richest one-fifth accounted for 83% of all economic activity, while the poorest one-fifth accounted for just 1.4%.

6 I use the word "elites" to refer to the most powerful people in any society, perhaps 1% of the population, that exercises substantial influence or control over the important cultural, political and economic institutions. In Australia, conservative commentators have recently taken to referring to substantial sections of the well-educated upper middle class that are concerned about larger social issues as elites. The assumptions behind this particularly cynical bit of spin-doctoring are that anyone able to spend energy and time focusing on wider issues must be rich, and that those concerned with hip-pocket issues are all "Aussie battlers".

7 Based on the ideas of Adam Smith and others.

8 There is good evidence that the poorest have got poorer through environmental degradation and loss of access to natural resources. Further, the energy slaves for the Third World majority may be in the form of Coca-Cola, blue jeans, TV, and other consumer goods while adequate housing, nutrition, health, education, meaningful work and social values are unattainable.

9 W. I. Thompson, Article in *Journal of the New Alchemy Institute* Stephen Green Press 1979.

10 See Article 13 "Permaculture as Development Aid for the North" in *David Holmgren: Collected Writings* 1978-2000.

11 "Background Briefing" ABC Radio National, 1 June 2001.

12 See Article 7 "Garden as Agriculture" in *David Holmgren: Collected Writings* 1978-2000.

13 For more current and detailed information about bushfire resistant design see B. Mollison, *Permaculture: A Designers' Manual* and D. Holmgren, *The Flywire House: A case study in design against bushfire* Nascimanere 1993

14 The CFA Community Fireguard strategy formally acknowledges the limits of professional management of fires and that only well-prepared and self-reliant households and communities will survive catastrophic fires.

15 J. Lovelock, *The Ages of Gaia: A Biography of Our Living Planet* Oxford University Press, 1989.

16 Some examples of what has been lost include: fewer jacks-of-all-trades in the community able to stand in when professional systems fail; fewer long-term employees with local and system-specific knowledge; loss of local food production around cities and on farms; just-in-time manufacturing with no storages of parts and materials; and general breakdown in community networks and contact between neighbours.

17 There has been major erosion of rural self-reliance around the world by economic forces and degradation of natural resources.

18 "Earthquake Lessons" *Whole Earth Review*, no. 68, 1990.

19 The total value of the city was reputed to be worth more than the whole state of California.

20 Since this was written, the crisis in the insurance industry has intensified considerably due to the events of September 11th 2001.

Use and **V**alue **R**enewable **R**esources and **S**ervices

Let nature take it's course

Renewable resources are those which are renewed and replaced by natural processes over reasonable periods without the need for major non-renewable inputs. In the language of business, renewable resources should be seen as our sources of income, while non-renewable resources can be thought of as capital assets. Spending our capital assets for day-to-day living is unsustainable in anyone's language. Permaculture design should aim to make best use of renewable natural resources to manage and maintain yields, even if some use of non-renewable resources is needed in establishing the system.

In restoring the balance between renewable and non-renewable resource use, it is often forgotten that these "new ideas" were the norm not so long ago. The joke about the environmentally aware person using a solar clothes dryer (washing line) is funny because it works on the very recent nature of much of this takeover of functions by technology and fossil fuels.

Renewable services (or passive functions) are those we gain from plants, animals and living soil and water without them being consumed. For example, when we use a tree for wood we are using a renewable resource, but when we use a tree for shade and shelter, we gain benefits from the living tree that are non-consuming and require no harvesting energy. This simple understanding is obvious and yet powerful in redesigning systems where many simple functions have become dependent on non-renewable and unsustainable resource use.

Permaculture design should make best use of non-consuming natural services to minimise our consumptive demands on resources and emphasise the harmonious possibilities of interaction between humans and nature. There is no more important example in history of human prosperity derived from non-consuming use of nature's services than our domestication and use of the horse for transport, soil cultivation and general power for a myriad of uses. Intimate relationships to domestic animals such as the horse also provide an empathetic context for the extension of human ethical concerns to include nature.

The proverb "let nature take its course" reminds us that human intervention and complication of processes can make things worse and that we should respect and value the wisdom in biological systems and processes.

Renewable Resources as Energy

The idea that all resources and even services are forms of energy of differing concentrations which drive the development of all self-organising systems was introduced and linked to more familiar financial concepts of wealth, power, capital and income in **Principle 2**: *Catch and Store Energy*.

In **Principle 3**: *Obtain a Yield*, the EMERGY (real wealth) measure of various biomass sources was related to the rates at which they are renewed, confirming that the slowly accumulating ones are more valuable.

The limited and often erratic flow rates of renewable energies are the reason they have tended to be displaced by fossil fuels which have very high and regular flow rates. As we go through the essential transition to declining energy, the limited and erratic nature of renewable energies provides a valuable negative feedback, reminding us that all natural resources must be used carefully and respectfully.[1]

Criteria for Using Renewable Resources

To develop a deeper sense of the appropriate (and inappropriate) use of renewable energies and resources, we need to understand both the broad patterns and specific requirements of renewable resources so that we:

- make the best of what renewable resources can offer
- ensure use is within the renewable limits of the resource.

What constitutes appropriate use will always depend on factors specific to the site and situation. The industrial mindset where resources are used in gross and standardised ways must be replaced by a great diversity of resources, each contributing in different ways.

While making greater use of renewable resources is a popular idea, the reality can be environmentally disastrous. Already controversy over the use of "waste" harvested from native forests for biomass energy is brewing as the next great fight over public forests in Australia as well as other countries. In less industrialised countries and regions where people largely depend on very depleted firewood and other potentially renewable resources, the local environmental impacts can be severe.

Resource replacement time and product half-life

To decide whether use of a renewable resource is reasonable, it is helpful to ask, "Will the function or product which the resource is being used for last at least as long as it took nature to generate the resource?"

Thus, it is appropriate to make relatively ephemeral daily use of the sun, tides, water and wind because they are daily or seasonally renewable energies. On the other hand, the use of timber resources from trees is more problematic because of the time the trees take to grow. Paper products have a half-life[2] of maybe a few years, but the trees being used to make the paper may have taken decades or even hundreds of years to grow. On the other hand, if the same tree is used to make a fine dining-room table, the table may have a half-life of hundreds of years and so this is a much more appropriate use. Half-life is a useful concept, but it tends to reflect the high turnover of materials in our throwaway culture. In a future low energy society where change is less rapid (see **Principle 9**: *Use Small and Slow Solutions*) and there is minimal waste and regular maintenance (see **Principle 6**: *Produce No Waste*), then the half-life of products could be greatly extended. A book made from high-quality paper could potentially last and be used as long as an heirloom dining table and so justify the use of slowly renewable natural resources (such as old-growth trees).

Multiple products from every resource

By-products with varying degrees of value are the inevitable outcome of any sensible use of a natural resource. We should be suspicious of any use that achieves 100% conversion to a single product because this probably represents a devaluing of many of the potential uses for the natural resource.

Some years ago a contractor was clearing the timber from a neighbouring residential block of land. As he began to cut the timber into firewood "foot blocks", I asked why he wasn't cutting fence posts from the straight and durable yellow box trees. He replied that a fence post was no good to him, so I offered to take the timber in uncut lengths for its firewood value. The same mentality on a much larger scale drives whole-of-tree harvesting for wood fibre or fuel, ignoring the potential of some of the trees for higher-value uses.

Total process evaluation

It is not possible to fully evaluate the appropriate use of natural resources in isolation from the wider industrial context. If the furniture manufacturing process demands the cutting and milling of large numbers of trees to generate enough high-quality timber to make tables, then are those tables any more appropriate than the low-value disposable uses, such as paper pulp or fuel wood, which use all the leftover wood? Alternatively, a craftsperson using wood left behind in the forest after logging to make a unique piece of furniture may be the best example of use of renewable resources.

A whole-systems approach that sees each resource and product as part of a larger industrial ecology reveals important cross-subsidies and links. The concepts of the Ecological Footprint[3] and EMERGY mentioned in previous sections are useful in providing quantitative answers to these questions.

Appropriate use

How well we use the products from natural resources is as important as the way those products are made. The dining table that is used each day to feed a large household is very different from the one used for the occasional dinner party in an otherwise empty house. One will become imbued with the memories and marks of living. The other will occupy space that is locked, insured, maintained and heated, doing little.

Capital asset size

As well as resource replacement time, the total size of the resource available is crucial because it can provide a buffer against the adverse effects of variations in rate of consumption. To some extent the time taken for growth and size (or area) factors can be multiplied together to gain a sense of how abundant a renewable resource really is. Historically, forests provided nations with a capital asset that accumulated in peacetime and was drawn down during wars. On a small rural property, a woodlot that could theoretically supply the fuel wood needs of a household may, in practice, be difficult to manage because a rotational thinning system that produces a constant year-to-year supply may be less than optimal management.

Investment of Non-Renewable Energies

Renewable resources generally supported the harvesting of renewable energies and resources in pre-industrial times. Human labour, animal power and many of the tools used in agriculture and forestry were all generated from renewable sources, frequently the same resources. Thus the horses that ploughed land to raise crops were generally fed from the same farms they worked. In traditional forestry operations the forest provided most of the materials for the structures and tools that were used in harvesting the forest. Sawmills were powered by steam engines fuelled by mill off-cuts.

Today, modern agriculture is the most pervasive and important example of increases in productivity from renewable resources by the use of additional non-renewable energies, materials and technology to assist in the management, harvesting and processing of natural resources. Although these processes have increased total yields, they have transformed agriculture from our prime means of harvesting renewable resources to one of our largest consumers of non-renewable resources. While teaching permaculture in Israel, I suggested that a glass of milk in Australia was perhaps 20% oil, while in Europe the figure was perhaps 50%; and that from what I had seen of Israeli dairy farming, 80% of a glass of milk must be oil.[4]

To harvest, store and use renewable energy sources requires varying inputs of high-quality (generally non-renewable) energy. EMERGY evaluations show the inputs for good-quality tidal and hydropower sites are very low because the landscape-forming processes have already created most of the infrastructure needed. In the case of biomass from forest waste or thinnings, the photosynthesis and the natural processes of the forest have already done most of the work.

In the case of wind power, the lower quality and erratic nature of the energy resource requires the much greater harvesting infrastructure. For solar electric power, the resource is most abundant but so low in quality that vast infrastructure is required relative to the quantity of harvested energy.

Solar cells: renewable salvation or technological diversion?

Despite these limitations, solar (photovoltaic) cells that convert sunlight to electricity have become the great symbol of renewable energy, out of all proportion in importance to a future with declining energy. They are seen by many as the environmentally clean source of power that will sustain a high-tech but energy-efficient society in the future. Although solar cells are a convenient way to generate electricity at locations remote from the power grid and they may be able to contribute during a transitional period from energy growth to energy decline, I believe that a high-tech society running on solar cells is the stuff of dreams.

The basis of this somewhat heretical view depends in part on methods for assessing the net energy yield of solar cells. In other words, do they collect more energy over their useful life than they took to manufacture and maintain? Studies have been done using a variety of embodied energy methodologies that show positive net energy results, but EMERGY analysis suggests the solar cells are a net loss.[5] The differences between this and the other methods are mostly due to differences in the evaluation of human labour and services.

In most other methods, human labour is rated by metabolic food energy use or, in some cases, national fuel share.[6] The EMERGY methodology measures human service input by national EMERGY use share. By these measures, human services in the United States are 200 times more valuable (and expensive) than metabolic energy and about double the simple national fuel share.[7] These figures are national averages, independent of what people actually use. Consequently, a semi-literate office cleaner "costs" the same as the most highly paid computer software engineer. This might be a nice idea, but it does not accord with economic or ecological realities. If we multiply the figures by income, we obtain (albeit crude) differential costs for human service inputs.

Manufacturing solar cells is a very high-tech process and requires highly skilled human input while the quantities of electricity generated to "pay" for that input are relatively small. However, it is the administration and operation costs that give such disappointing EMERGY yield ratios. These costs may be substantially reduced as solar power plants become more usual. The equivalent costs for small domestic solar systems may be much less.

Even with these improvements, the other important factor in discounting optimistic claims about solar cells is based on understanding the Maximum Power Law (see **Principle 3**: *Obtain a Yield*). Because solar energy has been the primary external energy available for life on earth, billions of years of evolution have probably already optimised the capture and conversion of solar energy. Technological alternatives are unlikely to match that efficiency when properly evaluated.

To quote Howard Odum:

> With research and manufacturing experience, the goods and services required to make the solar cells have been decreasing a little each year. Costs per unit power have been decreasing slowly. However, when this process of improving efficiency has gone as far as is thermodynamic, the efficiency of these cells may approach that of the green plant chloroplast which is nature's photovoltaic cell. Studies in biophysics providing curves of efficiency as a function of light intensity for isolated chloroplasts show them to be more efficient than hardware cells. It may be that the natural conversion of sunlight to electric charge that occurs in all green plant photosynthesis after 1 billion years of natural selection may already be the highest net EMERGY possible.[8]

Trees: Nature's Solar Power Plants

The ultimate development of the biological capture and storage of solar energy in forms useful to future generations of people is forest trees. Although trees do not yield electricity directly, they do most efficiently convert the very dilute solar energy into wood that we can more effectively substitute for many current uses of fossil fuel. For example, modern technology for wood gasifiers and micro-gas turbines appears to be a much more economical pathway for electricity production than solar cells.[9] EMERGY analysis of existing wood-burning power generation supports this comparison.[10] When this electricity production is combined with the following advantages, trees are clearly the best renewable energy source for the future.

Advantages of trees as renewable energy source include:

- they provide wood for decentralised production of transport fuels using intermediate technology (wood gas or methanol)[11]
- they provide structural timber products, fibre and wood chemical products to replace high-energy concrete, metal and synthetic materials
- forests produce honey and other products and environmental services
- wood-producing forests can grow sustainably on our poorest land unsuited to food production.

Appropriate Role of Solar Cells

Solar cells may be an appropriate way to make use of already embodied fossil fuel energy in human skill and industrial manufacturing capacity to contribute to electric power production during the transition to declining energy availability.

This is especially so:

- where small demand systems are remote from the grid
- in sunny dry climates, where low rainfall limits wood and other biomass production
- in sunny urban environments, where solar cells can double as a roofing material.

The greatest value from solar cells may be their role in forcing us to reassess how precious electricity is, and the inevitable conclusion that we should use it only for high-quality functions such as small electric motors, lighting and communications. *The increasing numbers of people living with autonomous solar power systems are the pioneers in a new culture of modest electricity use while continuing to participate in modern affluent society. The value of their actions may be far greater than any net energy gains or losses of solar cells.*

Nevertheless, in 200 years time the primacy of trees is likely to be entrenched, while solar cells may be one of the forgotten high technologies of the past that there is no way to manufacture.

This critique of solar power shows why permaculture, and this principle in particular, focus so much on rediscovering biological solutions as alternatives to our bias towards technology. The idea that "nature knows best" is often supported by the most rigorous scientific understandings.

Sustainable Use of Renewable Resources

If harvesting trees and other biological resources to provide for our energy and material needs is the way of the future, then the key issue is to ensure that harvest does not deplete the capacity of nature to continue to yield. Further, where natural resources have been depleted, we need to rebuild natural capital as described in **Principle 2**: *Catch and Store Energy*, so that in future times of need we can draw on these stores of energy and wealth.

Some environmentalists believe that the harvesting of natural forests can never be ecologically sound and that all forest products should be made from plantations on existing farmland. Others have suggested annual crops, such as hemp, as alternatives to trees as a source of wood fibre. Although hemp is an excellent substitute for cotton which needs high inputs of water, fertiliser and pesticides, hemp as a substitute for wood pulp does not make so much sense because it is an annual crop that requires arable land suited to food production. On the other hand, wood pulp can be a by-product of timber production on much less fertile and steeper land with greater environmental benefits.[12]

At Melliodora we use the wood harvested by thinning from well managed native forest[13] for structural uses and as fuel for cooking, water heating and back-up heating. I see this use as reflecting permaculture principles almost as much as the firewood and poles we have harvested from our plantings at Melliodora.

What constitutes sustainable harvesting of natural forests is complex and can only be proven over long periods of time with careful management and documentation. But these difficulties are not good reasons to avoid using natural resources in favour of more technologically mediated solutions; in the end we have to get our hands dirty if we are to adapt to energy descent. *The image of clean green technology where we do not need to mess with nature or kill anything to provide for our needs is, in the final analysis, an illusion. That illusion appears to have substance only because generations of the world's more affluent urbanites have been disconnected from nature.*

In a world of energy growth, it is natural for high yielding and intensively managed systems to replace low yielding and self-maintaining wild ones; but in descent, the efficiency of self-maintaining wild resources will again be critical even if human population pressures require greater demand for yield and greater care in harvesting. Sustainable yield can be defined as the surplus above that required for the system's own needs. Complexity of natural systems and fluctuations determined by seasons and other factors require us to be very conservative in what we harvest, relative to (what we believe is) the sustainable yield. How and when we harvest is just as important as how much.

Fluctuating and mobile resources

The importance of timing has already been discussed in **Principle** 3: *Obtain a Yield*. While the accumulation of wood in living trees and as slowly curing firewood allows considerable flexibility, in timing of harvest, many plant and animal resources disappear if we don't take the opportunity to harvest at appropriate times. The whole crop of a stone-fruit tree can disappear with birds or rot in a week. The timing of hay cutting to maximise the nutrient value before seed set is a critical skill for pastoral farmers in many regions.

An example of a wild system that can be harvested each season without adversely affecting future yields is yabbies living in a dam. Because there is a constant inflow of organic material in catchment run-off to feed them, and the reproductive capacity of these crustaceans is so great, they rapidly breed to replace harvested numbers. Without harvesting, the stock of yabbies cannot increase beyond a certain point; an unharvested dam will often contain many small yabbies, while harvesting allows others to grow big.

Many short-lived water birds, such as ducks, can be harvested in great numbers some years to avoid a population crash. But because birds migrate across regions and even continents, large numbers at one place does not necessarily indicate a surplus. A wider understanding of the movement and gathering patterns of wild birds is needed to ensure sustainability of harvest.

Regulation of wild harvesting in traditional societies

Because wild systems can be harvested without the need for specific work or other inputs (other than harvesting), it is natural to begin to regard these yields as free. Without any required payment or feedback, over-harvesting can result. Where these systems are part of a common rather than private resource, the incentive to over-exploit may be even greater.[14]

What constitutes a sustainable yield may only be shown after centuries of experience and mistakes. This knowledge then needs to become incorporated in a cultural tradition that can pass on the learnt wisdom and provide sanctions and taboos to prevent short-sighted behaviour, self-interest and greed from damaging the natural abundance. We are reminded of Native American traditions about the need to consider the effects of our actions for seven generations into the future and the European fable about the goose that laid the golden egg.

In traditional agricultural societies, natural systems beyond the boundaries of intensive agricultural land provided important and complementary yields and were almost universally owned in common; the origin of that much-reduced feature of the English village, "the common". In Nepal, one hectare of arable land used to be adequate to support a farmer and his family from rice production, vegetables, a few tree crops and livestock. However, this system required an additional seven hectares of common forest to provide the animal, fodder, fuel, construction materials and other special yields such as medicinal herbs.[15]

A Hare Krishna devotee[16] with considerable experience of farm management and permaculture design presented a permaculture zoning interpretation of the traditional Indian regulation of land use by the Brahmins through the caste system. He explained that everyone had the right to garden (Zones 1 and 2) to provide for family needs, while those with the technical ability and entrepreneurial drive were allowed access to the arable land (Zone 3) to produce staple crops for the market. The use of the forests and rangelands (Zone 4) by farmers was strictly regulated and, of course, there was always the sacred (Zone 5), beyond utilitarian use by anyone. The Brahmin elite understood that entrepreneurial drive of farmers which generated material wealth for society in a privatised Zone 3 was constrained by the need for inputs of human and animal labour. If unrestrained in Zone 4 commons, this wealth-producing but short-term focus would greedily over-exploit the apparently free resources of fertility, water, fodder, fuel and other yields that managed natural systems provided to the whole of society.

Sustainable harvesting of native forests

Given careful management, it is possible to harvest firewood, poles and sawlogs from native forests without reducing the capacity of the forest to continue its full range of ecological functions and to provide these yields to future generations. Despite the ups and downs of over-harvesting and mismanagement, some regrowth eucalypt forests in Australia show that timber harvesting can be a mechanism for restoration of forests,[17] even if the policies that provided those examples have since succumbed to more rapacious and short-term exploitation.

Like the yabbies in a dam, the total canopy cover in a forest is limited, but that canopy may exist as many thin saplings or a few large trees. Thinning of regrowth forest stands allows the remaining trees to grow faster and to a larger size. If the most desirable trees are removed, the value of the forest is degraded over time. If, in thinning, the least desirable trees are removed, the maturing forest will have increased values. In this case, by obtaining a yield of lower-quality, less useful wood (as firewood), we ensure there will be more useful wood in the future (as sawlogs). Put simply, "remove little trees to grow big trees." This is often counter-intuitive for people who have experience at reforestation (growing little trees) but none in managing established or existing forest.

Even if we understand very little about forests and forestry, we can go into the forest with the question "What does the forest have to give?" If we use observation skills, including the decision-making loop (see **Principle 1** *Observe and Interact*) it will become evident in most forests that small stunted individuals of the dominant canopy species are abundant and can be removed without much risk of adverse effects. The skill needed to fell such trees is not great and the effort needed to process them for use not onerous. If we make the best use of all the trees that we fell, we can think about the results and proceed in small incremental steps. The yields from thinning might only be firewood and mulch but could include posts or craftwood if the type of tree is suitable. In the process of obtaining a yield, we improve the capacity of the forest to catch and store energy in larger, more valuable trees.

In thinning a forest, we can use very simple and absolute criteria for what is a "good" tree to retain, or we can use complex and contextual criteria that take account of multiple values. In practice, for many forest owners, limitation on the capacity and cost of thinning is more of an issue than whether the selection criteria could be improved. Failure to thin regrowth forest because we are not sure it is a good idea will only encourage future owners and managers to pick the eyes out of the poor inheritance we have left them.

At Fryers Forest Eco-village,[18] we are thinning a 100-hectare native eucalypt forest of Box species in a way that obtains a yield adequate to cover the costs of the work while increasing future timber and other forest values. The continuing development of the forest management draws on best examples from the past and the latest practices in European nature-based forestry.[19]

Hunting for sustainable yield

Similar principles apply to the sensible management of wildlife as a source of meat and hides. Many herbivorous animals (both native and introduced) reproduce and proliferate to a degree that can damage the environment that supports them. Sensible harvesting by humane methods both provides a useful yield and keeps the population and its environment healthy. In Australia, we are more familiar with culling wild animals, whether kangaroos or rabbits, as pest control rather than harvesting. The intention has often been to inflict the maximum damage on the population for the minimum effort; killed animals are wasted or used for pet meat, for example, a very low-grade use.

Even when hunted animals are seen as a valued resource, dysfunctional macho killing of the largest dominant male animals (eg the stag with large antlers) tends to damage herd health and viability, and yields meat that is too rank and tough to eat. This hardly helps the argument that hunting can be a model of ecologically sound resource use.

On the other hand, when young, barely mature animals (especially males) are culled, the genetic vigour and health of the herd is maintained and even enhanced. In most species and environments, these young animals are likely to die from stress, competition or predation. In addition, they are generally easier to hunt because of their risk-taking behaviour. The meat from young animals is tender and with fewer hormones which make the meat of older animals rank.

Older proven breeding animals, especially dominant males, represent the selected genetics and (in long-lived, intelligent animals) knowledge and learned behaviour that can be passed on to younger animals. Like old-growth trees, dominant animals represent the essence of the population and environment. They should have the highest value in any managed wild system.

No farmer would kill his prize bull for meat, and yet this sort of behaviour has often prevailed in culling wild animals. Rabbit trappers who were the main control on rabbits in Australia before biological control through myxomatosis in the 1950s understood this wisdom. The trappers "farmed" rabbits by not setting traps on breeding mounds[20] and so made a living where the grazier struggled to do so.

Today, most indigenous wildlife species, along with native forests, have achieved the dubious distinction of being both worthless and "sacred". When wild resources have low value, the cheapest and most wasteful harvesting methods are used and there is little care or management to sustain the resource. On the other hand the emerging popular sanctity of indigenous animals and trees is very shallow because it is still founded on the premise that the resource has little monetary value, so we are not giving away anything by forgoing use.

In contrast, sanctity in indigenous cultures was founded on the premise of very high material value of wild resources generally, and the most valuable became sacred. We can aspire to redeveloping this respectful valuing of nature's gifts. As long as we live from the oil well and the coalfield, we would do well to pay homage to them rather than take them for granted like spoiled children who have everything but value nothing.

When we consciously obtain, and value, any harvested yield (from the fruits of the season to the fossil fuels of bygone aeons), we both provide for our needs and recognise our connection to and interdependence on the cycles of nature.

Renewable Services from Nature

While sustainable use of renewable resources is fundamental to energy descent, it is the non-consuming use of nature's services, which the horse icon illustrates, where we can experience a deeper harmony with nature. When we make use of the services of living nature (animals, plants, soil and self-purifying water systems) to provide for our needs, we reduce our demand for consuming both renewable and non-renewable resources.

Classic permaculture examples

In explaining how we can make better use of renewable services once more, "the chook tractor" is the classic permaculture example: the natural scratching action of the fowl is used to cultivate the soil instead of a tractor (and plough) or rotary hoe. This comparison of the fowl to the tractor reveals that the functions of a tractor are better understood than those of a fowl. At the same time as it cultivates, the fowl's foraging for insects replaces the use of toxic insecticides and its manure replaces use of energy-intensive fertilisers.

In a similar way we can think of grazing animals as mowers, and plants as water and nutrient pumps, shelter and living fences. Living soil can be thought of as a filter, purifier and store of water and nutrients. Streams, swamps and other waterways can be self-purifying water storages.

In **Principle** 2: *Catch and Store Energy*, I used the building of soil fertility by careful use of natural mineral fertilisers as an example of the appropriate use of non-renewable resources. Although mineral fertilisers are still relatively cheap, we should generally give priority to making soil nutrients more available through biological processes. Fostering the invisible action of soil organisms, the passive functions of plants and the traditional roles of livestock in building fertility is generally preferable to adding those nutrients out of a bag. This is especially true in established gardens and intensively managed farms, where past applications of fertilisers have often created a surplus of locked-up and unbalanced nutrients.

Pioneer plants that build soil fertility, with or without the help of soil microbes, are a feature of almost all permaculture-inspired gardens and farms. Legumes that support nitrogen-fixing microbes are the most common example. Symbiotic mycorrhizal fungi and similar organisms, which mine phosphate tied up in insoluble form in the soil, are another.[21] Through use of this microbiological service, we can save our nearly exhausted rock phosphate sources for use where there is a genuine deficiency.

Deep-rooted pasture plants such as chicory, that extract nutrients from and open poor clays are most valuable in slowly improving soil. Similarly, some long-lived forest trees, like oaks, have proven ability to thrive in poor soils in southern Australia, and slowly accumulate critically important calcium and boron in their leaves. Soil animals such as earthworms and termites physically mix and improve the soil layers.

We can think of countless examples of use of nature's renewable services, from the common to the novel. In all cases, design and management are the keys to getting a useful result. Often management is required to focus or contain the plants or animals providing the service. Without this, the solution becomes a new problem.

Examples from Melliodora include:

- design to reduce the chance of fowls scratching newly planted vegetables by keeping a breed of birds (Black Australorp) that can (generally) be contained by standard-height fences
- hard pruning of nitrogen-fixing tagasaste and wattle shrubs in the gardens and orchards (for goat fodder and tree mulch), to prevent excessive moisture competition and shading of vegetables and fruit trees.

However, in both cases, we have occasionally failed to effectively control the "biological service". It is understandable that, when non-renewable resources provide readily available and "cheap" solutions, the biological solutions tend to be abandoned.

Often a combination of the technological and biological elements is complementary. At Melliodora:

- we have used a motorised slasher as a powerful tool for conversion of brambles, weeds and rough pasture to mulch and worm food, but over time grazing animals are gradually replacing the slasher as converters of the seasonal pasture surplus
- we control the grazing of goats and the foraging of poultry in the orchard with movable electric mesh fencing (actual annual electricity use is less than that for a regularly used light globe).

In poorer countries there are huge opportunities to make use of biological services to solve what appear to be intractable problems. Of course, design and management remain critical. There are many stories about inappropriate introduction and use of exotic species, but these bad stories should not blind us to the fact that introduction of easily reproducible useful plants which fill a missing niche is one of the most simple, democratic and cheapest development solutions.

The gathering momentum of using leguminous trees and shrubs in African subsistence agriculture provides a dramatic and well-documented case study of use of biological services.

An African example

The yields of staple crops (especially corn) harvested by peasant farmers in Africa and other tropical regions have remained low or fallen as the land degrades and populations increase. The poorest farmers are completely dependent on natural fertility and rainfall. Cultivation is their only management option to maintain yields. However, repeated cultivation reduces organic matter in the soil and kills earthworms that maintain structure. These changes lead to a decline in soil structure that in turn leads to greater need for cultivation. Uncultivated common land once provided wild foods, fuel, animal fodder, and consequently, manure to improve cropping soil fertility. They are almost everywhere degraded to near desert or taken over for large-scale commercial pastoralism or cropping. Corn, a high-yielding "gift" from the New World, originally increased the food supply for Africans in the way the potato did for the Irish, but it is a soil-depleting crop.

During the 1990s the International Centre for Agroforestry (ICRAF)[22] in Nairobi researched the use of fast-growing leguminous shrubs and trees to allow the poorest farmers to rebuild soil fertility and provide fodder for dairy cows. These biological solutions have in turn produced more manure for soil fertility, and milk for needed protein and cash. The dramatic local success of these systems is leading to their adoption across Kenya and beyond. These classic permaculture solutions, using plants as substitutes for unavailable conventional fertilisers, are increasing yields of corn up to tenfold. It is ironic that some of the legumes used come from Mexico, the original home of corn which was brought to Africa centuries ago.

Such huge increases in productivity are a saviour, but they bring long-term hazards. The massive rise in available nitrogen and consequent yields will inevitably, over time, expose the next weakest link in the fertility chain. I would speculate that calcium is likely to be the next limiting factor in most soils, a problem that is accentuated by leaching of any unused nitrogen. This acidification process, which is perhaps the most widespread form of land degradation in southern Australia, has shown that the huge increases in pastoral productivity derived from the subterranean clover and superphosphate revolution in the 1950s were unsustainable.

In the African case, the fact that the legumes used are woody perennials and that further research and extension work is encouraging the planting of longer-lived, non-leguminous trees, including *Prunus africana*,[23] may prevent the acidification problem emerging sooner than it need.

Perhaps the key factors in these deeper applications of biological solutions will be social and ethical rather than technical. Will the minor boom in rural affluence created by the agroforestry revolution lead to dysfunctional consumer affluence in the next generation, or a deepening respect for the earth and its gifts?

Working animals

The horse icon that illustrates Principle 5 reminds us that the non-consuming services we gain from working relationships with domesticated animals have a much greater history and potency than simple examples such as the chook tractor. Before industrialisation, domesticated beasts of burden — especially horses, but also dogs, bullocks, donkeys, camels and elephants — were as important as the pyrotechnic skills in increasing human well-being and power to tap natural resources. Unlike fire, which appeared to come from human brilliance (or the gods), domesticated work animals required humans to care and respect nature in a way that could be as intimate and empathetic as the relationships between people. I don't want to romanticise those relationships, but instead point out that the value of a well-bred, well-trained working bullock or horse (let alone an elephant) could be greater than that of a human worker. Is it any wonder that so much of our mythologising about animals concerns these working companions?

Examples abound in history of the power associated with working animals. The value of dogs in increasing people's ability to capture wild game led to the proliferation of hunting breeds with very specific characteristics and temperaments, while a host of other functions drove the selection of other animal breeds.

The Australian dingo and its relatives in Thailand are now recognised as the ancestors of domesticated dogs. The dingo's relationship to aboriginal Australians since its arrival about 4000 years ago suggests that animal domestication was a co-evolutionary process: mutual benefits resulted in incrementally closer relationships between humans and wild animals, rather than the "capture and tame" model of domestication. Rather than seeing working animals as slaves, this perspective shows an evolutionary partnership.

The rapid integration of horses into the culture of the native Americans following their arrival with the Spanish is testament to the power that horses provided. It is not widely recognised that the use of the working horse as a measure of machine power since the Industrial Revolution has its roots in an expanding role for larger and faster working horses in Europe immediately prior to industrialisation. The shift from bullocks to horses for transport and farming in the lead-up to the Industrial Revolution in Europe represented a significant increase in speed, and therefore power, which were essential precursors to industrialisation.[24] Bullocks could work while being fed rough pastures and tree and shrub fodders; working horses capable of faster work pulling coaches or ploughs required more concentrated feed (primarily oats) produced from good cropland. The extensive oat monoculture[25] that developed to support the horse economy was itself possible because of the supplementation of local agricultural production with imported food and fibre (wool and cotton) from foreign colonies.

Even today, with cheap fossil fuels and high technology, working animals remain important.

- Despite cheap all-terrain motor bikes, sheepdogs remain an integral aspect of the labour efficiency of Australian wool production, where savage economies of scale demand that one man manage around 3000 sheep.
- Advances in robotics are yet to economically replace guide dogs for blind people or sniffer dogs in the customs and police services.
- In small-scale forestry, working horses remain competitive with the latest advances in small tractors. A well-trained horse snigging a log can be sent off to a loading ramp by itself and will stop and wait if the log is caught up. Navigating a wild forest might still be too great a challenge for even the latest advances in robotics. In Sweden, the combination of working horses with lightweight modern vehicles and hydraulics technology represents the best of both worlds.[26]

The impediments to the wider use of horses in small-scale forestry are the lack of people with the skills, and the cost of well-bred, well-trained animals. Both positive and negative stories about working farm animals abound. The productivity and safety of working animals depend as much on the empathetic, even loving, relationship between person and animal as it does on breeding and equipment. A purely rational comparison between the productivity of machines and working animals is not possible because the difference between the best and worst of working animals is the difference between heaven and hell, while farmers do not have to be devoted to their tractors to succeed in farming.

At Melliodora, cockatoos are very destructive of fruit and nut crops. Most deterrent strategies are of little use against these powerful, intelligent, long-lived flock birds, while netting is very expensive. Personally I am waiting for the Hepburn falconer as the long-term solution. A local falconer with suitable trained raptors could, for a fee, keep the valley free of cockatoos during the fruiting season.[27]

Another example, more speculative still, of creative possibilities concerns the brush turkey. These and other Australian megapodes are extraordinary birds in that they construct large compost heaps to incubate their eggs. In the early 1980s on the south coast of New South Wales, a permaculture colleague, Peter Brew and I were speculating that these birds from northern New South Wales which tame readily, could be domesticated. We had observed that in making compost heaps these birds dramatically reduced the fire hazard from ground fuels in moist forests. We wondered whether domesticated birds could be usefully added to large-scale food forest systems with predator-proof fencing. Such systems could be used to breed up local stock for reintroduction to local rainforests and over time compost heaps (in the domestic systems) could be harvested for use in intensive gardens and some limited egg and meat production.

As someone who doesn't have a natural empathy with animals, I recognise my own limits; I respect people who work effectively with animals as the conservators of the functional human intimacy with nature that will be an essential element of any culture adapted to energy descent. *Those working to combine traditional skills and knowledge with the latest understanding in non-violent training of animals, and who make creative use of readily available modern technology in combination with animals, exemplify permaculture in action as much as the horticulturalist saving heirloom varieties or maintaining the skills of grafting.*

Ecosystem Services

Specific non-consuming services that we gain directly from plants and animals are part of a larger field of environmental or ecosystem services. Most of these services operate at a large, even global, scale. They include atmospheric purification and climate maintenance, soil-building processes and purification of surface and underground water bodies. The degree to which we are dependent on the continued operation of these gifts of nature has been highlighted by global warming. Evaluations using various methods of environmental accounting show that the global value of environmental services is greater than total economic output.[28]

At a more local level, previously functioning and free ecological services are noticed when they break down due to overload or abuse. The cost of providing alternatives to these free services highlights the absurdity of the notion of an economy independent of nature.

Water purification services

Perhaps the most important of these free ecological services is (or was) the natural purification of water performed by flowing rivers. In pre-industrial times, people living by large rivers let them take away human and other wastes (which they do anyway during periodic floods). The energetic and biological purification processes of a flowing mass of water can make river water safe to drink for downstream communities. The religious purifying effects of the Ganges river in India epitomise the recognition of this aspect of great rivers. With rising populations and additions of more toxic industrial pollutants, rivers have lost their purifying capacity; a myriad of human health and other problems have resulted.

About a century ago in industrialised countries, surrogate river systems of reticulated sewerage dealt with the critical health problems, and in more recent decades an explosion of environmental technologies have been developed to replace or protect ecosystem services that were once free.

The latest development phase in environmental engineering has been the recognition that natural systems provide the best models for water purification. Artificial reed beds, living machines[29] and other such technologies are some of the fastest-growing aspects of environmental technology, especially in Europe and North America. Swamps are the natural models for much of this technology. It is ironic that some of the best and lowest-cost water purifying systems have evolved where sewerage outfalls have been fed into natural swamps at rates moderate enough to allow the adaptation of the swamp ecosystem.[30]

Beyond the design of simple earthworks to allow biological processes of water purification to work, the ecological succession of streamside vegetation can effectively reinforce these water-purifying functions. In a submission to government, a university limnologist and I advocated this approach as an alternative to current streamside weed destruction and planting of purely indigenous vegetation adapted to low water and nutrient flows.[31]

Willows have been found to be efficient and complementary to reeds in effluent purification in Europe,[32] but very little research was done in Australia before the large-scale, nationally funded willow removal programs in southern Australia. Local research[33] suggesting willows were 10 times more efficient than eucalypts in purifying water along local streams shows that willow removal programs are counterproductive for maintaining and improving water quality. Once again, we have free and invisible services being destroyed without recognition of their value.[34]

The current flurry of research of artificial wetlands has tended to obscure the extraordinary water-filtering capacity of deep, well-structured soils. A metre or two of free-draining, well-structured loam and clay loam soil can destroy pathogenic organisms in sewage water and free tied-up surplus minerals from clay and humus particles. Where such favourable soils exist, the much-maligned septic tank absorption systems work well. Improvements in soil structure and humus content increase the capacity of all soils to perform this water-purifying function (see the discussion of the role of mineral balance in soil structure in **Principle 2**: *Catch and Store Energy*).

Composting as microbial service

The flush toilet mixes large volumes of pure water with small volumes of human waste, and thus creates a problem. This never arises with the compost toilet, which utilises the services of ubiquitous microbes, and sometimes manure worms, to turn undesirable waste into safe and useful compost.

Composting on domestic and industrial scales is increasing, rapidly driven by demands to reduce landfill volumes. It provides a useful and saleable supply of organic fertiliser. Worm farms, small and large, are increasingly being used to process wet and high-nitrogen wastes from domestic kitchens and farm dairy sheds. Cuba has 176 large-scale vermi-compost centres which take urban waste to make organic fertiliser for the extensive urban agriculture program.[35]

Existing natural systems — forests, soils, rivers, swamps — can all provide water purification services so long as population densities are low and toxic substances are minimal. However, some natural agents such as the common white rot fungus break down even dangerous and persistent chemicals.

Environmental technology

Where existing wealth from non-renewable resources is high and development is intense, artificial and managed systems (such as reed beds) represent an improvement on using high-energy mechanical solutions. These current state-of-the-art environmental technologies may never become universal in poorer countries. Even in rich countries, the embodied energy required for some environmental technology probably means they contribute to more environment damage elsewhere than they prevent.

In a recent design project[36] I was involved in designing a dam and wetlands to catch urban stormwater for natural purification and reuse for irrigation in a city farm project. Below each stormwater outlet I had a small pond dug and planted to locally wild cumbungi reeds.[37] The reeds will catch much of the debris and sediment from the suburban street run-off and, over time, the pond will fill up with silt. A backhoe can dig out the pond and place the reeds and sediment to compost in surrounding urban forest plantings. Remaining pieces of reeds will re-establish the system; plastic and other non-compostable rubbish can be easily removed by hand at a later date from the matured compost.

At one major stormwater outlet where there was no space for a cumbungi pond, the supportive City Council offered to fund, construct and maintain one of the new award-winning "CDS" litter traps constructed of concrete and galvanised steel. These structures use a centrifuge action to separate rubbish from water. Although very effective, there is no doubt in my mind that the capital cost, let alone the embodied energy cost, of this technology[38] creates more environmental damage (in its manufacture) than it could possibly save in preventing plastic and beer cans spreading down the local creek.

Although I devote the next principle (**Principle 6**: *Produce No Waste*) to elimination of waste, this principle reminds us that nature often uses apparent wastes to expand and strengthen living systems, which in turn provides people with more resources. So long as we work within nature's limits, this cycle of abundance can be continually strengthened. (**Principle 12**: *Use Change Creatively* expands on this theme of adaptive change.)

Conclusion

The developing environmental orthodoxy is that we must increasingly separate all human support systems and ourselves from nature in order to prevent damage. This is a delusion created by generations of urban affluence separated from the cycles of nature. Although there are many positive and useful examples of environmental technology based on natural models, the underlying principles of separation of people from nature are philosophically and energetically flawed. We need to recognise that sustainable systems are more likely to emerge from an intimate partnership with nature, rather than application of natural design principles within the confines of the technosphere. Thus the slogan "nature knows best" is appropriate.

1 Mollison's 5 categories of resources also provide a perspective directly relevant to this principle.

1. Those which increase with modest use,
2. Those unaffected by use,
3. Those which degrade if not used,
4. Those reduced by use, and
5. Those which pollute if used,

See *Permaculture: A Designers' Manual* chapter 2.4

2 Half-life means the time taken for half of the quantity or value of a product to be degraded or lost.

3 See Redefining Progress web site http://www.rprogress.org/

4 The facts behind these guestimates are that in Australia the cows graze clover/grass pastures (all renewable processes) with smaller fossil fuel inputs in making phosphate fertiliser, fencing, dairy shed processing etc. In Europe the cows are kept in sheds and fed crops grown with fertilisers and harvested and transported with machines. In Israel, the situation is similar to Europe except that the crops are all grown using irrigation water pumped with fossil energy. For more interpretations of Israeli landscapes and land uses see Article 14 "Impressions Of Israel: A Permaculture Perspective" in *David Holmgren: Collected Writings* 1978-2000.

5 Two EMERGY studies of installed solar systems in Austin Texas and Nashville Tennessee in the 1990's showed EMERGY yield ratios of 0.41 and 0.36 respectively (-anything less than 1 is a net loss). For comparison, lignite coal produced electricity in Texas yields 2.1 and hydroelectric power in NZ yields 10. See H.T. Odum, *Environmental Accounting*.

6 Metabolic energy 2500 kilocalories per day, fuel share (USA) 229,000 kilocalories/day,

7 National share of EMERGY (USA) of 557,500 coal equivalent em (bodied) kilocalories/day from Odum 96.

8 H. T. Odum, *Environmental Accounting* chapter 8.

9 Figures from CSIRO research quoted in *Ecos* no. 98, March 1999.

10 See H.T. Odum, *Environmental Accounting* chapter 8.

11 These can also be easily converted to hydrogen should the much vaunted hydrogen economy manage to become established in the next decade.

12 For more detail on this argument see Article 17 "Hemp As A Wood Paper Pulp Substitute? Environmental solution or diversion from the search for sustainable forestry?" in *David Holmgren: Collected Writings* 1978-2000.

13 Fryers Forest eco-village community with 100 ha of managed box forest. See Fryers Forest web page http://www.holmgren.com.au/FF.html

14 American author Garret Hardin called this "The Tragedy of the Commons" and argued that over-exploitation of common resources was inevitable. However, diverse examples across cultures show effective and sustainable management of common resources has occurred often for centuries without over-exploitation.

15 Lee Harrison pers. comm.

16 Presentation by Gokula Dasa at Hepburn Permaculture Design Course 1995.

17 The 64,000 ha Wombat Forest in central Victoria was completely devastated in the gold rush between 1860 and 1890. A complete ban on cutting sawlogs until the 1930s and continuous thinning for firewood and later short lengths for hard board manufacture helped create a valuable and diverse forest. Unfortunately bad State government policies and management since the 1970s have resulted in over cutting for sawlogs and pulpwood from mature forest and created large stands of dense, unthinned regrowth. For my views on local public forest management see Article 15 "Wombat Forest Submission" in *David Holmgren: Collected Writings* 1978-2000.

18 See HDS web site http://www.holmgren.com.au/FF.html

19 The Prosilva Network of Nature based forestry See web site
http://ourworld.compuserve.com/homepages/J_Kuper/page1_E.htm

20 Bill Mollison, pers. comm.

21 Especially on the roots of Ericaceous and Proteaceous plants (eg Arbutus and Banksia). Although these species are typical of very low fertility environments, the potential to use these or similar species to mine previously applied but locked up phosphates in fertilised soil will be an important aspect of sustainable agriculture in the future. Of the most commonly applied mineral fertilisers, phosphate is most prone to being locked up (mostly as iron and aluminium phosphate). As a result most farm and garden soils can be considered as stores of phosphate which, given the right conditions, mycorrhizal fungi should be able to mine for use by crop plants. Surplus nitrogen and potassium on the other hand tend to gas off or leach from the soil.

22 See "Feeding the Soils of Africa" in *Ecos* 103, April-June 2000

23 A local indigenous tree in Kenya, threatened by over exploitation for the bark that is sold as a medicinal remedy for prostate disorders.

24 K.M. Dallas, *Horse Power* Fullers Bookshop Hobart 1968.

25 Monoculture means the growing of agricultural and forestry crops in large stands of a single species, sometimes of a single variety.

26 For an excellent and practical modern text on the subject see Sidback, Hans, *The Horse In The Forest: Caring, Training, Logging* Swedish University of Agricultural Sciences Research Information Centre 1993.

27 Experiments in the Yarra Valley with eagles have kept vineyards free of pest bird damage but wildlife laws in Australia are a considerable impediment to this form of pest bird deterrent becoming more widespread.

28 Costanza et al "The Value of Everything" in *Nature* 395, 430 (October 1998).

29 Ocean Arks International http://www.oceanarks.org/

30 An alpine swamp receiving ski resort effluent at Thredbo for more than a decade was extensively studied in 1981 and found to be highly efficient at removing both nitrogen and phosphorous. Attempts to improve the performance of the wetland showed it was a highly co-evolved system best left alone. See *Ecos* no. 60, Winter 89.

31 See Article "Submission to Inquiry in Pest Plants in Victoria" in *David Holmgren: Collected Writings* 1978-2000.

32 Peter Harper (Centre For Alternative Technology Wales), pers. comm.

33 Michael Wilson *Post gold rush Stream regeneration: implications for managing exotic and native vegetation* Centre for Environmental Management, University of Ballarat and presented at the Second Australian Stream Management Conference in February 1999.

34 For an overview of this major failure within the Landcare movement to recognise and value existing vegetation along urban and rural streams see Article 19 "Permaculture and Revegetation: Conflict or Synthesis" in *David Holmgren: Collected Writings* 1978-2000.

35 Cuba's extraordinary development of urban and rural organic agriculture is one of best large-scale models for transition to descent. The collapse of the Soviet Union's supply of subsidised fuel, fertilisers and pesticides initiated the proactive Government policies and programs, which are now providing social, economic and ecological benefits beyond those originally imagined. See following report www.projectcensored.org/intro.htm

36 Harcourt Park, Bendigo City in Central Victoria

37 Typhus species

38 CDS units cost around AUD$35,000-AUD$70,000. Servicing and maintenance costs are estimated to be in the order of 5%-12% of capital cost per annum. See web site www.cdstech.com.au/products.html

Produce No Waste

A stitch in time saves nine
Waste not, want not

This principle brings together traditional values of frugality and care for material goods, the mainstream concern about pollution, and the more radical perspective that sees wastes as resources and opportunities.

The industrial processes that support modern life can be characterised by an input–output model, in which the inputs are natural materials and energy while the outputs are useful things and services. However, when we step back from this process and take a long-term view, we can see all these useful things end up as wastes (mostly in rubbish tips) and that even the most ethereal of services required the degradation of energy and resources to wastes. This model might be better characterised as "consume–excrete". The view of people as simply consumers and excreters might be biological, but it is not ecological.

Bill Mollison defines a pollutant as "an output of any system component that is not being used productively by any other component of the system".[1] This definition encourages us to look for ways to minimise pollution and waste through designing systems to make use of all outputs. In response to a question about plagues of snails in gardens dominated by perennials, Mollison was in the habit of replying that there was not an excess of snails but a deficiency of ducks.

The earthworm is a suitable icon for this principle because it lives by consuming plant litter (wastes), which it converts into humus that improves the soil environment for itself, for soil micro-organisms and for the plants. Thus the earthworm, like all living things, is a part of web where the outputs of one are the inputs for another.

The proverb "waste not, want not" reminds us that it is easy to be wasteful when there is an abundance but that this waste can be the cause of later hardship. "A stitch in time saves nine" reminds us of the value of timely maintenance in preventing waste and work involved in major repair and restoration efforts.

Waste or Exchange in Nature

The issue of apparent waste in nature needs to be understood in terms of system and subsystem boundaries. The tripartite altruism discussed in **Principle 4**: *Apply Self-regulation and Accept Feedback* showed that apparently wasted energy and resources from one organism or species contribute to supporting lower-order system providers and higher-order system controllers. These feedback payments and taxes help maintain the flow of energy and environmental stability for the organism, so they are not wasteful.

For example, plants lose up to 10% of their primary chemical energy in carbohydrates through their roots, which appears at first to be wasteful.[2] In fact the "lost" carbohydrates feed symbiotic and free-living micro-organisms in the soil, which supply the plant with critically needed mineral nutrients. Thus, what appears to be waste is actually exchange.

Nutrients in shed leaves are processed by soil organisms and converted to humus that can then feed the plant. This also has the effect of stimulating a very rich soil ecosystem and other plants from which the original tree may benefit directly or indirectly.

On the other hand, there are more risks of loss to the plant in this process from fire, leaching, erosion and competition from other plants. Trees hedge their bets by extracting some of the mineral nutrients from old leaves before they are shed. This efficiency in nutrient recycling is internal to the plant organism but it requires metabolic energy and resources. Eucalypts are very efficient at extracting phosphorous from mature leaves because they have evolved in soils deficient in this critical nutrient. Because eucalypt foliage is so low in nutrients and contains oils and other substances toxic to soil life, eucalypts are poor at building humus-rich (high-energy) fertile soils. Deciduous trees tend to build humus-rich fertile soils more rapidly, at least partly because the quality of shed leaves is superior.[3]

Similarly, animal species vary in how efficiently they extract all the nutrients from their food. Carnivores and omnivores (such as dogs, fowls and people), which live on energy-rich and nutrient-rich food, are much less efficient than herbivores, which live on lower quality foods. Consequently, the manure from herbivores is a less concentrated source of mineral nutrients. In China in 1900 the farmers who bought nightsoil from the various European cantons in Shanghai paid the highest price for the German product because Germans ate more meat than other nationalities and therefore excreted more nitrogen and mineral nutrients.[4]

Thus *as a general rule, organisms and systems that are sustained by rich energy resources generally appear more wasteful, but they generally support richer co-evolved systems.*

This view of waste as a potential abundance can be seen everywhere in nature because over time any unused resources will, by system co-evolution, become the energy source for something else. Some of the factors that contribute to this process include limits on available energy, biodiversity and competitive pressure.

Similarly in traditional pre-industrial societies, waste is minimal due to limited resources; any wastes are readily absorbed and recycled by environmental systems because they are made from low-cost biodegradable materials.

Waste Minimisation

The slogan "refuse, reduce, reuse, repair and recycle" provides us with a hierarchy of strategies for dealing with waste. Refuse means to decide not to engage in the consuming action or task in the first place because it is not necessary. Reduce means to minimise the materials and energy required or the frequency of the consuming action. Reuse means either reuse for the same purpose or put to the next best use. Repair means to use skill and very limited additional resources to restore function. Recycle means to break down into more basic elements or materials before being reprocessed for the same or other uses.

Refuse and reduce

After generations of fossil fuel-based affluence, the opportunities for refusal and reduction in rich countries are so many that "conservation" is said to be the greatest new resource available to exploit. For example, in the electricity industry, promoting or subsidising energy-efficient light bulbs can be equivalent to a new power station in saved demand.

Introduction of compost toilets, rainwater tanks and other water-conserving strategies can replace the need for new reservoirs for town water supply.

The revolution in reduction has been greatest in industry and business because cost and competitive pressures drive rational decision-making. However, *the greatest opportunities for savings from waste refusal and reduction are at the household and personal level, where people often feel that extravagance and waste are elements of their sense of freedom and affluence. In affluent societies new extravagant expressions of consumption develop with each generation which, in succeeding generations degenerate into a habitual norm, and eventually an addictive necessity.* The need to buy new (but poor-quality) clothes, the habit of throwing away half the food on the plate or leaving the lights on at night so the kids don't get scared are all behaviours which show this pattern from extravagant consumption to habitual norm to addictive necessity. Addiction to wasteful habits is a factor that has been underestimated in driving consumption by the majority and in keeping the disadvantaged poorer than they need be.

Reductions in over-consumption of food, drugs, material goods, media and entertainment all have the potential to improve quality of life, and they are the most obvious examples we have of the concept of "do more with less".

Governments do not generally support major social changes away from addictive consumption, even though the social and environmental benefits would be great, because the growth economy is inextricably tied (that is, addicted) to dysfunctional over-consumption.

Container reuse

Fired pottery containers were one of the essential technological innovations necessary for the agrarian revolution. The opportunities to reuse industrially produced containers are enormous in the transition to energy descent. This "next best use" catches energy at each step of degradation before it degrades to the point of being unusable by a system. For example, if we reuse a food container that is far better than sending them away for recycling. Most official waste reduction strategies place little emphasis on reuse or confuse it with recycling.

Reuse for the same purpose was a normal unquestioned part of domestic and commercial activity less than 40 years ago. The returnable milk bottle worked well when systems of production and distribution of milk were smaller, domestic standards of care in cleaning bottles were high, and there was little fear of litigation.[5] The more minimalist system of decanting into everyone's private milk billy also worked reasonably well in its social context. With declining energy, it is unlikely that we will redevelop these systems in an orderly fashion.

Free or cheap industrially produced containers, which are rubbish for the average consumer household, are a bonus for those involved in self-reliant living. This abundance can be seen as a subsidy to the minority who adopts the lifestyle of energy descent. Some argue that if it's not possible for everyone to behave this way then it is not sustainable.[6] I see these advantages to "early adopters" of any innovation in behaviour as a perfect mechanism for accelerating adaptive change.

Reusable containers provide enormous opportunities for reuse in permaculture systems. Some examples of container reuse and next best use at Melliodora include:

- Spring Valley and other seal-top bottles for hot preserves and pickles
- plastic shopping bags for food storage, as rubbish bags and other uses

- old-style beer bottles for making beer and juice
- egg cartons for egg storage and sale
- cardboard and polystyrene boxes for fruit storage and sale
- plastic pots and plant tubes
- glass flagons (with the bottoms cut off) as cloches, or joined together to form glass bricks
- film containers for seed storage or pill containers when taken to the local vet
- milk cartons as plant pots and tree guards
- 20-litre olive oil drum for a drum fire for roasting chestnuts
- 200-litre drums (purchased) with clip lids for grain storage
- 200-litre polyethylene pickle barrel (purchased) for liquid manure drum
- 20-litre nylon plastic buckets for animal feed and other storage, general farm use and pan toilet buckets
- stackable take-away food containers for nails, screws and other fasteners.

Although a few of these are purchased from commercial "recycling" (that is, reuse) enterprises, many we collect from other householders because we do not get enough due to our minimal purchasing of commercial food and other goods.

Food and water waste

This is another area where there are great opportunities for waste reduction. The obvious and most important behaviour change is to eat leftovers made into delicious dishes. What is left after that is better fed to fowls than composted—the fowls convert the scraps into eggs and also provide manure for soil fertility. We collect local restaurant waste, as our fowls would starve to death on what leaves our kitchen.[7]

Similarly, fresh hot water can be used to bath the family, wash the clothes, and then rinse the nappies before irrigating the fruit trees.

These simple hierarchies of next best use were self-evident to my parents' generation, but they often need to be explained to younger environmentalists who do not have an intuitive understanding of the energy quality hierarchy.

Limitations to reuse

The opportunities for reuse of unwanted or discarded materials are so vast in affluent societies that two traps for the enthusiastic reuser and recycler need mentioning.

First, it is possible to collect so many things that one cannot reuse them before they degrade (due to weather, termites etc.) or are mislaid. This is an example where the focus on efficiency of potential reuse decreases useful outcomes (maximum power) as explained in **Principle** 3: *Obtain a Yield*.

Second, general design strategies can become so dependent on the availability of complex manufactured items at no cost that, when the supply dries up, the design solution is no longer appropriate. For example:

- the "Eltham style" mud brick house using large reused bridge timbers was an appropriate design when these timbers were being burnt on the roadside; today, when many of these timbers are re-sawn for fine furniture, the 300-mm square post may be more of an extravagance

- sheet mulching is an excellent way to quickly establish new food gardens with readily available organic wastes, but the transport of first-quality lucerne hay and other good-quality animal feeds to manage weed growth over large areas may be an inappropriate use of resources.

Repair

In self-organising biological systems from cells to landscapes, processes for repairing damage and restoring function operate to prevent waste of existing structures and to conserve energy and essential materials required to build new ones. For example, the process of wound healing involves a whole series of complex repair processes, from stemming the loss of blood to scar tissue formation. Similarly, wounds in the earth that destroy vegetation and topsoil are repaired by rapidly colonising pioneer plants which grow on exposed subsoil and begin the process of building new soil organic matter.

In traditional societies, tools and other heirloom possessions were lovingly repaired as soon as damage occurred. For my parents' generation that lived through the Great Depression and World War II, the skills and commitment to repair were fundamental to their ethos of proud frugality. Today some of these values survive condensed in dedication of a few, generally older, people taking the time to darn a much-loved jumper or spending endless hours in restoring old furniture or vintage motor cars.

The permaculture approach is to balance the obsessive focus on restoration of one valued item with more pragmatic and timely repairs of anything of value, following the proverb "a stitch in time saves nine". But it is hard for even the most dedicated repairer to apply this proverb literally when near new clothes can be bought from opportunity shops for a few dollars.

Maintenance, especially of buildings and infrastructure, is a more fundamental and cyclical aspect of repair, which is considered in more detail later in this principle.

Recycling

Recycling is the most generally overemphasised of the strategies for preventing waste. Recycling suggests some input of energy to actively degrade a material to its more basic constituents. For example, recycling a glass bottle requires energy to melt and remould the glass into a new bottle. Reusing the existing bottle is generally preferable.

In all ecosystems decomposer organisms such as earthworms recycle organic materials to humus and minerals in the soil for later uptake by plants. In the process, they gain energy from the decomposition process needed for survival and reproduction. This is the perfect model for recycling materials once we have exhausted reuse and repair options. The widespread innovation in industry to replace toxic and durable materials with non-toxic and biodegradable ones is one of the most successful aspects of industrial innovation using ecological principles.

Industrial Models

Industrial recycling as a transitional strategy

Increasing efficiency and complexity in industrial recycling have made inroads into the rubbish mountains accumulating at landfills in recent times, but new opportunities for centralised recycling of small and widely distributed source materials will decline as the costs of transport and energy rise. At the same time household recycling, such as composting for food production, will reduce the scale of source materials available for collection.

More and more in industry, we see retrofit design to make better use of what was previously wasted, along with design of more integrated systems.

New technologies[8] are being used to burn green waste (from gardens and urban landscapes) to generate electricity, reducing both landfill and greenhouse gas generation. A local intensive piggery[9] has received awards for its environmentally progressive redesign to make use of pig manure for methane to run the piggery and produce saleable compost.

These examples of reducing waste and pollution while achieving economic gains are commendable, but they are still only transitional technologies making use of current wastes. As contraction becomes the norm, these opportunities might not be available, similar to the bridge timbers and the sheet-mulched garden mentioned before.

The current green waste stream from cities is available because urban affluence supports the management of gardens and landscape to export biomass as waste. Urban landscape managed according to permaculture principles would generate little, if any, green waste. The technology to efficiently burn green forest waste might still be useful, but the environmental and energy gains now being made will decline, as waste is not created in the first place.

In the case of pig husbandry, intensive piggeries will be far too energy and resource intense to be economic in the future, so what is seen as state-of-the-art environmental technology may actually be no more than bandaids on a flawed system. The latest research[10] on natural free-range pig raising (in Denmark) has shown that, when the pig is allowed to behave naturally in a free-range rotational environment and still access concentrate feed, it grows just as fast as when kept in costly sheds. Managers of intensive piggeries find it difficult to believe such evidence because it suggests that, even with current cheap energy and resources, intensive piggeries may simply be a system for keeping farmers in debt by investing in expensive technology and products.

These examples show how upstream redesign can cut off the resources that feed downstream waste-based industries. Further, they illustrate the more general need for fundamental change to avoid design cul de sacs. We can avoid some of these dead ends by considering, at least qualitatively, the benefits and losses to larger-scale systems (such as society or the wider environment) caused by a design strategy or solution. We can also take a fresh look at the intrinsic nature of the elements (such as gardens or pigs) that are generating the waste, to see if our understanding of the element is constraining our design solution.

A second Industrial Revolution?

The Industrial Revolution increased the efficiency in the use of human labour. This was most dramatic in the textile industry, where the use of technology and cheap energy and materials massively expanded labour productivity.

Today, with fuels and non-renewable resources in decline and our renewable resources of soil, forests, fisheries and so on in a poor state, the imperative has shifted to increasing the efficiency in use of every kilowatt of electricity, tonne of steel, cubic metre of wood or water. At the same time, human labour and skill in science, business and technology are abundant. Industrial ecology is a field of study that has emerged over the last decade. It provides some of the conceptual framework for radical examples of industrial redesign.[11] Green technology optimists like Amory Lovins[12] argue that redesign of manufacturing processes driven by market forces is rapidly producing examples of "Factor 4" improvements (twice the value of output for half the consumption of energy and resources) and even of "Factor 10" improvements. Essentially, Lovins believes that the application of natural design principles (read permaculture principles) of integration, feedback, no waste and so on to industrial and business processes is creating a second industrial revolution, which will increase material well-being while reducing environmental impact and depletion.

On the other hand, after some decades of doing more with less, the industrialised world is still highly dependent on consuming energy and resources for economic growth. The fact that the United States has mirrored Lovins' soft energy path of decline in the growth of energy and resource consumption while growing economically (despite unfavourable public policies) can be attributed, in large measure, to the extremely wasteful nature of American industry prior to the 1970s energy crises. There was lots of fat to be trimmed or slashed. Industrial and economic efficiency has a longer history in other technologically advanced countries, such as the Netherlands or Japan, so there is less fat to cut.

Other factors that contribute to the apparent success of the American economy in doing more with less include:

- large-scale migration of energy-intensive and polluting heavy industries to Mexico and other trading partners
- global dominance of information technology, especially the explosion in e-commerce
- rapid growth of economic activity in areas widely regarded as producing "bads and vices" rather than "goods and services", such as crime, the security industry, litigation, interventionist medical services.

The huge improvements in productivity that the green tech optimists envisage depend on more fundamental design patterns and capacity for integration. These are discussed in **Principle 7**: *Design from Patterns to Details* and **Principle 8**: *Integrate Rather than Segregate*.

The prospects for Factor 4 and Factor 10 improvements in energy and resource use are exciting but I believe Lovins' apparent faith in the highly centralised capitalist mode of production ignores analysis and history. Capitalism has throughout its history, been the very engine of consumption and waste, which has been periodically constrained by broader notions of the public good. That unfettered and centralised capitalism will somehow undergo some fundamental transformation to become capitalism based on minimising waste and benefiting the environment seems far-fetched.

New high-tech examples of "doing more with less" can be expected to proliferate as the costs of fuel rise. But this will generate enormous demands for capital and new energy investment at the very time when the large energy-getting industries (coal, oil, gas) will be placing the same demands on capital for a decreasing net energy return. The electricity crisis in California in the late 1990s shows some of these dynamics; it may be a foretaste of the struggle for capital and energy resources over meeting immediate demands or investing in transition.

Industrial efficiency and human ingenuity

If large high-technology industries controlled by corporations make major increases in energy and resource efficiency, this could conceivably provide a stepping-stone out of a high-energy society and put in place some of the conceptual frameworks needed for a more enduring, low-technology, low-consumption society. There is controversy over accounting methods for demonstrating whether or not these massive improvements are real; this is the same issue that underlies the debate over renewable energies (see discussion of EMERGY and other methods of accounting for the efficiency of solar cells in **Principle** 5: *Use and Value Renewable Resources and Services*).

It is widely believed that human ingenuity, design skill and culture are the keys to the second industrial revolution, but EMERGY analysis suggests these less concrete forms of human and social capital are themselves the product of past embodied energy from fossil sources. Although this informational infrastructure is more flexible and enduring than physical infrastructure, like other forms of embodied energy it is subject to gradual depreciation over time. Thus the current rash of brilliant breakthroughs in industrial redesign and engineering can be seen as the natural products of half a century of social democratic politics, education, welfare and other social products of affluence, all refined and honed by twenty years of more laissez-faire capitalism and individualism. (See **Principle** 12: *Creatively Use and Respond to Change.*)

While I am not suggesting that the bonanza of technological innovation is almost spent, there are plenty of indications that huge resources will need to be invested in coming decades to rebuild the depleted social capital which has been the source of current successes.[13]

From a permaculture perspective, most of the existing human and social capital is configured to solve large-scale technological and industrial problems within a framework of market capitalism. Even when more socially and environmental valuable outcomes are mandated, our cultural bias in training and culture causes us to continue to reinvent the old problems in new forms.

The severity of this difficulty is most clear when we think of the huge proportion of scientists and engineers working on military research. What useful job can you give to an aerospace engineer who has spent his career working on the design of some small part of swing-wing supersonic fighter-bombers? Beyond these immediate redeployment issues, society needs to invest its wealth to create new skills and ways of thinking that can help us design and manage natural systems which will continue to provide our sustenance long after we have exhausted the mine of industrial efficiency.

Durability and Maintenance

The prosaic subject of the maintenance of buildings and other physical infrastructure may be just as important in the transition to energetic and economic contraction as new high-tech ways to do more with less.

Maintenance must be one of the least romantic and least loved of everyday activities but it is a critical function in all systems. While repair suggests episodic response to accidental damage, maintenance is the pre-emptive response to the predictable and incremental depreciation of value that (reflecting the Second Law of Energy) affects all stores of embodied energy. These depreciation issues are most obvious in the built environment which does not have the self-maintaining aspects of biological systems.

European cultural traditions placed great value on the durable and the permanent, especially in the built environment. The fable of the three pigs and the varied value of their houses in defence against the wolf suggest these values are deeply embedded in our culture. The attention that is paid to good workmanship, maintenance and durability in the Germanic and Scandinavian cultures makes Australian attitudes seem more akin to those of tribal nomads. Norway has the oldest wooden building in the world. Throughout rural Scandinavia, the seasonal ritual of applying home-distilled pine pitch to the external timber of farm houses and barns was regarded as more of a sacred ritual than a necessary chore.[14]

To take examples from other parts of the world, mud bricks are hardly the most durable building material, but a 900-year-old building in New Mexico at Pueblo de Taos is reputed to be the oldest continuously inhabited building in the world. Such buildings are great testaments to the power of maintenance. Throughout the Mediterranean, stone terraced mountainsides appear as timeless landscapes; but as the last generation of peasant farmers ages, this marginal agriculture is being abandoned. Without maintenance, these landscapes may undergo large-scale erosion before they grow back to forest. The consequences of cessation of maintenance on the extraordinary and extensive rice paddy terraces of South-East Asia would be even more catastrophic.

One of the modern delusions about the built environment is that it is — or at least should be — possible to build things to be maintenance-free. This delusion expresses itself not just in the slackness of a new generation of homeowners, but also in our great public buildings.

Maintenance engineering

Maintenance engineering is a profession with a great future. Commitment of more resources and the most innovative and creative solutions will be required if we are not to be overwhelmed by the cost of maintenance and replacement in our built environment and service infrastructure.[15]

With declining energy, we will find that the resources devoted to maintaining the great stock of built assets and infrastructure created when energy and resources were cheap will become a huge burden, but the labour-intensive aspects of maintenance will be relatively cheaper.

However, if our assets are already in bad repair then labour inputs in maintenance can do little to recover the situation. The graphs[16] in Figure 16 show how the decline in any built system value responds to input of maintenance effort and resources. The third graph shows the effect of lack of maintenance where only large inputs can remedy the situation.

Overlaying this somewhat alarming situation has been the privatisation and corporatisation of almost all publicly owned infrastructure, and the widespread downgrading of funds and staff for maintenance, along with the value placed on high standards of maintenance. Increasing numbers of service failures, and even fatal accidents, in many countries have been attributed to this decline in maintenance standards. These have fuelled public criticism of privatisation, but they are only the visible tip of the problems created by lack of maintenance, which future generations of ratepayers, shareholders and customers will inherit.

These large-scale problems of failures of maintenance under the onslaught of economic rationalism are mirrored at the domestic scale by a lack of interest and effort in home maintenance. In contrast renovation remains an attractive option for the expenditure of surplus capital by Australian homeowners.

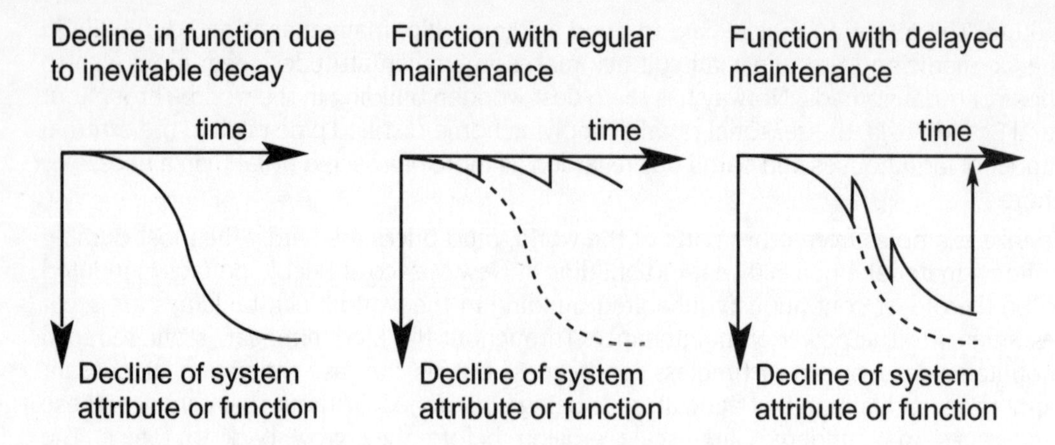

Figure 16: Infrastructure decay and maintenance

The decline in basic maintenance can be attributed to several factors:

- increased demands and opportunities to spend time on work and leisure activities away from home
- high mobility, undermining long-term thinking and values
- continual rises in real estate values, almost irrespective of condition
- decline in the work ethic and the traditional values associated with maintenance.

We realise that houses built in the 1960s, 1970s and 1980s are not lasting well. This failure in durability is often attributed to modern building standards, but it must be at least partly caused by lack of maintenance.

In *How Buildings Learn*, Stewart Brand used archival and current photography to chart the changes made to buildings by owners, occupiers and the natural elements. Brand shows that maintenance, or the lack of it, is one of the critical factors in the evolution and survival of buildings over time. He cites the Pompidou Centre (1979) in Paris as an example of a maintenance nightmare:

> considered a landmark in architectural history, and it is a major tourist attraction, competing with the Eiffel Tower. Where the Eiffel Tower (1889) exposed its structure in an elegant and monumental way, the Pompidou Centre tries to do the same with its services. But iron structures can withstand the elements much better than brightly painted ducts and pipes. The Eiffel Tower's lasting message to architecture is: exposed structure can be gorgeous. Pompidou's lasting message: never expose services.

Rediscovering a commitment to maintenance is one of the most important aspects of this principle. The graphs in Figure 16 make it abundantly clear that slack maintenance is a hidden generator of enormous waste through loss of value and performance in the built environment. **Principle** 12: *Creatively Use and Respond to Change* discusses the alternative strategy of using temporary and easily renewable materials.

Pest Plants and Animals as Wasted Resources

It is ironic that, at the very time in history when concern about the decline in nature's self-renewing abundance has reached a peak, there is an almost equal concern about unwanted abundances and plagues of plant and animal life. Pest plant and animal problems are more prevalent in rich countries for two reasons.

Available water, nutrients and land

The intensification of agriculture has (perhaps counter-intuitively) decreased the amount of marginal land that is managed. This has allowed redevelopment of nature, both native and exotic, on a scale not seen for centuries. Increased use of water and nutrients in intensive agriculture, as well as ornamental and recreational land uses, have stimulated the proliferation of the wild species that are best able to use these surplus resources.

Affluence

A complex set of economic and social factors has also contributed to the spread of unwanted plants and animals. Urban lifestyles, the social welfare net, low prices for food and other natural resources, a lack of people actively involved in agriculture all contribute to the lack of management, harvest and utilisation of wild species.

The prevailing response of land managers, environmentalists and society is to regard proliferating species as new forms of biological pollution when they are in reality, unused resources. Working out more creative and effective ways to use wild resources is a constant theme in permaculture design (see **Principle 3**: *Obtain a Yield*). The excessive surpluses of pest plants and animals represent a special opportunity. In these cases, the more we use the better.

The slogan "the problem is the solution" challenges us to look for creative ways to use what is otherwise regarded as worthless. The commercial harvesting of pest carp from Australian inland rivers for fertiliser is on the right track; but in the long term, higher-value uses, such as poultry feed, should displace use for fertiliser. Eventually, human consumption will be the natural and best control on carp in Australia but it may take a decline in affluence (or available fish stocks of preferred species) for that to be fully realised. Carp themselves are a response to inland waterways overloaded with algal growth. The algae are stimulated by excessive nutrients from farm fertilisers and town sewage entering the streams, as well as increased sediment loads.[17]

Beyond recognising our own role in contributing to nature's plagues, learning how to accept plant and animal plagues as opportunities for us to creatively provide for our needs while restoring a greater degree of balance epitomises our harmonious place in nature. Historically, the disenfranchised and the despised, such as the rabbit trapper, have filled this role. We need to turn this view on its head. In an environmentally aware society, people who manage to earn a living from using pests should be respected and acknowledged as models rather than misfits.

Wasted Human Resources

The ingenuity of a formally well-educated population, predominantly in rich countries, is a well-recognised form of social capital, but it pales in comparison to the knowledge and skill of traditional, mostly rural people, predominantly in poorer countries. Our current global economy devalues these skills; it destroys people's capacity to apply knowledge productively and, at best, treats them as uneducated and unskilled factory fodder.

The majority of people on the planet still have some personal or family experience of a culture of place and living from renewable local resources. Most of these people are struggling to climb aboard the train of industrial affluence. This primarily occurs through migration from rural areas to cities. In the process monetary incomes rise, but the people lose access to unmeasured wealth and have to discard their most useful technical skills and social values.

I do not want to romanticise the rural poor of the Third World. In some ways, they are like the aerospace engineer at the other end of the scale, unprepared by their traditions or experience to understand energy descent and carrying a baggage of prejudices unrealistic expectations. But to classify these people as the most destitute and incapable people on the planet is wrong. I believe *the world's poor, especially the rural poor, represent a vast pool of human resource and capability which is being debased and denigrated at the very time in history when it represents our greatest asset in the transition to reduced energy availability*. *Not only is this a great injustice, but it is also a stupid and grossly inefficient use of human resources*.

Many of the most progressive development projects by small non-government organisations working with the world's poor are consciously or unconsciously applying permaculture principles:

- acknowledging and supporting existing local systems which provide for people's needs, maintain cultural knowledge and foster husbandry of natural resources
- identifying local human and natural resources that are currently undervalued or wasted
- introducing very limited numbers of new species, tools and materials which have been found useful in similar scale traditional systems and which can be maintained using local skills and resources.

One of the most important side-effects of some of these projects is the re-evaluation of traditional knowledge that was being devalued as old-fashioned and ineffective. This devaluation occurs due to a complex of interlocking factors:

- sometimes the useful knowledge is bound within traditional integrated systems that no longer work as they once did because of overpopulation, war, resource degradation or social change
- the knowledge is only understood by old people, sometimes in a language that is no longer spoken by the young
- advertising and government propaganda promotes modern methods and denigrates local knowledge
- the keepers of traditional knowledge perceive that the rest of the world is affluent and uses modern methods.

Many permaculture designers who have worked in development projects in the Third World admit that they learnt more than they were able to contribute. Bill Mollison made the provocative suggestion that we need to get those with knowledge of traditional sustainable systems to the West to teach us how to feed ourselves;[18] this emphasises the fact that conservation and reinvigoration of aspects of traditional sustainable systems is not charity, but a path to survival for the global community.

I have no personal experience in development project work, but I remember being embarrassed when I was asked to present a slide show about Melliodora to a Bedouin community in the Negev desert during a visit to Israel in 1994.[19] The community was struggling to maintain some aspects of traditional life on the fringes of modern Israeli society. Most people worked in the local towns and were on the low rungs of the ladder to hopeful affluence, pulled and pushed by the usual factors, including government obstruction (and at times destruction) of their self-built houses on traditional land. I was a "famous ecologist" from Australia, and presumably rich. At the first slide of our mud brick house, an old man gesticulated excitedly. He said, "Look, a Bedouin house" because earth was a traditional local building material. Perhaps my status as a foreign expert living in an earth house has encouraged some of the younger men to see the value in traditional, rather than industrial, building materials.

In reality, the ability of outsiders to exercise much useful influence in these situations is very limited, unless working relationships based on mutual trust and respect are developed over a long period. Self-help development projects and fair trade co-operatives that allow people to maintain traditional skills and gain modest incomes to supplement self-reliant household and community economies are the natural complement to greater self-reliance and disconnection from the corporate version of the global economy in rich countries.

For those of us from rich countries not directly involved in helping these projects, the most powerful thing we can do to help is engage in the same process in our own homes and communities. With a much greater degree of self-reliance we:

- reduce demand for corporate-controlled exploitative export development of Third World resources
- increase the status of self-reliance in the emerging global culture
- free up capital to flow into underdeveloped countries to gradually correct the mismatch between overdeveloped and underdeveloped economies, which is indicated by the wineglass model of global economic activity.

Although the wineglass model shows a staggering imbalance of measured economic wealth, if the value of free environmental services were added the disparity would not be

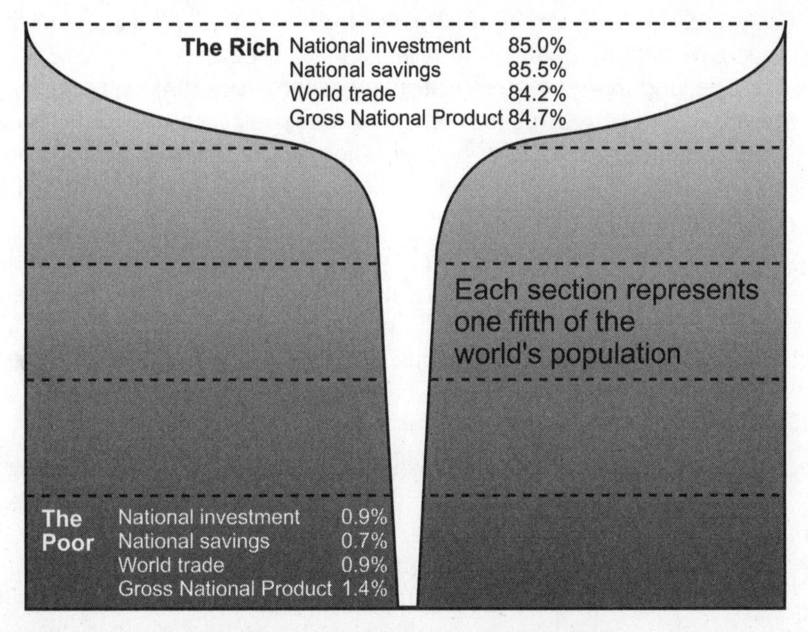

Figure 17: Wineglass model of (measured) global economic activity

quite so extreme. Poor rural people often still have access to relatively more of these unmeasured goods and services (through subsistence agriculture, hunting, etc.), although these sources of unmeasured wealth are diminishing rapidly. EMERGY evaluations of nations give a more realistic measure of real wealth. Although it is not directly comparable to the data for the wineglass model because it measures whole countries, the following table of EMERGY use per person per year paints a more realistic picture of the real wealth circulating in selected countries.

Country	Per person EMERGY use*
India	1
World average	6
Brazil	15
USA	29
Australia	59

*(x 1015 solar emjoules/year)

Table EMERGY use per person in the world and selected countries

The surprisingly high EMERGY per person for Brazil and the extraordinary figure for Australia reflect the very high free services of nature, as well as mining of non-renewable resources in these less-developed countries and, in the case of Australia, the low population. The view of most Australians who travel overseas — that Australia is the best place to live — possibly reflects more than simple national chauvinism and emphasises the immense opportunities in this country to make better use of that real wealth.

Rubbish as Nature

Visiting Israel in 1994, I was shocked by the prevalence of rubbish discarded in the streets and the countryside, the lack of maintenance of buildings and vehicles, and the cavalier approach to waste management and pollution. Israel is, of course, a Middle Eastern nation not a European one, and many Western European sensibilities that we take for granted do not apply. While it is tempting to regard these attitudes (very common in newly developing nations) as environmentally unenlightened, another perspective was pointed out by more than one Israeli, "If modern living generates rubbish, pollution and waste, then it is childish and naive to try to hide and ignore it."

From this perspective, the Israelis or Greeks having coffee on the street cafe as close to the choking car fumes as possible, may be more at one with their world than affluent Swedes living in beautiful, clean and green Stockholm, worrying about Strontium 90, synthetic oestrogens or genetically modified organisms getting into their organically certified food.

Conclusion

The permaculture principle "Produce No Waste" suggests a middle path to living lightly on the planet, while accepting unwanted outputs when they are an abundance that can be used by nature and others as sustenance. In a Gi Gong breathing exercise, we are encouraged to imagine ourselves breathing in clear white light and breathing out toxic black smoke: an idea a little confronting to modern environmental sensibilities. The wholistic view of waste is completed by a celestial scorpion floating above us which takes in the black smoke as nourishment.

1 B. Mollison, *Permaculture: A Designers' Manual.*

2 A. Smith, "Living Soil" *Permaculture Journal* no. 7, 1981.

3 This appears to be true even in southern Australia, where deciduous trees form wild stands on low-fertility bush soils.

4 F.H. King, *Farmers of Forty Centuries.*

5 Concerns over what milk bottles might have used for led to energy-expensive cleaning systems prior to refilling, which in turn led to claims that milk cartons were environmentally preferable.

6 Of course, this argument is based on the assumption that all workable solutions are mass solutions, whereas the diversity principle suggests descent is characterised by a shift from global mass solutions to diverse local solutions.

7 Large quantities of carbohydrate-rich food waste can be unbalanced for egg-laying birds, pigs were the traditional consumers of these wastes. In many domestic situations, a worm farm producing a surplus of protein-rich worms fed to fowls combined with high-quality compost is an optimal system.

8 Fluidised bed combustion.

9 See Future Harvest website, www.mov.vic.gov.au/FutureHarvest/fffuture.html

10 Bent Anderson (of the Danish National Institute of Animal Research) in *Proceedings International Federation of Organic Agriculture Movements conference* Lincoln University, New Zealand, December 1994.

11 See H. Tibbs, "Human Ecostructure" *Whole Earth Review* no. 93, Summer 1998, for an introduction and sources, including J.M. Benyus, *Biomimicry: Innovation Inspired by Nature* William Morrow & Co. 1997.

12 See Rocky Mountain Institute website http://www.rmi.org/

13 The parlous state of Australia's investment in education and research and development is one more obvious example of this problem.

14 Bertil Thermaenius, pers. comm.

15 D. Brett, "Infrastructure: Going, Going, Where?" *Ecos*, no. 61, Spring 1989, CSIRO.

16 After Dr Lex Blakey, reproduced in *Ecos*, no. 61, Spring 1989, CSIRO.

17 Sediment in rivers, especially the Murray, is primarily from bank erosion generated by running the river as a massive summer irrigation channel.

18 In the video *Bill Mollison The Permaculture Concept: In Grave Danger of Falling Food.*

19 See Article 14 "Impressions of Israel" in *David Holmgren: Collected Writings 1978-2000.*

Design from Patterns to Details

Can't see the wood for the trees

The first six principles tend to consider systems from the bottom-up perspective of elements, organisms, and individuals. The second six principles tend to emphasise the top-down perspective of the patterns and relationships that tend to emerge by system self-organisation and co-evolution. The commonality of patterns observable in nature and society allows us to not only make sense of what we see but to use a pattern from one context and scale to design in another. Pattern recognition, discussed in **Principle 1**: *Observe and Interact*, is the necessary precursor to the process of design.

The spider on its web, with its concentric and radial design, evokes zone and sector site planning, the best-known and perhaps most widely applied aspect of permaculture design. The design pattern of the web is clear, but the details always vary.

Modernity has tended to scramble any systemic common sense or intuition that can order the jumble of design possibilities and options that confront us in all fields. This problem of focus on detailed complexity leads to the design of white elephants that are large and impressive but do not work, or juggernauts that consume all our energy and resources while always threatening to run out of control. *Complex systems that work tend to evolve from simple ones that work, so finding the appropriate pattern for that design is more important than understanding all the details of the elements in the system.*

The proverb "can't see the wood (forest) for the trees" reminds us that the details tend to distract our awareness of the nature of the system; the closer we get the less we are able to comprehend the larger picture.

Pattern Thinking

Bill Mollison's introduction to patterns in nature[1] provides a broad, inspiring picture of a great field of potential applications in permaculture design. This search for spatial and temporal patterns in nature, which take us beyond the Euclidian geometries that dominate our educated thinking, is important for designers in every field. Mollison says "Learning to master a pattern is very like learning a principle; it may be applicable over a wide range of phenomena, some complex and some simple." Although it is important to understand the relevance of the organic and apparently irregular patterns of nature to human systems, often our attempts to apply these patterns are arbitrary and inappropriate.

Christopher Alexander's Pattern Language[2] was a milestone in the recognition and organisation of classic patterns of human-scale built environments. Developing a similar pattern language for the much broader scope of permaculture design is a need that several permaculture designers[3] have recognised. However, in developing such a pattern language there are two problems.

- The processes of biological growth and organisation represent a far greater and more diverse field than the built environment.
- The implications of the rise and fall of the energy base of humanity must be understood and expressed through design principles before we have the framework needed to systematically identify and organise appropriate patterns.

Whether we are designing a garden, a village, or an organisation, we need a broad repertoire of familiar patterns of relative scale, timing and geometry that tend to recur in natural and sustainable human systems.

In this principle I want to contribute to development of the pattern language of permaculture design by focusing on examples of structures and organisation that seem to illustrate the balanced use of energy and resources that we call good design. As explained in **Principle 3**: *Obtain a Yield*, this balance is actually one that achieves maximum power, but as energy availability and quality decline, our common sense and intuition about what is optimal design often fails us.

Further, we need to relearn pattern recognition because cultural innovation, especially media technologies, have scrambled the pattern thinking that was common in pre-industrial societies.[4] This loss of ability to see, hear and otherwise recognise the patterns of nature may be our greatest impediment in our attempt to adapt to realities of energy descent.

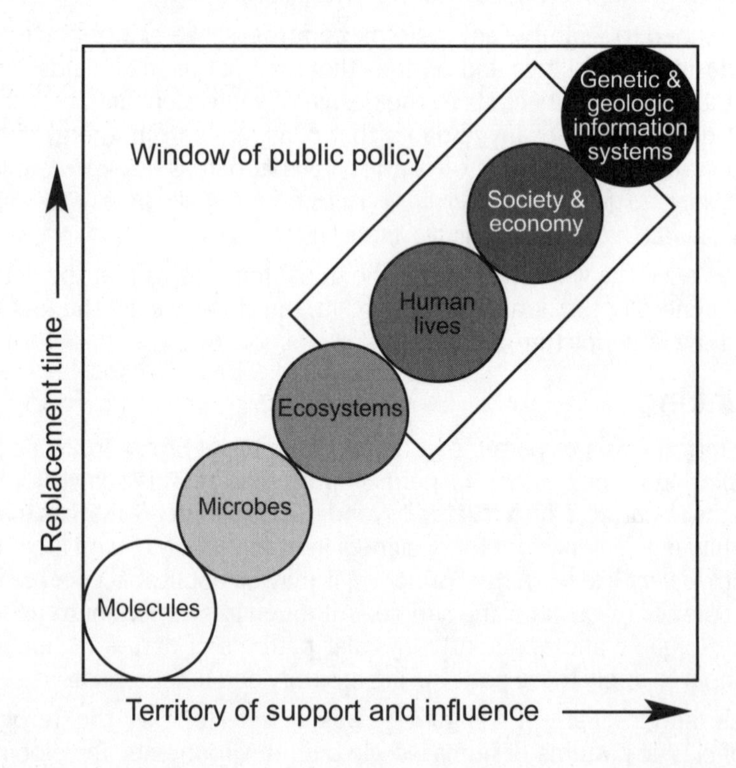

Figure 18: Hierarchy of systems graphed in time and space (after Odum and Odum, 2001)

Scales in Space and Time

Many issues and strategies in permaculture design relate to selection of appropriate scales of systems, decision-making and action. Spatial scales can be thought of as ranging from perceivable with the eyes (human scale), down to the microscopic, and up the global and beyond. Scales in time can be thought of as extending from the human daily life and lifetime, down to the fast and short-lived atomic processes, and up through ecological, historical and beyond to geological time.

Principle 9: *Use Small and Slow Solutions* considers the imperative to reduce size and speed in the transition from energy growth to descent. Here, I focus on a more general understanding of scale in both space and time as patterns for permaculture design that are as important as the zone and sector concepts are for site design.

We can readily see change in people, other living things and systems of a similar scale to ourselves, but we find it difficult to perceive — let alone comprehend — changes in macro or micro systems at scales very different from those that our senses and memory can access. Tools and technology, stories, history and culture have extended this range of comprehension to varying degrees, but human nature, to a surprising extent, is bounded by the human scale of the senses and personal memory. Science and technology, especially medical technology, are widely acclaimed for their power over micro scale systems that affect our well-being, but scientific understandings of macro systems have not received the same acclaim, perhaps because large-order systems are not so amenable to control and any benefits are slow to emerge. Similarly, spiritual teachings that focus on macro scale understandings and wisdom do not have wide appeal compared with ideas dealing with the here and now. This bias towards short-term thinking within our own lifetimes has been described as an evolutionary weakness of humans, which we must overcome if we are to survive.[5]

To effectively grapple with sustainability questions, we must develop a rich and contextual understanding of how scales in space and time shape the design and evolution of systems. Further, we need to develop a shared language for talking about scale because it is a constant source of error and misunderstanding in any collective action.

Figure 18 shows how systems across a large range can be graphed according to scale in space and time. Physically small, short-lived systems can be described as drawing from, and affecting, a limited territory and having a rapid replacement time.[6] The graph shows the window of public policy slightly expanded from the traditional human-only focus to include issues of natural system management and resource depletion. Because this general pattern of linkage between space and time in systems is so widespread in nature, it is useful for thinking about issues of appropriate scales in design for energy descent.

A more intuitive way to portray the same hierarchy of systems is shown in Figure 19. where large, slow-changing systems are composed of small, rapidly changing ones.

Physical scale or functional scale

Although physical scale is often a good indicator of functional scale in systems, this is not always the case. For example, predators are often a similar size to, or even smaller than, their prey, but they occupy much more territory and are less numerous. Thus Odum's description of "territory of influence and support", while not an immediately obvious characteristic of any element or system, is a more accurate description of their relative power than physical size.

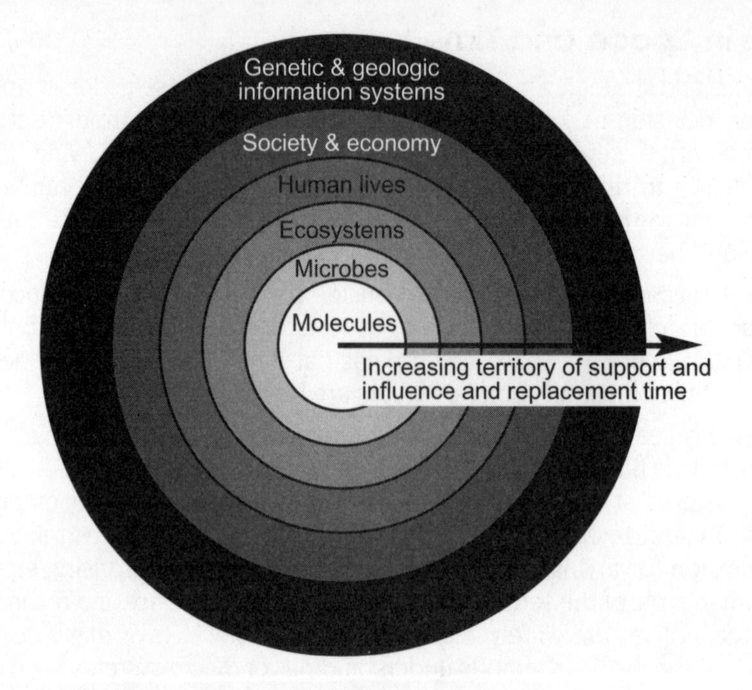

Figure 19: Hierarchy of systems nested in time and space

The difference in scale between related systems is often an order of magnitude. For example, predators typically occupy 10 to 100 times (1 to 2 orders of magnitude) more territory than their prey and are proportionally less numerous. These order-of-magnitude jumps in functional scale seem to also apply in the design of human systems (see discussion below of Permaculture Zones and Scales of Production and Growth).

Clustering of Energy in Space and over Time

A simple interpretation of the Second Law of Energy suggests that energy, matter and resultant activity and structure tend to become more diffuse and randomly distributed. Thus, concentrations were seen as random statistical phenomena. One of the most important insights of Chaos theory is that the concentration of energy, mass, activity and structure created by living processes is not an aberration of the Second Law of Energy but an inherent characteristic of the universe. Thus, *wherever (and whenever) in the universe available energy and matter are abundant, self-organisation leads to increasing complexity of activity and structure.*

Landscape patterns

The evolution of landscapes reflects the capture and storage of energy from the sun and from earth uplift, described in **Principle 2**: *Catch and Store Energy*. Although the primary sources of energy vary from place to place, the conversion of these primary sources by large-scale environmental systems of earth movement and climate further concentrates and structures the energetic potential and fertility available to plants, animals and people. This concentration operates in time as well as in a physical location, so much of the critical activity in nature occurs in short events. For example, most of the erosion that happens along stream courses occurs in very short-lived peak floods. Between these events, nothing happens. This pattern has enormous implications for sustainable land use, where adverse effects may only show up in occasional, short-lived events. In environmental monitoring, measurements that do not run continuously over long periods can lead to false conclusions because they miss critical events that describe and determine environmental outcomes. (See **Principle 12**: *Creatively Use and Respond to Change* for extensive discussion of the pulsing model of ecosystems.)

In the same way that the cellular architecture of organisms concentrates energy and functions in cell nuclei, landscape and ecosystems have nodes of concentrated energy, which support special, highly productive systems or elements. The implications of this tendency to clustering opens up many possibilities for ecological design, but it also warns us of pitfalls and unpredicted outcomes. The concentrations of fertility and harvestable resources are perhaps the most fundamental factors that have determined the patterns of human land use and settlement through time.

Rainforest versus eucalypt woodland.

The balance between rainforest and eucalypt woodland in many Australian indigenous landscapes reflected an ecological and cultural optimum, based on a clustering of energy and resources. High rainfall, high fertility landscapes may sustainably support an almost uniform cover of rainforest. Drier, low fertility landscapes can also be colonised by rainforest in the absence of fire, but catastrophic fire during severe droughts could wipe out rainforest from a whole catchment and lead to major losses of soil in erosion following the fire.

These dry, low-fertility landscapes regularly lose water, nutrients and organic matter at higher rates than rainforest (mostly after small fires and floods). These sources of energy can be caught by denser and more vigorous vegetation growing along streams. This concentration of energy may be enough to support a narrow corridor of richer, denser riparian rainforest, with high species diversity and high biological productivity.[7]

The indigenous landscapes of Australia frequently reflected this general pattern of site enrichment within landscape depletion. This pattern was at least in part the result of Aboriginal management[8] that reinforced these natural tendencies inherent in all low-energy landscapes.

While loss of energy and nutrients in one part of a catchment may be seen as local degradation, the effective capture and concentration of that energy in another part of the system is a master pattern for dealing with low or declining availability of energy. By allowing surplus resources from one system to concentrate at a particular site, a more energy-demanding, richer system can establish, which in turn captures more free energy.

Soil patterns

My own observations of skeletal and duplex soils in Central Victorian eucalypt forests show a mosaic pattern of soils at scales as small as one square metre. Hard-setting surfaces devoid of life and organic matter shed almost all water (and leaf litter), while areas of accumulation behind logs and other debris absorb this water, and create friable, humus-rich topsoils. As well as supporting herbaceous groundcover species, these sites allow deep absorption of water into the subsoil and fractured rock aquifers, which support the impressive drought-hardy Box eucalypts that dominate these forests.

Excavation in these soils shows a mosaic of buried organic matter, clays, silts, gravels and stones; these reflect long-term cycles of cut-and-fill erosion over thousands of years, creating localised degradation and enrichment. Similar large-scale patterns have been described as characteristic of arid ecosystems. These examples show how nature responds to degradation and loss of energy by concentrating efforts on capturing and using the mobilised energy and resources at points of natural accumulation which landscapes provide. The Middle Eastern and North African oasis, with its concentrated fertility and abundance amid the barren desert landscape, is the iconic representation of this phenomenon, which was both alien and inspiring to the European imagination. The ancient Nabbatean run-off agricultural systems in the Negev desert were modelled on patterns observable in the local landscapes.[9]

This self-organisation evident in wild landscapes and accentuated by indigenous and traditional land use should be seen as a general pattern for sustainable land uses in a low-energy future. It provides a framework for the location and scale of more fertile systems able to supply human food and other special needs, as well as dealing with land degradation across catchment landscapes.

Structural Patterns of Vegetation

Natural ecosystems are, almost by definition, at maximum power and optimal efficiency. This state results from long periods of co-evolution within the constraints of available energy. The ecosystem structure and organisation as well as the species present are important factors in getting the most energy for the least effort[10]

Forests as models for agriculture

The permaculture concept began with the idea of natural ecosystems, especially forests, as models for agriculture. In many parts of the world with moderate to high rainfall, especially steeper upland areas, forests are the predominant ecosystems.

These ecosystems:

- are dominated by large trees which grow tall through competition for light
- include understorey species that can use the filtered light and stable microclimate created by the canopy
- have diverse habitats for both small and large animals
- are very effective at holding soil against landslips and other forms of instability.

Many traditional agricultural societies recognised the value of forests for catchment protection and other long-term values. For example, the origin of the forest conservation ethic in Japan[11] dates from the 14th century, when only 30% of that mountainous country was clothed in forest. Floods devastated the lowland rice-growing areas. Since the 17th century Japan has retained more than 70% of its area under forest.

In some places, forests actually provided many of people's basic food needs. Corsica's "rural civilisation" was supported by chestnut forests which provided, as well as the usual resources of fuel, timber and animal forage, the staple food of the people. This and other examples documented by Russell Smith in the 1930s and 1940s[12] showed that food forests have been more productive and sustainable than grain agriculture in many parts of the world. I believe that the paucity of surviving examples of food forests that provide staple sustenance may be due to the effects of war, rather than any lack of agricultural productivity from food forests (especially in the "old world" of the Mediterranean and Middle East).[13]

Permaculture food forests

The permaculture strategy of establishing "food forests" — which are composed of a diversity of species that provide for peoples' needs, and yet have many of the characteristics of natural forests — is the best-known application of this principle. This idea has been enthusiastically and widely applied by permaculturalists. These systems, especially in moist subtropical and tropical areas, have been productive and to a degree self-maintaining, but they have also been criticised as inappropriate from more traditional organic[14] and biodynamic[15] perspectives.

In cool temperate climates most of the productive fruit and nut trees have evolved to flower and fruit and resist fungal diseases in more open environments than dense forests, whereas many subtropical species bear fruit under a shaded canopy and are well suited to forest systems. On a large scale, dense forests are only possible in high-rainfall areas or along streams and sources of abundant moisture. In low-rainfall regions, trees become more widely spaced, and a woodland structure is the norm. Most of Australia, prior to European arrival, was some sort of woodland. Woodlands are still multi-layered but the plants in those layers are more widely spaced than in forests to reflect the limitations of rainfall.

These caveats help us refine, rather than reject, the relevance of the food forest model to sustainable agriculture and encourage us to look for other ecosystem models.

Agroforestry and analogue forestry

The agroforestry concept[16] integrates productive use of trees into conventional agricultural systems. It developed in parallel with permaculture over the last three decades of the 20th century, with considerable intellectual cross-fertilisation. In its most limited sense, agroforestry consists of widely spaced timber trees within cropping and grazing paddocks.

The more recent concept of analogue forestry[17] mimics the indigenous climax ecosystem; it recreates the structural and functional interactions of the natural forest, using both indigenous and exotic plants. In many ways it is a restatement of the original permaculture concept with an academic lineage from systems ecology, but its cultural roots are in traditional tropical systems of subsistence mixed tree cropping.

Landcare and Redesigning Australian Agriculture

Perhaps the greatest vindication of the concepts of permaculture, agroforesty and analogue forestry has been the recognition that it is the lack of perennial and woody plants in Australian agricultural landscapes that is the prime cause of salinity and a range of other land degradation symptoms affecting our rural landscapes. Annual crops and pastures are simply not efficient at using rainfall. This unused water leaches nutrients and causes acidification, and contributes to a rising saline groundwater. The Landcare movement[18] has spread these understandings across rural Australia.

The structure of ecosystems reflects annual rainfall more than sunshine levels. The period of active growth of vegetation also closely follows the seasonality of rainfall. Thus in southern Australia winter-active vegetation predominates, while in northern Australia the reverse is true. Although irrigation allows us to bypass these climatic limitations to some degree in garden agriculture, large-scale systems need to be designed to reflect and use rainfall where and when it falls. Efficient use of rainfall is now one of the catchcries of the Landcare movement.

However, many scientists and farmers now accept that broadacre agriculture requires fundamental redesign. Systems ecologist Ted Lefroy[19] studied the use of water and nutrients in Western Australian wheat belt agriculture as measures of ecosystem efficiency. His work was one of the starting points for a large-scale multimillion-dollar research and development project involving the Land and Water Resources Research and Development Corporation and the CSIRO, called Redesigning Agriculture for Australian Landscapes. Whether such projects make a major difference to the design of agriculture on the ground remains to be seen.

Optimum Plants and Animals

Despite the enormous variation in the characteristics of individual tree species, some of the general patterns of global climate and biogeography are useful in understanding which systems and species are likely to be most adapted to our local environment. Bill Mollison devotes a whole chapter in *Permaculture: A Designers' Manual* to explaining the broad patterns of climate and microclimate because of the importance of this basic geographic knowledge in comparing and integrating information for similar environments.

Deciduous and evergreen trees

One of the most noticeable differences between forests around the world is the predominance of evergreen trees in most climates, except in the temperate zones of the northern hemisphere where deciduous species predominate. Deciduous species are best able to use the seasonal pattern of strong alternation between cold winters and warm summers, so it seems unusual that deciduous species are so rare in the southern hemisphere temperate zones. The explanation of more subtle but fundamental differences between the northern and southern hemisphere temperate climates has major implications for permaculture design of tree crop and orchard systems.

Because of the predominance of ocean in the southern hemisphere, the alternation between summer and winter is less extreme. This milder climate favours the possibility of growth in winter and, in winter-rainfall areas, the leaching of soil nutrients.[20] Thus evergreen species are generally better adapted because they take advantage of the more erratic growing conditions. The lesser seasonal variation also leads to erratic cold changes in spring and even summer. Deciduous trees, which replace their whole canopy each spring, are more vulnerable to frosts during the growing season than evergreen species, in which the majority of foliage is always mature and therefore less vulnerable.

Most of the fruit and nut trees that can be grown in Australian frosty cool temperate climates are deciduous, and yet the winters are just warm enough for plant growth (by a limited range of species that are resistant to cold and frost damage). To make efficient use of winter rain at Melliodora, we make use of as many frost-hardy evergreen fruiting species as possible (feijoa, olive and loquat), but our large deciduous orchard is interplanted with tagasaste and wattles, which are evergreen, winter-active; they fix nitrogen, attract bees, and provide mulch, poultry and goat fodder. Between the trees we have diverse mixed pastures of grasses, legumes and herbs, which are also winter-active, and under the trees we have spring bulbs that are also active well before the trees are in leaf. This system yields less fruit than a traditional orchard of the same size but, as well as providing other yields, it is more efficient in growing organic matter, using natural rainfall, recycling nutrients and preventing leaching.

In milder coastal areas of Victoria a wide range of subtropical evergreen species, such as avocado, citrus, macadamia and sapote, can be grown. This astounds visitors from the northern hemisphere, where at similar latitudes winter cold precludes such species. In these areas high-density evergreen food forests become a design option.

Fertility-loving and infertility-loving plants

The adaptation of plants to fertile (high-energy) and infertile (low-energy) soils also illustrates subtle but fundamental patterns important for design. It is often assumed that plants that are found naturally growing on infertile soils prefer these conditions. In fact, it is their relative efficiency and therefore competitive advantage on poor soils that accounts for their natural occurrence. Most plants grow better the higher the fertility, but only plants evolved (or bred) to high-fertility soils are reasonably efficient in these situations.

Although our aim may be to increase productivity, matching plants and animals to suit current conditions is also an important strategy. The hardy nature of local indigenous and other Australian native plant species makes them ideal for providing shelter and other functions in unirrigated and unfertilised soils. When we plant a new garden, the soil fertility may support only a limited range of vegetables, but over time we can introduce more demanding ones. The use of pioneer plants that grow in and improve poor soils is a strategy which is considered further in **Principle 12**: *Creatively Use and Respond to Change.*

Plant and animal biomass as indicators of fertility

The relative amounts and types of plant and animal biomass in natural and managed ecosystems can be used as a general indicator of mineral fertility. Although the permaculture (and now mainstream scientific) view is that substantial perennial plant biomass is an important characteristic of sustainable agriculture, there are understandable reasons why humans have tended to reduce woody and perennial biomass in managed systems.

Ecosystems with large amounts of woody plant biomass often predominate on infertile, leached soils in high-rainfall areas, whereas the archetypical human habitats tended to be drier regions with more grassy vegetation. These regions have vegetation with a higher mineral content which supported large numbers of animals, and heavy seed-producing grasses (parents of modern grains).

North American landscapes

The ecological and soil mineral balances in the short-grass prairie, tall-grass prairie and eastern deciduous forests of the American mid-west were closely studied by William Albrecht.[21] He showed that the balance of minerals in the lower biomass and climatically drier prairies favoured large herbivores, food crops, and thus people; while the deciduous forests to the east with higher rainfall had more leached soils, less ideal for crops and naturally supporting fewer large herbivores. These deciduous forests with abundant seed crops of acorns, beechnuts and hickories that feed squirrels and turkeys are replaced on lower-fertility soils by conifer forests. The conifer forests have abundant woody biomass but few nutrients, large seeds or animals.

Victorian landscapes

More familiar to me are the natural patterns of Victoria. Here the drier and more fertile western volcanic plains were covered in grasslands and open grassy woodlands, while the higher-rainfall but more leached sedimentary soils to the east supported tall eucalypt forests. As a general pattern, the grasslands supported more large herbivores (kangaroos) and Aboriginal people. Within the native grasslands the most nutrient-dense plants were small legumes, tuberous wildflowers and other relatively delicate herbaceous species that were mostly eliminated by intensive sheep grazing, leaving the larger, more fibrous tussock grasses.

This pattern was widespread in pre-European Australia and explained the distribution of Aboriginal and early European settlement.[22] It partially explains our ancestral antipathy towards dense forests.

Sydney landscapes

For every recognisable pattern there are exceptions, but increasingly sophisticated contextual pattern recognition can often incorporate these exceptions into a broader understanding. For example, on the north shore of Sydney Harbour, the moderately fertile Ashfield shale loam soils supported tall wet eucalypt forest, while the infertile sands of the Hawkesbury sandstone supported more stunted forest or low heath land.

It is frequently reported in historical interpretations of the early settlement of New South Wales that settlers cleared the tall forests in the mistaken view the soils would be more fertile. It may have been that the experience around Sydney led them astray. While tall forests tend to occupy less fertile land than the best agricultural soils, the extreme mineral imbalances and lack of moisture-holding capacity of poor sandy soils create physiological drought stress and support only short, stunted woody vegetation. Although a health land may superficially look like grassland, it has few soft herbaceous species palatable to grazing animals and a high proportion of hard-leaved and toxic vegetation, which tends to burn in fires even when green. This ecological gradient from fertile grassland and open woodlands on minerally rich soil through tall forest to stunted heath land on infertile soil is illustrated in Figure 20.

Understanding these general and similar, but often local, patterns allows us to comprehend the mosaics of landscape diversity, their limitations and their potential for broad-scale sustainable land uses, as well as their application to cultivated diversity in smaller, more intensive systems.

These observations can temper the tendency of some permaculturalists to become "biomass junkies", regarding total plant biomass as the sole measure of good land management. The permaculture strategy of tree crops and food forests is not about creating massive biomass forests; instead, it focuses on trees that have the maximum potential to feed people and livestock. These tend to be smaller trees, often deciduous, with soft leaves that decompose readily and create non-acidic compost soils; they have large fruits, seeds, nuts or pods with a high content of carbohydrate, protein or oil; and they are often fire retardant. Many people make the mistake of including tall forest species, especially eucalypts and conifers, which can easily dominate these systems without contributing much of value. Especially in high-rainfall with leached soils, the excitement of fast growth is quickly overtaken by the frustration of biomass that nothing can eat and that takes sunlight, water and nutrients from plants that do yield food and fodder.

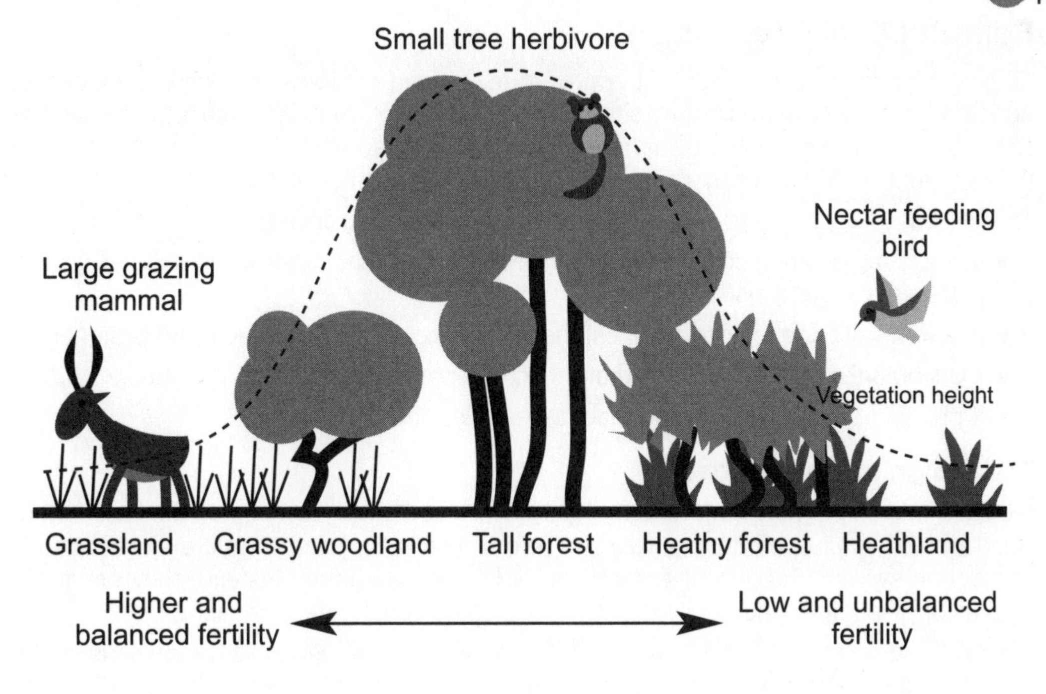

Figure 20: Gradient of vegetation type, canopy height, and characteristic herbivores reflecting soil fertility in temperate mid-rainfall regions

On the other hand, if we need hardy species for shelter, fuel, timber and other less demanding yields than human food, these fast-growing, maximum biomass species are very useful. At Melliodora we even find an appropriate place for heath land shrubs that attract native birds, growing on the cut shale bank of the house site that is suited to few other species.

Grazing animals

The idea that energetically optimal systems follow broad patterns can be extended from plants to animals. For example, in comparing cows and goats as milk producers, cows produce more milk on good quality grass and clover pasture (high-quality energy), but on poorer forage, including woody shrubs (low-quality energy), goats are relatively more productive. While cows may be the best animals in high-rainfall cool climates to maintain healthy pasture, goats are the best animals to create such pastures in the first place. At Melliodora, goats have helped convert rough pasture and woody weeds to better-quality pasture, which is now also supporting other animals such as geese, while our abundance of tree fodder continues to provide an ideal balance to lush pasture for the goats.

In comparing cattle and sheep with kangaroos in their efficiency in converting forage on semi-arid rangeland into meat, it is now widely acknowledged that kangaroos are far more efficient.[23] I don't know of any attempt to compare cattle and sheep with kangaroos on high-quality fertilised pastures, but theory would suggest that domestic animals would be more efficient in this situation. When available energy is rising, it is natural to search for higher-yielding solutions; when energy availability is falling, efficiency of energy conversion is more important.

Permaculture Site Design

The most common application of permaculture design principles has been in the design and development of permaculture sites. The "site" concept used in permaculture is similar to the use of the term by the design professions[24] to mean a limited parcel of land, often focused on a central point that is generally a dwelling or other building.

The site-based approach to permaculture design has several advantages:

- it encourages small-scale, nucleated development that mimics the cellular growth patterns of organisms
- it is well suited to residential sites, the primary concern of permaculture designers
- it encourages distinct systems that reflect the particular nature of the site
- it provides a focal point for permaculture zones and sectors.

Site design as cellular design

We can think of a permaculture-designed garden (Zones 1 and 2) as a human rural settlement cell. There is a limit to efficient garden size before we have to jump up into a more complex production system. Successful gardens do not keep expanding. Instead, they provide a surplus of plant stock and knowledge that help to establish new gardens.

The dwelling and the human household it contains are analogous to the nucleus of the cell, providing control, management, and the information for cell reproduction. Households can be thought of as "reproducing" when children leave home and establish their own. Again, it makes sense that there is an upper limit beyond which household size doesn't work well. Ironically in the Western world, our households are too small to be efficient in food production and preparation, as well as other functions. The more traditional extended family household of between five and fifteen members was large enough to efficiently support many of the functions we focus on in permaculture self-reliance.

Despite the great challenges in recreating community, the expanding interest in eco-villages and co-housing as part of the permaculture vision is implicit recognition of the problem that the nuclear family is too small in scale for many aspects of ecological living. This problem is considered from another perspective in **Principle 10**: *Use and Value Diversity*.

Zones, sectors and slopes

The site design tools of zone, sector and slope allow us to organise information about the site into useful patterns and provide a starting point for an overall concept plan. They also help answer questions about the placement of new elements progressively introduced to a site.

Permaculture zones are more-or-less concentric areas of intensity of use, which describe the power and efficiency of people working from the focal point (a dwelling). The closer to the centre, the more efficient and intensive is our use of the land; the further away we go, the more we must rely on self-maintaining elements that require little input from us, and generally yield less for us. "Starting at the back door" reminds us not to fall into the common mistake of "overreach" when developing a site. If we extend our activities too far and fast while the immediate territory is not organised and working well, we find our energy dissipated.

In most situations, scattering a few vegetable seeds, a little water and some compost across a grass paddock gives no yield; the same effort and resources can create a small but productive food garden in one corner of the paddock.

Although the zones are conceived of as concentric, this is never so in practice. Slope, soil, aspect and infrastructure all cause particular zones to shrink or expand. Even the idea that each zone is a continuous band enclosing the inner zone does not necessarily work on the ground. The development of Melliodora makes specific use of the zoning concept, but it is not concentric due to the nature of the site and effects of title boundaries.

It is useful to think of each zone as characterised by a particular set of plants and animals, management strategies and structures. This is useful within a bioregional and cultural context, but it may need to be varied considerably in other contexts. It is a mistake to turn this simple design concept for organising a site into a blueprint.

Figure 21 shows the permaculture zones described in ways relevant to southern Australian regions with moderate rainfall (450–1000 millimetres).

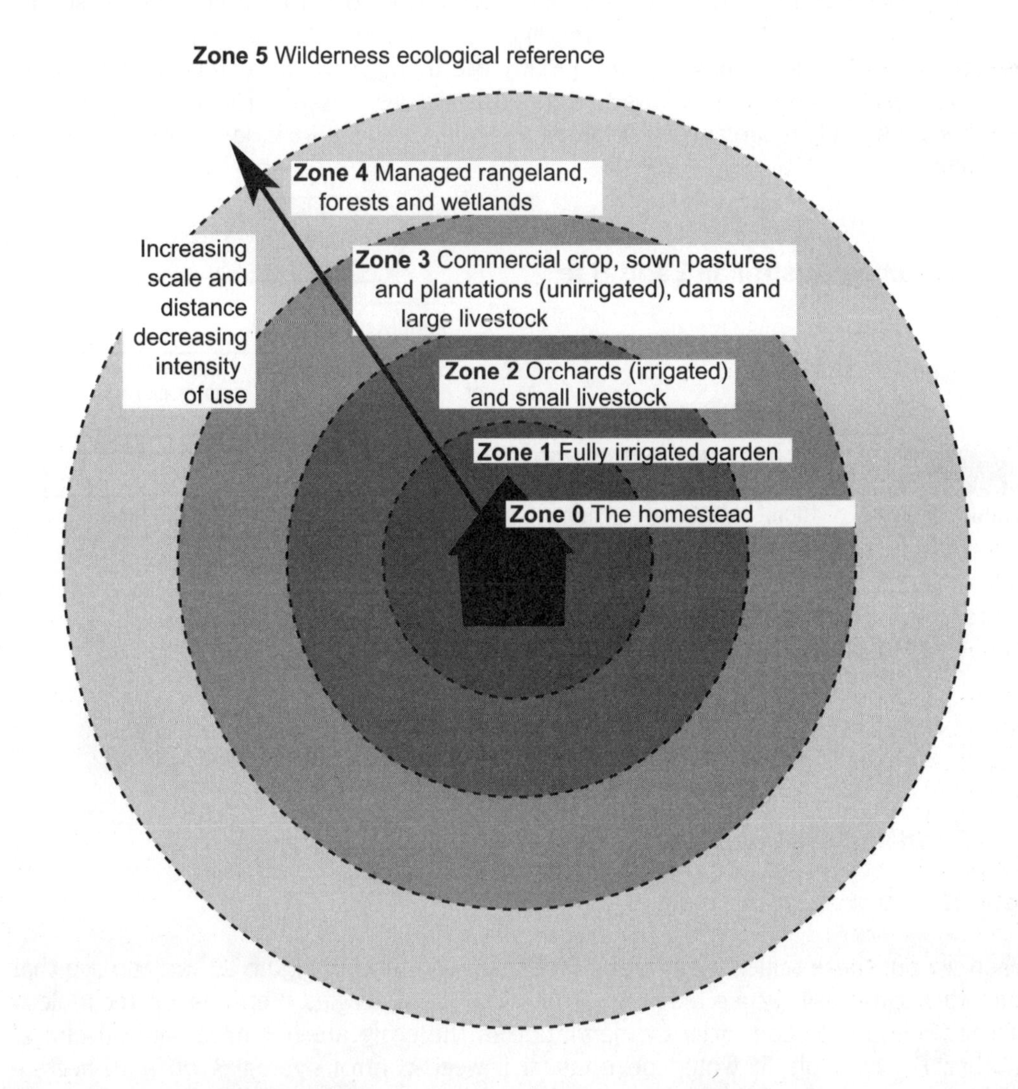

Figure 21: Conceptual zones of intensity of use around a rural homestead

Zones and scale

Each consecutive permaculture zone is not only further from the house, but by implication also larger in area. The question of how large these zones are, like their specific nature, is a matter of design methods that will vary with environment and context rather than principle. However, it is useful to explore how this widely used and variously applied permaculture design tool can be better understood as a step towards a pattern language for permaculture site design.

Some permaculture designers divide any given property, even an urban block, into five notional permaculture zones. Although in a general sense it may be relevant to divide properties into different zones of some sort, I find it more useful to think of the permaculture zones as land use zones that relate to particular physical scales. Zone 1 food gardens vary substantially in size. A small, intensive food garden might be 10 square metres, while a large, intensive system might be as much as 1000 square metres. Similarly, the size of commercial farms varies greatly, from intensive horticultural enterprises growing specialty herbs as small as 1 hectare through to massive cropping properties of 1000 hectares, while rangeland pastoralism and forestry might include another two orders of magnitude.

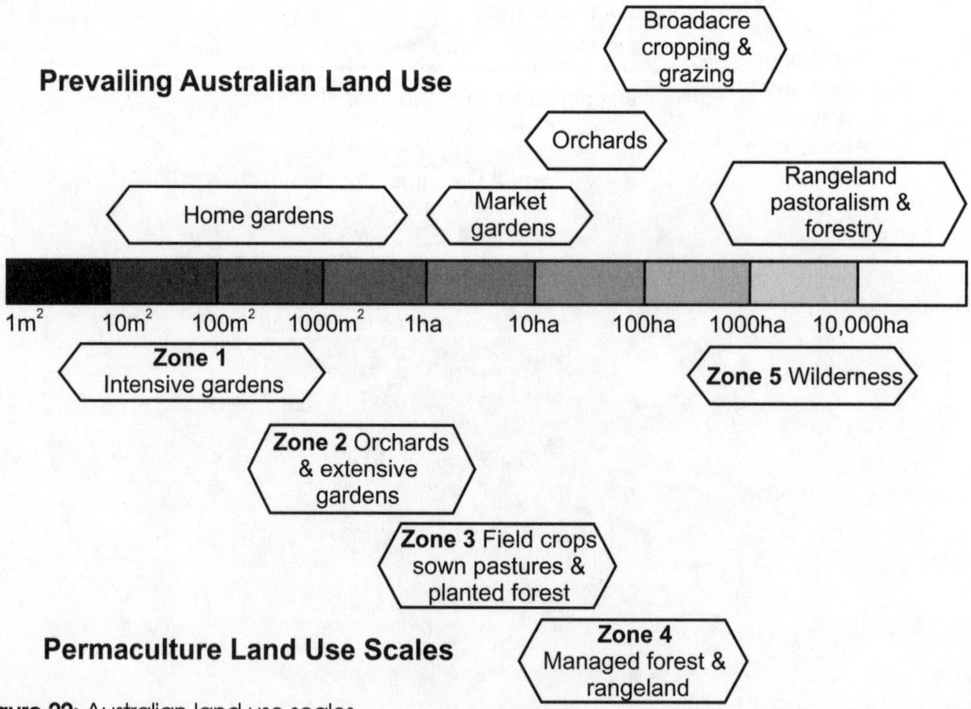

Figure 22: Australian land-use scales

When we put these scales together by order of magnitude in Figure 22, we can see that gardening covers as large a scale range as farming. The strategies, methods and technology that are relevant at one order of magnitude in the scale are not necessarily useful at another. For example, it would be unusual if weed-control strategies on a 10-hectare market garden applied to a 100-hectare cropping and grazing property. In fact, the farmer's crops and pastures may well be the market gardener's weeds.

Most of the food grown in Australia is produced on properties from 10 to 1000 hectares. While permaculture principles can be applied to land management at all scales, one of the overall effects of applying the principles is to highlight the huge opportunities at the scale of gardening rather than farming (see "Permaculture Scale" in **Principle** 9: *Use Small and Slow Solutions*).

It is common for disagreements about gardening and land management strategies and techniques to be based on misunderstandings about scale. As a consultant, I am regularly asked how to control blackberry or how to plant trees, as if the appropriate techniques are independent of context and scale. In considering the plethora of gardening methods and techniques, even the difference between 10 square metres and 100 square metres can be critical to success or failure. In this context, the permaculture zones provide a useful scaling tool to group not only species, but also strategies and techniques, according to the intensity of use and area of land required. Over time it should be possible to develop some common understanding within a particular bioregion (and possibly cultural context) of what strategies and techniques are appropriate in the various permaculture zones.

Linear versus aerial thinking

Perhaps the lack of awareness of scale is partly due to the tendency for us to consider scale as linear rather than area dependent. A garden of 10 by10 metres is only three times longer and wider than a garden of 3.3 by 3.3 metres and yet it is 10 times larger in area. In most production systems it is the total area that is the critical determinant of the energy and resources required for management and the yields that are likely to result.[25]

On large scales (1000 square metres and larger), it is rare for the whole garden to be managed intensively. Traditional orchards, animal runs and many systems of low-input gardening should be considered as Zone 2 systems, which can comfortably push out the scale of garden agriculture another order of magnitude to 10,000 square metres (1 hectare). At Melliodora we have an intensively managed area of 1000 square metres around the house, while most of the rest of the property (8000 square metres) is an extensive Zone 2 drip-irrigated orchard system with free-ranging livestock rotated through it (see Figure 24).

Optimum scales of production and growth

For any particular production system and set of techniques, there is an optimum scale for best productivity and stability as well as simplicity of management. A significant increase (or decrease) in the scale of operations will require a jump to a different production system and set of techniques. Once the investment in the new production system is established, growth in scale is rapid up to the new optimum size. Beyond this optimum, growth is again slow.

Consider an enthusiastic and capable gardener selling surplus from an intensive garden, which is kept fertile with compost heaps made by hand. Early success leads to a rapid expansion in the number of beds and compost heaps and the sale of good organic vegies at a local market. However, beyond some optimum scale it gets harder to make enough compost. Research suggests a few possible strategies:

- get a small truck and bobcat to collect and turn large compost heaps
- rent a much larger piece of land, grow green manure crops with the aid of rock minerals, and rotate the vegetable beds around the land (with the aid of a mechanical cultivator or slasher)
- develop a large rotational chook tractor, using electric mesh temporary fencing.

Whichever strategy is chosen, to help pay for and make full use of the new equipment, the gardener progressively expands the garden to five times the previous size (which may be ten times larger than the small system was at its optimum scale). After a few teething problems, the new system is very productive; with increasing demand, growth is again possible until a new set of problems appears— perhaps weed or insect control or marketing of the produce. Another set of changes and investment are necessary to allow expansion. Maybe this time it is all too difficult and less sustainable; after paying off the equipment, the gardener decides to scale back to the optimum size for current strategy and technology.

This development process is like a bike journey, heavy going up the hills followed by a breezy ride down into the next valley, as illustrated in Figure 23. Some hills prove too steep, and we turn around to roll back down the hill into the last valley. This roller coaster or pulsing pattern of growth can be observed throughout natural and human systems wherever there is the surplus energy to get to the next valley. It is also analogous to the tendency in evolution for species to gravitate around a limited number of design solutions. (See **Principle 12**: *Creatively Use and Respond to Change.*)

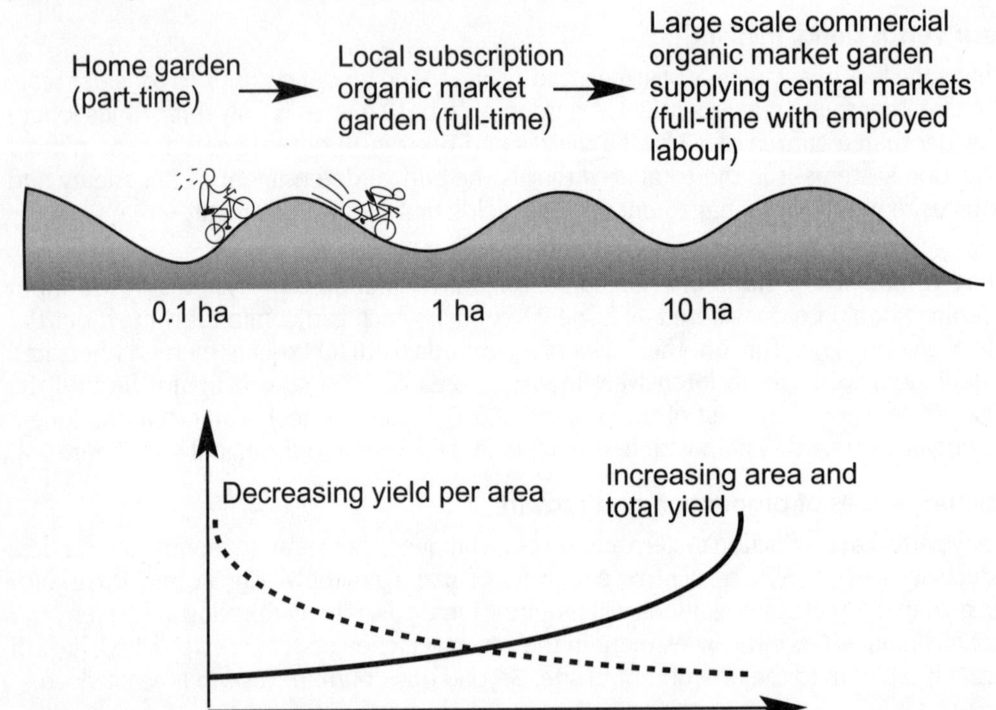

Figure 23: Optimal vegetable production scales with development hills that must be climbed to reach the next optimum valley

Sectors

Sectors radiate from the site focal point, defining the direction from which external wild energies enter the site. By far the most important of these, in temperate latitudes, is the sun sectors, which vary greatly between summer and winter but can be precisely defined (from solar charts). Other sectors - wind, fire, rain, and flood - are less precise. They are based on local observation, skills in reading landscapes, and regional data.

Each bioregion has a generalised sector layout, which is modified by topography, microclimate and land use for each site. Using the sector concept to understand and take account of the wild energies that influence a site allows us to make best use of those energies and diffuse or deflect their occasional destructive nature.

Once we are familiar with using the zone and sector concepts in a range of design situations, they become integrated as a mental map that allows us to filter a large number of possible location and relationship options. When everyone has this understanding, the mandala-like pattern of zones and sectors becomes a key building block in a bioregional culture of place.

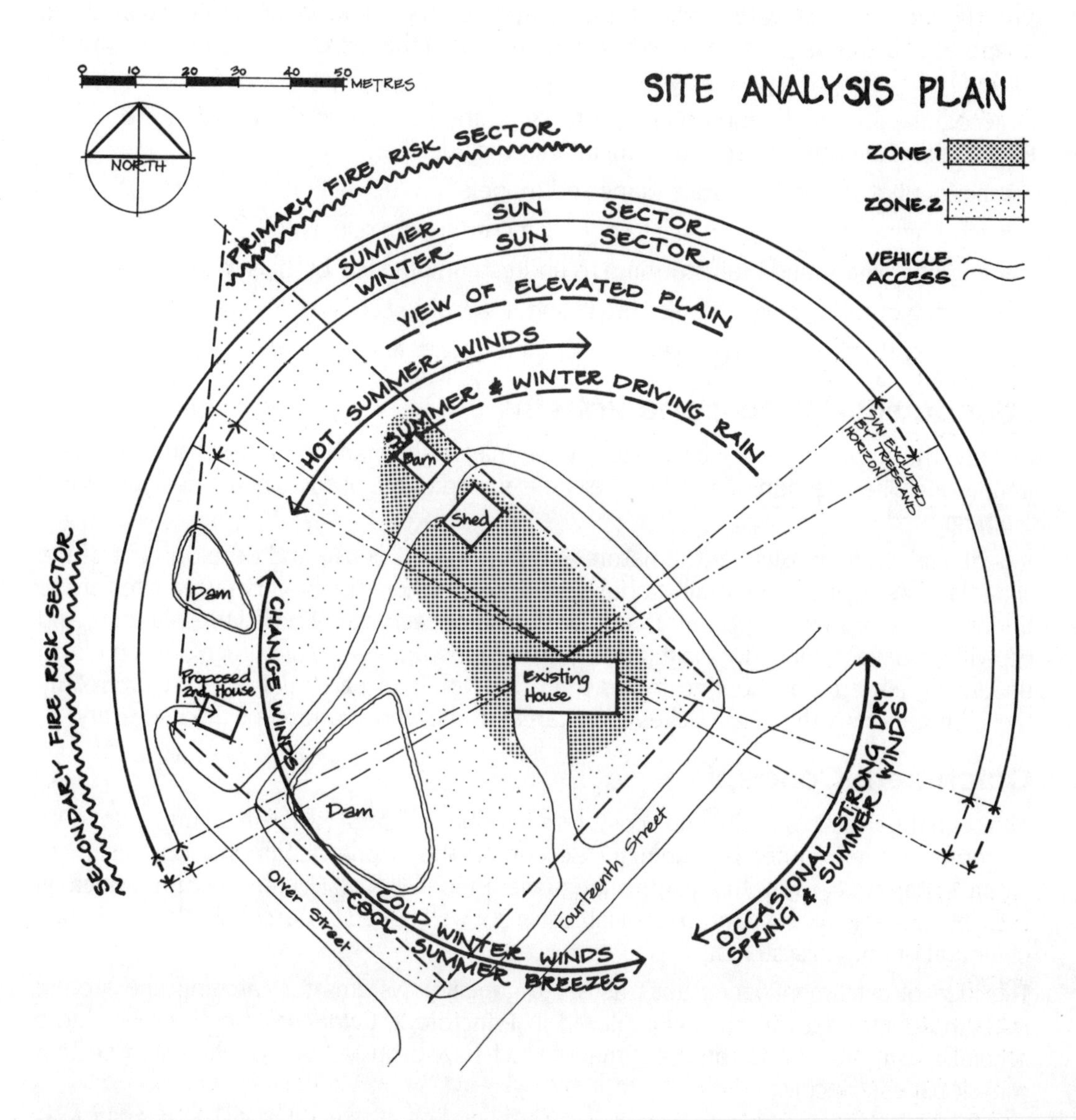

Figure 24: Zone and sector analysis of Melliodora

Slope

On a perfectly flat site, the force of gravity works neither to our advantage or disadvantage. On a slope, even a very gentle one, part of the gravitational force is acting to move things down the slope. This is elementary, but if we design a site to take advantage of this force and minimise the degree to which we are working against it, then we have a more energy-efficient system.

For example, by siting tanks and dams (when practical) high on hills, we can gravity-feed water without needing to pump. If the primary access to a site is from above, the importation of bulk materials such as mulch and firewood is easier to manage. We can minimise erosion by use of terraces, swales, contour access tracks and other structures that slow the movement of water and soil down slope. At Melliodora we have applied all these examples of use of slope, some of which can be seen in the site cross-section in Figure 25.

Besides the concepts of zone, sector and slope, a range of design tools can be used to help us recognise and apply appropriate patterns. Some of the site design tools we use in Permaculture Design Courses and consultancy include:

- aerial photos, overlays, scale plans and models
- mud maps and concept plans to quickly record design concepts
- cognitive maps and brainstorming to understand our view of the site
- scoring or rating of key development options or siting choices
- flow charts to explore complex development processes and works.

Permaculture Landscape Patterns

Despite the importance of site design in permaculture practice, it has some inherent limitations. The question of whether the site is suited to the use it is being designed for is often ignored.

It is difficult to fully consider the nature and implication of use and development of the wider landscape, including multi-nodal development where the links between sites are as important as, or more important than, the details of each site. Whole farm planning and eco-village development, for example, require a landscape-based approach that builds on underlying patterns of the land rather than the details of particular sites. To overcome these limitations, other approaches that start with the whole landscape are necessary.

Catchment Concepts

Although the concepts of site and ecosystem have been central to a general understanding of permaculture, neither is helpful in describing the strong geographic dimension of nature's patterns. Even the energy language modelling of Howard Odum, which is excellent at depicting systemic relationships, is not well-suited to describing the geography of natural or human systems.

The idea of catchment landscapes as self-organising systems for catching and storing water, nutrients and carbon was explored in **Principle 2**: *Catch and Store Energy*.[26] More general insights about rivers and catchments which have been part of a "catchment protection consciousness", include:

- river catchments cover the whole landscape and generally include several different ecosystem types

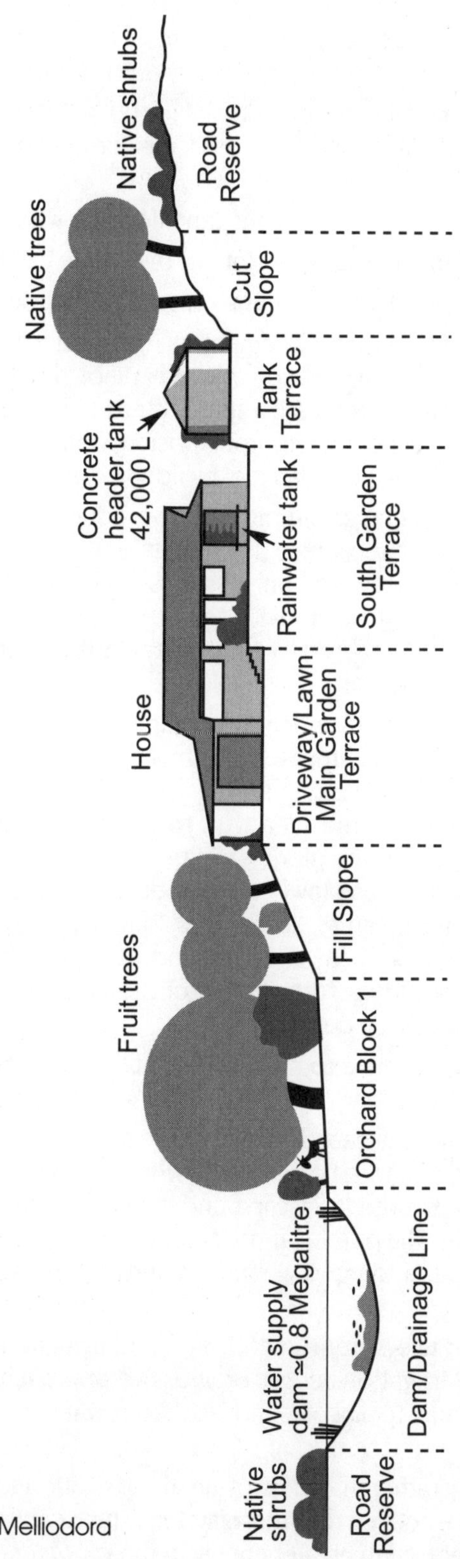

Figure 25: Use of slope at Melliodora

- their general patterns reflect the flow and form of energy, especially in rainfall and water flow
- rivers are the lifeblood of landscapes and it is no accident that their waters and associated fertile alluvial flats have been the focus for human development down the ages.
- what happens at the head of a catchment can affect the whole river system
- the health of river systems reflects the state of the whole catchment
- headwaters are pure, but infertile; estuaries are fertile, but accumulate toxins.

These ideas have been an important element of the Landcare movement but were articulated earlier in the Soil Conservation movement of the 1930s and 1940s. These movements have promoted a general catchment pattern of land use: headwater regions are forested to protect these steeper regions from erosion, while at the same time protecting the valuable agricultural lowlands from damage by flooding, siltation and salinity.

Catchment protection can be seen as simply one by-product of applying this principle to catchment landscapes. Once we recognise that forests must be the predominant land use in the uplands of catchments, then "the only issue is what sort of forest".[27] As discussed above, the permaculture concept began with the question of how to create forests that were agriculturally productive, rather than simply seeing the need for less farmland and more forest for catchment protection.

Small-scale food forests are the best-understood example of the concept. Other broadacre examples relevant to stabilising catchments and farm economies have also been influenced by the permaculture idea of making better use of shrubs and trees because of their energy efficiencies. Since the mid 1980s the idea that trees and forests planted on farms for essential catchment protection (erosion, salinity, etc.) could be equally productive elements in the farm economy has slowly taken root right across Australia. Agroforestry combinations — timber and pasture, timber and fodder tree species, seed and honey production, dry-land tree crops such as olives and carobs — are being developed as alternatives to annual crops and pastures. Some of the innovative examples in this wide field have come from permaculture designers.[28]

Other permaculture strategies relate to making our catchments more energy-efficient. For instance:

- keyline and other catchment-based permaculture strategies can be seen as reflecting the broad design rules of natural catchments when they seek to slow the flow of both water and nutrients down the landscape and cycle them through soil and plants as many times as possible. The primary function of surplus run-off water stored in keyline dams is to irrigate pasture to rapidly build soil fertility and structure (which can then store more water)[29]
- strategies to catch and use water run-off from buildings and hard surfaces for household use and garden irrigation reflect progressive approaches to urban catchment management, which aim to absorb and use stormwater as close to the source as possible
- strategies for managing rather than destroying mixed exotic and native vegetation along streams are based on evidence that these systems represent the natural successional adaptation to catchments with greater energy in run-off water and nutrients.

All these strategies can be incorporated in a catchment master pattern, as shown in Figure 26. Catchment drainage patterns tend to be dendritic, or tree-like; the river is the trunk, the tributaries are branches and twigs, and the permanent forests of the headwaters are the canopy. This pattern is a revegetation metaphor, giving us the priority locations for trees and forest within a catchment landscape. This river catchment master pattern is useful, but we need to integrate it into other spatial patterns before we can construct a complete pattern language for permaculture design across landscapes.

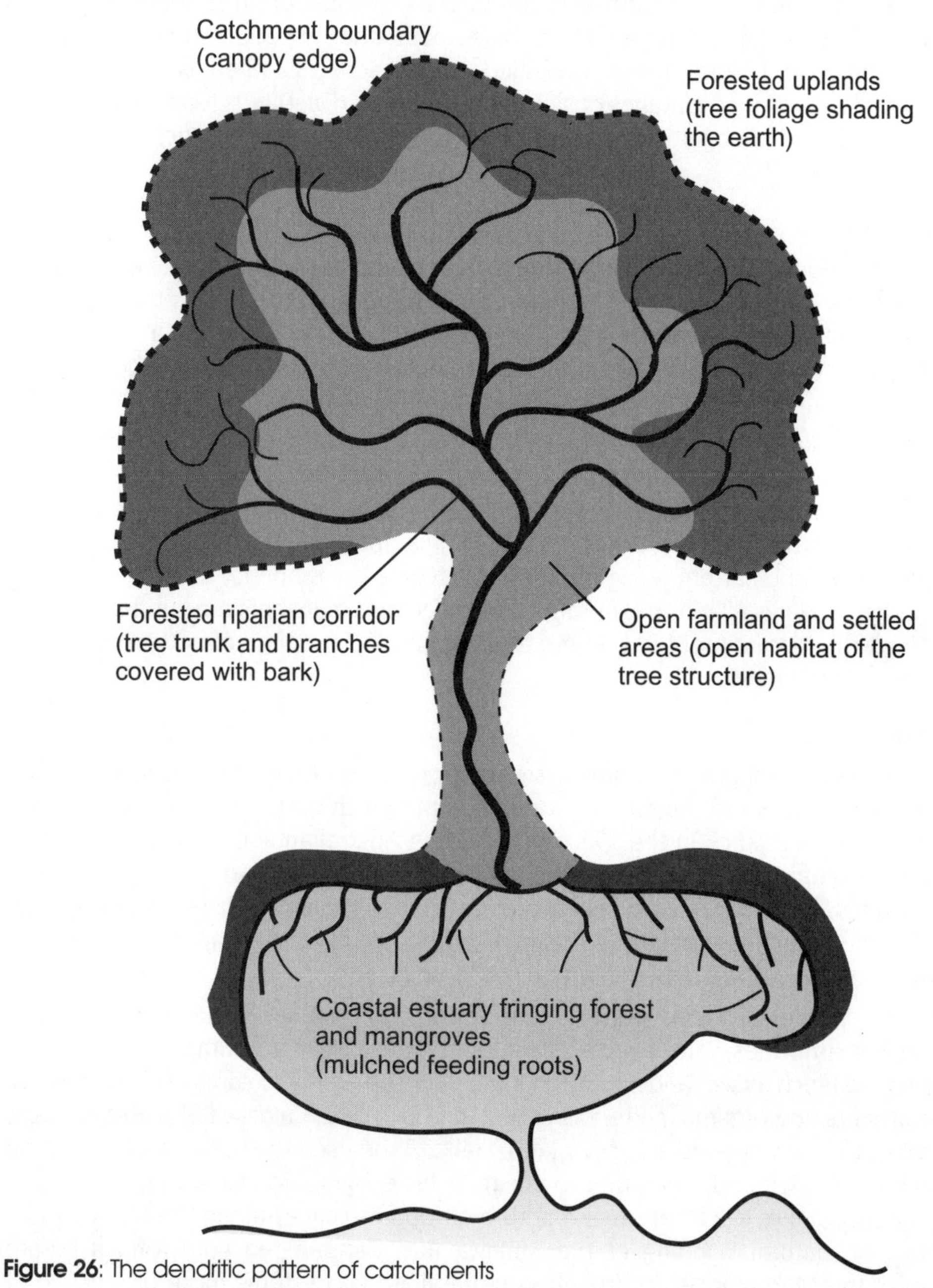

Figure 26: The dendritic pattern of catchments

Land Evaluation and Mapping

During the second half of the 20th century, various methods of land capability assessment were developed in Australia and elsewhere to pre-empt the long-term degradation from uses and management incompatible with the nature of the land. The simplest methods use slope and soil type to distinguish land that is subject to erosion and therefore not suitable for arable farming or urban development. While of some use, these methods have generally failed to predict long-term and complex land degradation problems, such as salinity. Also, they tend to reflect a cultural bias about what is good and poor land. For example, sloping land tends to be classified as problematic, whereas permaculture strategies emphasise the advantages of slope. Poorly drained wetlands tend to be seen as unproductive, whereas recognition of aquaculture as one of the most productive land uses can reverse this view.[30]

Land Systems

More sophisticated methods, such as the Land Systems concept, incorporate ecosystems in a geographical framework. The Australian development of the concept is generally credited to Christian and Stewart[31] who defined a land system as "an area or group of areas throughout which there is a recurring pattern of topography, soils and vegetation".[32]

The method identifies whole landscapes as systems (which often cut across adjacent catchments in bands from headwaters to the sea) reflecting climate, geology and topography. Within these land systems there is a characteristic pattern of repeating land "components", which have more uniform microclimate, slope, aspect, soil type and natural vegetation. Individual sites can be seen as providing a finer level of detail. This top down approach to landscape pattern recognition can overcome the tendency to a blinkered and reactive understanding of land that farmers and other land managers are subject to when only dealing with specific problems within existing paddocks; a case of not being able to see the wood for the trees.

Applications

An indication of the strength of the land systems approach is provided by the studies done of Kangaroo Island in South Australia prior to clearing much of the island for agriculture. As a result of those studies in the 1950s, the leading Australian soil scientist Northcote warned the government that salinity would emerge following the clearing of specific land systems as a result of hydrological changes, even though there was little direct evidence of salinity in the natural landscape.[33] That advice was ignored and today landholders and taxpayers are paying for the remedial revegetation work.

The Land Systems concept also helps us to describe and evaluate land independently of current land use priorities, values and economics in order to take advantage of biologically productive land (such as wetlands mentioned above). The ability to catch and store water, mineral nutrients and organic matter described in **Principle 2**: *Catch and Store Energy* helps us reorient our notion of good and poor land. Haikai Tané used the patterns of biological productivity in aboriginal dot painting to illustrate his ecographic model of rivers (Figure 8). This painting shows the location and nature of food resources along the Murray River floodplain. The natural flooding of the Millewa and Barmah red gum forests before regulation of the Murray River for irrigation maintained ecosystems more productive of animal protein than the unsustainable flood irrigated dairy farms that followed.[34] In arid regions Alan Newsome has noted a correlation between aboriginal mythology and land systems.[35]

Land Systems describe those self organising systems as a series of nested geographic systems which can be projected up to encompass the bioregion and down to include individual sites. This and similar technical languages and mapping tools can be used by landscape ecologists[36] to provide more concrete ways to understand landscapes and apply ecological principles to their management.

Most Land Systems studies published by government[37] are at scales which are mostly useful to planners and broadacre farmers but can also be of use to consultants, real estate agents and those searching for the right land to buy (to improve on more piecemeal understandings of the diversity of land within any bioregion).

In Whole Farm and Landscape Planning[38] I further explore the use of Land Systems and other planning concepts and tools especially as a foundation for whole farm planning. Some examples of the application of Land Systems in my design work have been published.[39]

Limitations

The Land Systems method is little understood or used outside professional and scientific circles, and its relevance to permaculture is not generally recognised. Training in Land Systems identification and mapping has mostly been by the apprenticeship model within government agencies charged with land capability assessment.

However, considerable difficulties remain in distilling landscape patterns that can be widely recognised and applied in design. First, the patterns and classifications within one bioregion do not necessarily translate to another. Attempts to standardise the survey methods and terminology,[40] although useful, tend to ignore more informal "reading" of the landscape as discussed in **Principle** 1: *Observe and Interact*. From a permaculture perspective, this more wholistic and flexible approach to recognition of landscape patterns is as important, and complementary to understanding formal surveys.

Second, the description of changes to land as "improvements" or "degradation" remains culturally loaded. As well as deciding whether an induced change in land use is good or bad, there is the problem of deciding how fundamental and permanent it is. Many improvements to land productivity turn out to be short-lived; but, on the other hand, so do some forms of degradation. Ecological processes do not always return land to some past equilibrium; just as often they transform it into some new state, requiring us to describe novel and evolving land systems (see **Principle** 12: *Creatively Use and Respond to Change*).

It is only after we have entered a low-energy future that landscape evolution processes and bioregional economies and culture will have settled enough for patterns to be a readily recognised and widely used pattern language. Given the geological scale and novel nature of change wrought by industrial culture, reading landscape is destined to generate localised and contingent patterns. Some may see this as a prognosis for frustration; I see it as exciting detective work, sifting the diversity of place and process to inform design and management more in tune with nature.

Bioregional Architecture

In pre-industrial times, architecture tended to reflect the bioregional climate and local availability of materials. The industrial era has broken those patterns. In searching for more sustainable building solutions there is a tendency to regard the ecological building methods as a supermarket of possibilities, a chaotic diversity. Like the introduction of new species of cultivated plants, the ecological building field is still one of experimentation with old and new ideas.

While this may be functional in providing examples of what works and what doesn't over time, ecological architecture will come to reflect bioregional patterns. For instance, state-of-the-art ecological building in northern Europe has largely returned to traditional materials, such as walls of rough-sawn vertical boards of naturally durable local conifers and sod roofs. These designs are driven not by romantic desires to return to the past but by the latest research on sick building syndrome and animal livestock housing. This has shown that materials that are permeable to water vapour, have some thermal mass and high insulation qualities that make the best combination in cold climates.

Lightweight houses on poles, and even exploded designs with rooms as separate buildings, are suitable in humid tropical and subtropical climates, where cooling air movement is essential and termites are a great problem. In south eastern Australia, such designs are expensive to insulate and heat and are almost impossible to make safe against bushfires. Here, more compact designs, generally on the ground, with greater internal thermal mass make sense. The colder the climate, the greater the benefits from the reduced surface to volume ratio of compact and integrated buildings. At Melliodora, the house includes a greenhouse, and garage/workshop within a single building envelope.

The great thermal stability of underground buildings has to be balanced against the engineering complexity and cost of the strength and damp-proofing required. In continental climates such as that of central Australia, and even more so in North America, that balance is favourable. The touted superior insulation value of straw bale walls is of greatest benefit in extreme climates, but it must be matched with top-quality roof insulation. Compact two-storey buildings make sense in cities, but in rural areas large lightweight metal roofs over single-storey buildings are more useful for rainwater collection.

Some architects and owner-builders try novel design solutions, especially with roofs, in search of a sculpturally unique design. Almost always these solutions cost more and waste materials, and they frequently don't work in the long term to exclude the weather. Despite his genuinely great achievements as a designer, Frank Lloyd Wright was cursed by many of his clients for roofs that leaked. It's sometimes hard to accept that most of the solutions for keeping out the rain have already been worked out and that all we need to do is copy them.

I am not denying that innovation and novelty can lead to breakthroughs in design or that cultural norms are already optimal, but I argue that prizing design novelty is no better than rigid conservatism.

Owner building experience

One of my first experiences in house building was helping some friends build a small octagonal house in the mountains of Tasmania in the early 1970s. My friends and I were more capable carpenters than many novice owner-builders, but we found that the construction of a dormer window in one of the triangular roof sections of the little house demonstrated some of the reasons for conventional design. After carefully measuring the angle cuts for the jack rafters using a rafter square, we were puzzled that none of the cuts fitted to the ridge rafters of the main roof. Sitting up on the bank above the house eating lunch, we had an end-view of our dormer, and suddenly my friend realised the cause of our problems. The sides of the dormer roof we were attempting to frame were not simple planes but complex three-dimensional curved surfaces.[41]

That and other experiences made me understand some of the wisdom in the conventional rectangular timber framing of house design. Although it was possible to do things differently, this frequently involves not only more time and skill but often greater waste in off-cuts of (rectangular) materials. The proverb "don't reinvent the wheel" seems to apply in building structures more often than not.

This raises the tricky subject of aesthetics in architecture, which, I admit, I am studiously ignorant of, since my focus at design school was on ecological functionality, which led to the permaculture concept. Caution about this intellectual minefield would suggest leaving well alone, but I cannot resist outlining a few thoughts on the value of aesthetics.

Aesthetics of Descent

I think the rejection of landscape aesthetics in permaculture was necessary to counter the assumption that only aesthetic design issues are worthy of debate, which leaves functionality as simply a matter of technical detail. Only vigorous debate about the nature of functional design in energy descent can expose the dysfunctional forces driving design of our inhabited landscapes and built environments (see **Principle** 3: *Obtain a Yield*). However, any wholistic design will include an aesthetic aspect. The creative tension between function and aesthetics can be seen as analogous to, or even a subset of, that between materialism and spirituality illustrated in Figure 5 (see **Ethical Principles**).

The Industrial Revolution made a nonsense of many of architecture's truths that were distilled in art and ornamentation. The maxim "form follows function" characterised modernist architecture.

Single, unsustainable functions of industrial scale have dominated the built environment, accentuating alienation from the modernist functionalism and producing a plethora of aesthetic responses, collectively called postmodernism, which attempt to give a multi-faceted aesthetic veneer to alienating modern environments. This role of aesthetics in building design must be seen as one of the primary diversions from ecological realities, akin to role of ornamental horticulture in disguising unsustainable gardens and land-scapes.

The 1970s oil crises coincided with new approaches to less energy intensive building, partly arising out of the counterculture owner-builder movement and awareness of indigenous architecture (without architects). There was a surge of interest in more sustainable building concepts and methods in architecture schools. Some of this overflowed into professional and building industry experimentation with energy-efficient buildings and revival of natural materials, but the impetus was lost for a whole host of reasons. Unfortunately, the passionate aesthetic debates that have consumed the attention of the architectural profession in recent decades can be seen as a case of "fiddling while Rome burns".

Just because permaculture emphasises ecological and human scale functionalism does not mean that aesthetic appreciation has no place in permaculture. A positive view of the ecological role of aesthetics suggests that it represents the distillation of the essence or truths of a design culture in forms that have a sensory and inner (spiritual) resonance. Once the bioregional patterns of modest ecological design become established, we will see the re-emergence of bioregional aesthetics that act as a type of design shorthand or signature. Aesthetics in this context can be seen as distilled patterns specifically attuned to human sensory response, which reinforces the recognition of appropriate patterns. In the transition from growth to descent, much of our task is to find the appropriate patterns. Over time, opportunities to be decisively innovative will decline, but the potential for every artisan and craftsperson to add their own unique touch in building and construction can provide a much more democratic expression of aesthetics by a wider range of participants than is possible in current society with its division between function and aesthetics. In the low-energy future, the older architectural maxim "god is in the details" may again be true.

1 See B. Mollison, *Permaculture: A Designers' Manual* chapter 4.

2 Christopher Alexander et al, *A Pattern Language: Towns, Buildings, Construction* Oxford University Press 1977, is being used by some permaculture designers as a model framework for developing a pattern language for productive landscapes using permaculture principles and strategies.

3 For example, Warwick Rowell from Western Australia. See Rosneath Eco-village website http://www.rosneath.com.au/

4 For further exploration of how media technologies may have reduced our pattern recognition ability, see Article 24 "Do Media Technologies Scramble Young Minds?" in *David Holmgren: Collected Writings* 1978-2000.

5 See Robert Ornstein and Paul Erlich, *New World New Mind: Moving toward Conscious Evolution* New York Doubleday 1989.

6 Although small organisms may be short-lived, their function is often continued by their offspring. Thus in systems thinking, small systems are characterised as having a "rapid replacement time" rather than being "short-lived", which suggests a momentary event, not repeated.

7 Ecologists use various indicators of biological productivity, including total photosynthesis and respiration, standing plant biomass, standing animal biomass, and length of food chains.

8 See Article 4 "Aboriginal Land Use" in *David Holmgren: Collected Writings* 1978-2000.

9 For a brief description of these systems and the research on their modern adaptation for sustainable agriculture; see Article 14 "Impressions of Israel: A Permaculture Perspective" *in David Holmgren: Collected Writings* 1978-2000. For the full story, see M. Evenari et al, *The Negev: The Challenge of the Desert* Harvard University Press 1971.

10 "Effort", in a system ecology sense, is the feedback of high-quality energy that an ecosystem contributes to maintain its energy supply from the large-scale environmental sources.

11 Haikai Tané, pers. comm.

12 See Russell Smith, *Tree Crops: A Permanent Agriculture* Devin-Adair 1953.

13 See Article 6 "Historical Precedent for Permaculture" in *David Holmgren: Collected Writings* 1978-2000.

14 Peter Harper, biologist and organic gardener and teacher at Centre for Alternative Technology, has been critical of the forest garden model as promoted and applied by British permaculturalists.

15 John Bradshaw of the Victorian Biodynamic Gardeners Association has criticised the dense planting of fruit trees as encouraging disease.

16 For an early overview see R. Reid and G. Wilson, *Agroforestry in Australia and New Zealand* Goddard & Dobson 1985.

17 See R. Senanayake and J. Jack, "Analogue Forestry: An Introduction" in *Monash Publications in Geography* no. 49, Monash University Melbourne 1998.

18 See Article 16 "The Landcare Movement: Community-Based Design and Action on a Scale to Match the Continent" in *David Holmgren: Collected Writings* 1978-2000.

19 Ted Lefroy of the Co-operative Research Centre for Legumes in Mediterranean Agriculture at the University of Western Australia. Lefroy is one of the few researchers in Australia using EMERGY accounting to take this modelling to a more integrated level. It is not surprising that the total EMERGY in rainfall is much greater than that of sunshine in all but arid regions. Thus efficient use of rainfall is efficient use of embodied solar energy.

20 In the more continental climates of the northern hemisphere temperate zone, the cold temperatures (and freezing of the soil) limit leaching.

21 See C. Walters Jr, (ed.), *The Albrecht Papers*.

22 See Article 4 "Aboriginal Land Use" in *David Holmgren: Collected Writings* 1978-2000.

23 T. Flannery, *The Future Eaters* Reed Books 1994, discusses the metabolic efficiency of Australian animals.

24 See K. Lynch, *Site Planning* MIT Press 1971, a standard text, and C. Alexander, *A Pattern Language: Town, Buildings, Construction*, an inspired work, for a better understanding of the site concept.

25 You can even garden on 1 square metre, which adds another order of magnitude to the scale range. I would consider gardening at this scale to be a Zone 0 system, because it is so small and intensive that the built environment (in the form of containers, paving, supporting balconies, adjacent walls and enclosing greenhouses) is the prime determinant of system productivity.

26 For further exploration of this idea, see Article 25 "Why Catchment Landscapes Catch and Store Water, Nutrients and Carbon" in *David Holmgren: Collected Writings* 1978-2000.

27 To quote Bill Mollison on the video *Global Gardener: Gardening the World Back to Life* .

28 Kym Kingdon's designs for alley farming the semi-arid Victorian mallee as implemented by grain and sheep farmer Anthony Sheldon. See the Museum of Victoria's *Future Harvest* Exhibition case study, http://www.museum.vic.gov.au/FutureHarvest/case15.html. Darren Doherty has designed and implemented mixed timber plantations and tree crop systems integrated with keyline concepts for clients especially in central Victoria. See Australia Felix website http://www.australiafelix.com.au/

29 See P.A.Yeomans, *Water For Every Farm* first published Murray Books, 1965 Revised by Ken Yeomans See http://www.keyline.com.au

30 The permaculture insights about the potential of land types traditionally seen as poor have led some towards a view of ecological equivalence, which sees all land as different but equally valuable, ignoring inherent biological productive potential. The influence of the ecological restoration movement on thinking has led to the view that less productive land is more valuable because of its ability to resist environmental weed invasion. I commonly hear clients say that their land is good because it has no weeds in the bush, which, as often as not, reflects very low fertility. Conversely, I hear land colonised by vigorous soil-building weedscapes described as "highly degraded" when it might be catching water, minerals and organic matter from more "natural" environments that are losing these same resources.

31 See *Technology In Australia* 1788 - 1988 A *condensed history of Australian technological innovation adaptation during the first two hundred years* Compiled by Fellows of the Australian Academy of Technological Sciences and Engineering Australian Academy of Technological Sciences and Engineering 1988 (On-line edition 2000 http://www.austehc.unimelb.edu.au/tia/037.htm

32 C.S. Christian and G.A. Stewart 1953, *General report on survey of Katherine-Darwin Region* 1946 Commonwealth Scientific and Industrial Research Organisation, Australian Land Research Service 1. Canberra: CSIRO.

33 Sabina Douglas-Hill, pers. comm.

34 For an explanation of the partial reflooding of these floodplain forests see J. Murphy "Watering the Millewa Forest" in N. Mackay & D. Eastburn *The Murray* Murray Darling Basin Commission 1990.

35 A. E. Newsome "The eco-mythology of the red kangaroo in central Australia" *Mankind* 12, 327-33 1981

36 For an overview of the field see T.T. Forman and M. Godro, *Landscape ecology* John Wiley and Sons 1986. For an example of the North American application of these concepts see *Proceedings Land Type Associations Conference: Development and Use in Natural Resources Management, Planning and Research* April 24 – 26, 2001 University of Wisconsin Madison, Wisconsin Department of Agriculture Forest Service Northeastern Research Station General Technical Report NE-294
http://www.fs.fed.us/ne/newtown_square/publications/technical_reports/pdfs/2002/gtrne294.pdf

37 For example, the study covering my home catchment, N.R. Schoknecht, *Land Inventory of the Loddon River Catchment* Land Protection Division, Department of Conservation, Forest and Lands Victoria 1988.

38 Article 9 in *David Holmgren: Collected Writings* 1978-2000.

39 The case studies D. Holmgren, *Melliodora* (*Hepburn Permaculture Gardens*), and D. Holmgren, *Permaculture in the Bush* Nascimanere 1992, show how the pattern language of land systems has informed my design of small permaculture properties. D. Holmgren, *Trees on the Treeless Plains: Revegetation Manual for the Volcanic Landscapes of Central Victoria* provides examples of large-scale landscape planning informed by Land Systems concepts.

40 The Australian standard reference is R. C. McDonald et al, *Australian Soil and Land Survey: Field Handbook* Inkata Press 1984.

41 They were hyperbolic paraboloids. If we had included valley rafters in the framing this would have eliminated the curve, but not greatly reduced the overall work and complexity.

Integrate Rather than Segregate

Many hands make light work

In every aspect of nature, from the internal workings of organisms to whole ecosystems, we find the connections between things are as important as the things themselves. Thus "the purpose of a functional and self-regulating design is to place elements in such a way that each serves the needs and accepts the products of other elements."[1]

Our cultural bias toward focus on the complexity of details tends to ignore the complexity of relationships. We tend to opt for segregation of elements as a default design strategy for reducing relationship complexity. These solutions arise partly from our reductionist scientific method that separates elements to study them in isolation. Any consideration of how they work as parts of an integrated system is based on their nature in isolation.

Principle 3: *Obtain a Yield* and **Principle 4**: *Apply Self-regulation and Feedback* cover some aspects of integrated systems from the bottom-up perspective of the elements or individuals. How does each element (or person) look after its own needs through self-reliance and contribute to the larger system? **Principle 7**: *Design from Patterns to Details* emphasised the top-down perspective in order to identify and apply appropriate patterns that guide the design and self-organisational growth of elements and relationships (the details). This principle focuses more closely on the different types of relationships that draw elements together in more closely integrated systems, and on improved methods of designing communities of plants, animals and people to gain benefits from these relationships.

The ability of the designer to create systems that are closely integrated depends on a broad view of the range of jigsaw-like lock-and-key relationships that characterise ecological and social communities. As well as deliberate design, we need to foresee and allow for effective ecological and social relationships that develop from self-organisation and growth.

The icon of this principle can be seen as a top-down view of a circle of people or elements forming an integrated system. The apparently empty hole represents the abstract whole system that both arises from the organisation of the elements and also gives them form and character.

In developing an awareness of the importance of relationships in the design of self-reliant systems, two statements in permaculture literature and teaching have been central:

- each element performs many functions
- each important function is supported by many elements.

The connections or relationships between elements of an integrated system can vary greatly. Some may be predatory or competitive; others are co-operative, or even symbiotic. All these types of relationships can be beneficial in building a strong integrated system or community, but permaculture strongly emphasises building mutually beneficial and symbiotic relationships. This is based on two beliefs:

- we have a cultural disposition to see and believe in predatory and competitive relationships, and discount co-operative and symbiotic relationships, in nature and culture
- co-operative and symbiotic relationships will be more adaptive in a future of declining energy.

Permaculture can be seen as part of a long tradition of concepts that emphasise mutualistic and symbiotic relationships over competitive and predatory ones.[2] *Declining energy availability will shift the general perception of these concepts from romantic idealism to practical necessity.*

The proverb "many hands make light work" reminds us of the intangible benefits from collective rather than solitary action as well as the more general synergistic nature of integrated systems in which the whole is greater than the sum of the parts.

Integration in Nature

Perhaps the most startling example of integration is the complementary character of the two most important biochemical processes on earth, photosynthesis and respiration.

Photosynthesis is the process by which plants use solar energy to power the conversion of carbon dioxide and water into carbohydrate and oxygen. The carbohydrate so created is the beginning of the whole life-support system on the planet. Over 1000 million years of evolution have already optimised the efficiency of the process. For a discussion of the relative efficiencies of green plants and solar cells, see **Principle 5**: *Use and Value Renewable Resources and Services*.

Respiration in both plants and animals is the "slow combustion" process that releases the chemical energy in carbohydrates to support metabolic activity, growth and reproduction. But the by-products of respiration, carbon dioxide and water, are the raw materials for photosynthesis. The perfect balance between these essential life processes could be taken as the ideal model of an integrated system.

In a similar way, plants provide the food for animals, which in turn provide the fertiliser needed by plants. Even plagues of leaf-eating insects in forests are now recognised by ecologists as integrated elements of the ecosystem: they cull sick trees and recycle foliage as fertiliser.

The flows of materials in these processes are cyclical rather than linear; they circulate in closed loops. The recycling consciousness of recent decades has made this one of the better-understood aspects of natural systems and sustainable design.

Types of Ecological Relationships

Ecological relationships can be thought of as on a scale from destructive or consuming to constructive or creative. I characterise these relationships as follows.

Predatory relationships

In a predatory relationship, one organism lives by the death of another; but within this harsh reality, a strong interdependence develops. The predator depends on the continued reproductive health and vigour of the prey population, and the predated species depends on the predator to maintain health by weeding out the sick and weak individuals. Animals or people harvesting bulbs and other tuberous plants can benefit the plant by thinning the clump and loosening the soil; a most benign example of predation.

Parasitic relationships

Parasites are lifeforms that feed off larger, more powerful and longer-lived host organisms. Parasitism is one of the most common relationships in nature, although many are microscopic and invisible. Parasites generally reduce the health of the host, generally without killing it. This delicate balance tends to favour mildly debilitating parasites and suppress the most lethal ones. Parasites often need to move from one host species to another, and have been shown to change the behaviour of their host to facilitate that transfer.[3]

Competitive relationships

Competitive relationships occur where living organisms have the same needs and struggle by growth or behaviour to gain those needs from available resources before the competitors do. Competition is more common between individuals of the same species, such as a regenerating stand of forest trees, but also occurs between plant species and animal species occupying the same ecological niche.

Avoidance relationships

Even where plants and animals appear to be competing for apparently identical resources, diversity and specialisation tend to allow more efficient and complete use of resources by avoiding competition. For example:

- in a forest of mixed species — some with shallow surface roots and some with deep taproots — the trees can generally grow closer together than in a forest of trees with the same type of root system
- by scheduling times of grazing, preferred feeding grounds and trails, wallabies avoid conflicts over feeding territory[4]
- very small birds (such as pardalotes) that eat insects on the outer canopy of forest eucalypts do not compete with larger birds (such as tree creepers) that work the bark of the same trees for insects.

Such species are said to occupy different niches formed from slightly different combinations of habitat and forage resources within the ecosystem. Avoidance relationships are neutral, with needs, timing or spacing providing a form of segregation.

Mutualism relationships

Where living organisms have differing needs and in the process of meeting those needs they provide benefits to each other; the relationship is mutual or co-operative. The examples from nature are innumerable. The flocking behaviour of migratory animals provides mutual benefits in avoidance of predators, navigation and mutual assistance (as in herds of deer crossing fast-flowing rivers). In rainforests, trees provide a shaded, cool, moist understorey environment for ferns and other species, which in turn further enhances the moist decomposing environment so that it more rapidly returns litter to compost and nutrients for the tree. Mutualisms often link several species in a web of interdependence. I remember sitting around the fire one evening with Bill Mollison when he identified a bird-call as a sign that the masked owl had made a kill. Since the owl eats only the neck meat of its prey, several other species feed from the kill; no doubt for them the sentry birdcall was the sign that dinner was ready.

Symbiotic relationships

Symbiotic relationships go beyond mutualism to the point where the organisms become so interdependent that they cannot live without the other. Lichen, a symbiotic combination of an alga and a fungus, is the classic example from nature.[5] Symbiotic nitrogen-fixing bacteria on the roots of legumes are one of the best-known examples, critical for organic and low-input agriculture.

Symbiotic integration is now thought to be one of the important mechanisms that make possible sudden evolutionary jumps. For example, the nucleated cell, which is the building block of all complex organisms, is thought to have originated from a symbiosis between two previously free-living primitive life forms. (See **Principle 12**: *Creatively Use and Respond to Change*.)

Polar opposites or emergent union

The differences between symbiosis and predation may seem great, but both involve closer integration. Indigenous hunter-gatherer cultures generally regarded the process of killing and consuming animals, and even plants, as one of integration. The maxim that "we are what we eat" is true at many levels. On the other hand, symbiotic relationships may appear constructive, but they also represent a complete loss of autonomy for both organisms.

These apparently opposite tendencies in nature can be seen as different paths to integration, as shown in Figure 27. This top-down or whole-system overview of ecological relationships allows us to make sense of the diverse and apparently contradictory evidence about how ecosystems work. In addition, it provides a more integrated model for thinking about relationships both in practical system design and in wider human society.

Each Element Performs Many Functions

Multi-function and complexity

The idea that elements in systems are simple and serve a single function is part of the dysfunctional mechanistic view of the world that still dominates our culture nearly a century after science began to recognise (in theory at least) that the universe was not just a giant clockwork mechanism. Every element, especially living plants and animals, is in itself a complex system with many different characteristics, requirements, outputs, and potential uses.

In nature, multi-functionality is the norm. The trunk and branches of a tree hold up the leaves for efficient collection of solar energy, and the sapwood conveys water and nutrients to the canopy, and carbohydrates to the roots. In addition, the trunk and branches provide habitat in bark crevices, faults and hollows for insects, birds, mammals and other lifeforms, that benefit the wider ecosystem, which in turn may provide benefits to the tree.

This multiple functionality at multiple levels within systems provides a deep and broad structure to integration within ecosystems. It challenges us to extend our thinking about natural systems and their elements beyond the obvious. Bill Mollison[6] has spoken of the relationship between the forest and the river as an exchange system: the forest gives energy-rich leaf litter to the river, while fish migrating upstream are nutrient-rich material (from the sea), which can spread into the forest (via predatory birds). This relationship is analogous to the sap flow in a tree, exchanging nutrients and energy between the roots

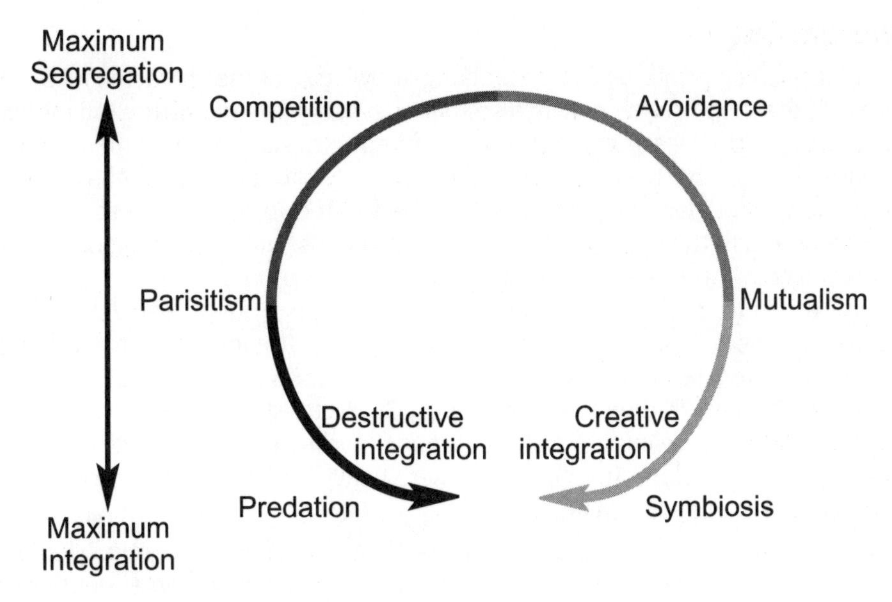

Figure 27: Integrated and segregated relationships in nature

and the leaves. For indigenous hunter-gatherers, this multi-functional nature of everything was self-evident. Today it is so lacking that a system as complex as a proposed new city was named the Multi Function Polis,[7] as if it were possible to imagine a city that had a single function.

Sometimes our fundamental concepts of structures prevent us from seeing the complexity of functions they integrate. Howard Odum modelled New Zealand beech forest using the energy circuit language (described in **Principle 2**: *Catch and Store Energy*). He separated the canopy of the forest (a producer system), from the trunks, flowers, fruits and roots (consumer and storage systems). These various parts of the forest tree structure were each seen as important as whole groups of animal consumers, such as forest insects. At this level of detail, Odum placed all the birds, mammals and reptiles in one consumer group with about the same importance as decomposers and soil animals.

The maximum yield fallacy

Principle 3: *Obtain a Yield* discussed the valid tendency to focus on the most productive yields and functions as reflecting the Maximum Power Law. Nevertheless, this tendency has become extreme in modern society, where a "monoculture of mind"[8] has simplified our thinking. As with other ways of thinking that permaculture is attempting to overturn, the ignoring of secondary yields or benefits in favour of a single yield or benefit, especially among farmers, was a natural response to readily available energy. The problems of extra work and pollution — the result of a failure to design to use all the outputs and benefits of any plant or animal — were covered by the great profit from the primary yield, which was itself made possible by the fossil energy subsidy. But without increasing energy availability, this approach is highly dysfunctional (see **Principle 6**: *Produce No Waste* and **Principle 11**: *Use Edges and Value the Marginal*). The multiple yields and functions of plants, animals, buildings, infrastructure and human activity itself now need to be reassessed so we can design and rebuild beneficial links between elements.

Functional analysis

To design integrated systems, we need a working knowledge of the characteristics, needs and outputs of the potential elements in a system. In teaching permaculture Bill Mollison[9] has used the fowl as the classic example to illustrate the maxim "each element performs many functions".[9] He lists the "intrinsic characteristics, products and behaviours and needs" of the fowl. The element (fowl) needs to be located in the system so that it can behave naturally, and in the process gain access to its needs, while providing outputs that can be directly used by other elements in the system. By this process we minimise our own work, and also minimise pollution (unused outputs). Of course, our primary interest in the fowl is eggs and meat, but often the neglected outputs (manure, heat, feathers) or characteristics (foraging, scratching) offer special opportunities for design that incorporate these factors as benefits rather than problems. Characteristics of different breeds of poultry are also factors in refining our design understanding of the fowl. For example, at Melliodora the purebred Black Australorp fowls are heavy birds and can be contained by one-metre fences, a critical design factor.

For plants, a similar functional analysis shows there are many potential yields and functions for any element. At Melliodora we have pruned many of our larger nut trees to a single trunk, between 1.5 and 1.8 metres tall, partly to increase the harvest of valuable furniture wood possible from these trees at the end of their useful (nut-bearing) life.

Sometimes, obtaining secondary yields actually contributes to primary yield. For example, at Fryers Forest[10] the grandfather of the previous owner thinned the regrowth forest over the first half of the 20th century to increase the grass for his sheep. The retained yellow and grey box trees developed large canopies, which in turn provided increased nectar yield for his honeybees. When they grew to a mature size they provided his grandson with a substantial inheritance of valuable timber. By resisting the prevailing practice of that era to remove all the trees in order to create more sheep pasture (single function), the landowner created greater total value while leaving the land in good shape and conserving biodiversity.

This multi-functional use of plants and animals often involves compromises, because varieties and breeds suited to one use may not be so good for another. In industrial agriculture there is the power to maintain conditions that favour single-function varieties and breeds, and so pursue the highest single yield. With declining energy, flexibility to changing conditions and needs is more useful than the apparent gains possible from specialisation.

Each Important Function is Supported by Many Elements

"Each important function is supported by many elements" is another maxim about integrated systems used in permaculture teaching. Although it can be seen as simply a corollary of "each element performs many functions", this key permaculture concept introduces other ways of thinking about how to design and develop integrated systems.

Back-up system

Back-up elements, systems and methods provide an important function or yield when the usual source or method fails. Obvious examples are the back-up electric power generators and batteries in hospitals, telephone exchanges and other essential services. The redundancy, cost and apparent inefficiency of having back-up systems that sit doing nothing for most of time used to be accepted as essential in order to maintain important system functions in almost all circumstances.

The word "redundant" has come to mean unnecessary or excessive, but older meanings include "characterised by superabundance"; it derives from the Latin redundans, meaning overflowing. In electronics, it refers to elements in a system that are not normally used, but come into operation if an active element fails. This essential design strategy in electronic engineering derives from the understandings in systems theory of the design of complex self-regulating, self-maintaining natural systems. In nature, a myriad of back-up systems exists for every important function. For example, in the human brain there is a huge redundancy and flexibility; it is possible for people with substantial brain damage to learn or relearn tasks that would have been processed by the damaged section.

In natural systems, the relative cost of maintaining back-up is not necessarily high. First, the elements are often self-replicating cells or organisms; and second, unlike the standby generator, they perform other functions instead of doing nothing.

Complementary contribution

Another aspect of this maxim is the idea that different elements contribute in varying ways to an important function. For example, the human eye is a highly complex, integrated system, in which the rod and cone cells in the retina respond to differing light intensity and wavelength. The exquisitely sharp and colourful images that we see are a result of these and other elements working together. Specialisation of elements within the eye produces a better image, but there is also a degree of redundancy. Vision is possible, if not as good, with only the rod cells, or only the cone cells, functioning.

In soil ecology, a vast diversity of organisms and micro-organisms contribute to the recycling of organic matter. Earthworms are the most easily observed; they digest organic matter and turn it into soil humus. In healthy soil, a dynamic and complementary balance of fungi and bacteria complements the macro scale activities of the worms. Fungi are most efficient in processing tough, carbon-rich plant cellulose, while bacteria are more effective in processing soft tissue, proteins and other nutrient-rich compounds.

Fodder shrubs

Fodder shrubs and trees are an excellent example of complementarity. They are a useful element in animal grazing systems, but their value has been consistently underestimated because conventional reductionist research has focused on yields of edible dry matter and the weight gain of animals on pure diets. But in almost all situations, it is the complementary nature of tree and shrub fodders that is the greatest value. By accumulating edible feed without drying off and going to seed like herbaceous pastures, trees and shrubs can function as living haystacks that can be grazed when pasture fodder is scarce. In addition, the different nutritional characteristics of fodder are often complementary. For example, the high protein content of leguminous shrubs complements bulky dry grass fodder.

Although trials of plant and animal productivity are useful, the key design issue is how to integrate appropriate species of tree and shrub fodders into existing farming systems so these complementary benefits can all contribute to the sustainability of the farm and the landscape.[11]

Water supply design

Household water

In designing human settlements few functions are more important than water supply. In affluent countries the reliability and low cost of reticulated town supply has, for generations, led us to take it for granted. What else can you have reliably and constantly delivered to your property for less than one dollar per tonne? What other resource do we need in small quantities on a daily basis (drinking) and yet at times in vast quantities (irrigation)?

On a rural property, the design and construction of a reliable and efficient water supply system is generally the most important design issue and the greatest expense after house design and construction. There is no specific profession or trade that routinely designs and installs integrated, self-reliant systems for water supply. In my experience as a permaculture consultant, owners, plumbers and equipment suppliers often work without any grasp of the integrated design issues involved. Specialised engineers who are familiar with the design issues are rarely hired because of cost; even if they are hired, they frequently have experience only with large, expensive systems that are over-engineered for domestic purposes. Architects generally have little knowledge of water systems and are unaware that self-reliant water supplies are not a plug-in addition to a house, but need to be considered at the beginning of the design process.

The most common (but undesirable) water supply for Australian rural households is an electric pressure pump drawing from a rainwater tank at ground level. The house is plumbed as if it were connected to mains water; every time a tap is turned on, the pressure pump starts. The convenience of high-pressure water encourages waste when, in fact, supply is limited to the store in the rain tank. Leaking taps cause the pump to cut in and out, increasing noise, electricity use, and wear of the switch mechanism. In case of a power failure, with a tank at ground level and no gravity-fed outlet, no water is available. Power failures are common in bushfires, so life and property can be at risk with a tank full of unusable water. A back-up petrol-driven fire-fighting pump is the add-on solution to salvage this disastrous situation. Unfortunately, such pumps tend to sit unused for long periods; as a result, people forget how to use them, the fuel goes stale, and wasps' nests can block the exhaust.

A more integrated system includes a header tank on a hill or a high stand, providing a moderate pressure supply. The header tank is fed by an electric pump that can still be automatically switched, but only needs to come on when the tank is, say, half full. In case of a power failure, a minimum of half a header tank is available. A small additional tank on a low stand, catching water directly from the roof, can be gravity-fed to an extra tap at the kitchen sink at low pressure without any mechanical aids. If rainwater for this tank is collected from a section of the roof without flue pipes, and a first-flush diverter is installed, such a system can give very clean water for drinking, as well as providing back-up water supply to the main system in case of a power failure. We use this system at Melliodora as an alternative (and back-up) to mains water.

Irrigation water

Our primary irrigation supply at Melliodora is from dam water that is pumped to a header tank by both a windmill and a petrol-driven pump. The windmill is slow but provides adequate water for some of the year. The petrol pump is fast and is needed during the main irrigation season. Thus, the two pumps are complementary. The gravity pressure from the header tank is adequate for most uses, but the petrol pump makes it possible to pressurise the whole reticulation system for irrigation at the driest time of year, also the peak bushfire hazard period. Because the petrol pump is in regular use for irrigation during this time, it is more likely to be reliable in case of a bushfire. We have roof sprayers for bushfire safety on our shed and barn; these depend on the pump for full pressure, but they will still work to some extent on the gravity feed from the header tank.

These examples illustrate a fine balance between the complementary and back-up aspects of design redundancy, which reflect the robust and reliable character of natural systems.

Simplification and Segregation

Simplification is perhaps the human default response to systemic problems (see **Principle 10**: *Use and Value Diversity*). By eliminating the apparently less important elements involved in a conflict, we reduce the complexity of management. When the elements involved in conflict are essential or too powerful to eliminate, we often resort to a strategy of segregation. Simplification and segregation are fundamental methods of avoiding excessive conflict and competition between elements or sub-systems; they tend to go hand in hand. Use of one increases the opportunities to use the other, as we pare the system down to its manageable bones. These are valid but overused strategies for dealing with dysfunctional system overgrowth and excessive complexity.

Segregation in the garden

By spacing vegetables widely, the conventional gardener prevents competition for water, sunlight and nutrients between plants. This allows all plants to grow to their maximum size, even though it requires more land and more work in weed control. By eliminating weed growth in the garden, we reduce competition with our crops. These and similar efforts at reducing competition by segregation make our systems biologically simpler, increase valued yields and are easier to manage. But inadvertently they contribute to other problems, such as a breakdown in the free environmental services of maintaining soil fertility and controlling pests and diseases.

A fundamental segregation tool is the garden fence, which excludes animals that would consume, scratch out, or trample our food plants. Although permaculture gardening strategies focus on how to reintegrate animals, especially poultry, in gardens and orchards, fences and enclosures are generally needed to maintain segregation at appropriate times.

Segregated homestead design

The segregated approach can also be seen in design strategies for the built environment. For example, on the traditional Australian farm, space is not a limited resource; homesteads, with their various buildings and structures, often spread over a large flat site with little connection or integration between the elements. When a new building is planned, the major design criterion is often plenty of space all around so there is no interference with other functions and elements. This segregated design is not completely illogical, because it makes use of abundant space to ensure easy access for vehicles and machines. The fact that it inhibits use of many of the secondary benefits or yields from buildings is seen as a minor problem or not relevant. From a permaculture perspective, these segregated systems are highly inefficient and wasteful.[12]

Zoning controls to segregate land uses

In the late 19th and early 20th centuries, the problems created by polluting industries led to land use planning and zoning controls. These sought first to segregate noxious industries from residential areas, and eventually to separate all land uses under the logic of avoiding conflict. Zoning controls and the resultant segregated land uses have become so accepted that for many in local government they are synonymous with urban and regional planning. It is only more recently that segregation of land use by zoning has been questioned, partly because of sustainability concepts such as permaculture.

Ethnic segregation and apartheid

Under a similar logic of avoiding conflict, segregation of social and ethnic groups has been tried in various societies throughout history. Perhaps the most spectacular attempt in the 20th century was the apartheid policy in South Africa. Social and ethnic segregation has been almost completely discredited as an ethical or workable social policy, although in extreme cases such as Palestine it still has some currency as the only possible way to reduce conflict and violence.

Separation of church and state

The separation of church and state as a result of the European enlightenment, and the progressively widening gulf between scientific materialism and spirituality, are traditionally credited with reducing the social conflict, oppression and superstition which dominated many societies where these apparently incompatible aspects of life were integrated. Segregation of spirituality and materialism in the modern world has provided humanity with an opportunity to take stock of conflict between these opposite parts of us. To that extent, modernity has been a useful, but not sustainable, state for humanity.

In Figure 5 (Ethical Principles) I depicted this polarisation as generating an emergent union that has two aspects, one creative and the other destructive.

Co-operation and Integration

While it is reasonable to understand segregation as a valid but overused design strategy, the opportunities to use co-operative relationships to build more integrated systems are everywhere around us. Permaculture includes many examples of application of this principle in the garden and on the farm, but integrated design has a much wider currency.

Industrial integration

The redesign of manufacturing processes to achieve greater energy efficiency and reduced waste (see **Principle 6**: *Produce No Waste*) is taking hold, and with it comes the imperative to integrate previously segregated elements and sub-system. Sometimes the natural response to industrial good news stories, such as the piggery that uses manure to generate methane gas, is to ask why the system wasn't designed that way in the first place. The answer is often that the new way is more integrated, and therefore more complex either in its physical design or its organisation and planning. This complexity of integrated systems provides the rationale for segregated systems when resources are abundant.

Amory Lovins[13] argues that large-scale redesign of industrial processes, driven by market forces, will lead to integration more akin to ecosystems. His promotion of the hypercar concept and the hydrogen economy rest more on fitting things together to create complementary and integrated systems than on the invention of new technologies. *Whether low-tech and socially accessible permaculture models or the high-tech corporate ways prevail, integration of previously segregated systems appears to be a fundamental principle driving post-industrial design.*

Companion planting and permaculture guilds

The emphasis on building more mutual and co-operative relationships while reducing the impact of predatory and competitive relationships is a key permaculture strategy for more effective integration within and between systems. Companion planting of vegetables and herbs, originally based on observations of mutualistic effects by biodynamic researchers, has popularised the idea that plants do not necessarily compete and may have beneficial effects on one another.

In the teaching of permaculture, the ecological concept of guilds provides a wider view of mutual relationships by identifying groups of plants, and even animals, which can provide mutual benefits.[14] Guilds, which can be reliably demonstrated at a bioregional and local level, provide a pathway to condense substantial complexity in design of polycultures, especially those involving long-lived perennials and trees. However, much more research is needed in this field because what appears to work well in one environment and soil may not work in another.

Integrated land uses

Integrated rural land uses, where every farm is to some degree a forest, were perhaps more central to the original permaculture vision than companion planting and guilds. Agroforestry, analog forestry, alley farming and other models for integrating trees with traditional farm land uses are examples of the ways that this vision is coming to fruition.

However, the technical issues involved in integrating trees with crops and pasture are only part of the picture. The prevailing system of freehold land ownership remains one of the greatest impediments to the development of integrated land use in rural Australia. Almost every large farm has some potential to include enterprises as diverse as livestock husbandry, cropping, horticulture, aquaculture, apiculture and forestry in ways that increase the productivity of all the enterprises. Unfortunately, it is uncommon for one farming family to have the skills, capital, or even the cultural disposition to manage this diversity.

Rural landscapes with a patchwork of pastoral, cropping, horticultural and even forestry land use on separate freehold titles may be quite diverse, but the level of integration is always impeded by the tendency of each owner-operator to maximise primary enterprise. For example, the grazier sees trees as taking up space that could be growing more grass, while the forester sees grass as simply a fire hazard and "unstocked" forest.

Urban land tenure models

Ironically, examples of reintegration of previously segregated land uses have emerged in urban areas driven by quite different forces. The control or elimination of many of the noxious aspects of industry, the development of new technology and the expansion of the service sector have again made it possible to live close to commerce and industry. The adverse social consequences of segregated urban land uses have also become obvious to urban planners. Although new inner city, mixed developments of residential, commercial and even industrial uses hardly represent models of sustainability, they demonstrate that the principle of integrated systems is gaining on a number of fronts.

My own realisation of the importance of land tenure to the vision of broadacre permaculture landscapes came when working briefly with New Zealand planner and resource ecologist Haikai Tané in 1979 and 1984. Haikai told me how the relatively recent innovation, strata or body corporate title was being used in some shopping centres and industrial estates. As co-owners of a shopping centre, small businesses gained economies of scale to employ marketing, accounting and other shared services, as well as more bargaining power with the large chain stores that shared the centre. In industrial estates, the sharing of waste management services provided economies of scale for small businesses dealing with stricter pollution control regulations. We saw the potential for these industrial estates to develop more complex examples of industrial ecology, where the output of one business was the input of another.

Further, we saw the obvious opportunities to use this modern form of land tenure to allow the integration of forestry, horticulture, aquaculture and apiculture into existing broadacre pastoral farms without the need to subdivide the land. We also recognised that, within the current affluent society, the desire to live in a rural environment provided the economic opportunity for reorganising farms into rural residential communities, while expanding and integrating their potential for productive land use. I outlined these opportunities for broadacre integrated rural land use in 1984,[15] and later tried to get planners to accept this form of creative land tenure as a solution to the ongoing loss of prime agricultural land to rural residential development.[16]

Rural eco-villages with body corporate land tenure provide a more sustainable way — economic, ecological and social — for people to live in attractive rural environments than conventional subdivision. However, the development and proliferation of this alternative in Australia has been slow. This is due to a number of factors including:

- planning schemes give no incentives for groups or developers who initiate this type of rural resettlement and, to some degree, actively discourage it
- rural land is relatively cheap, due to low commodity prices and poor economic viability of agriculture
- there is a reluctance to deal with the complexities and constraints of community (see discussion of intentional communities below).

Even when rural residential communities have been developed specifically with permaculture in mind, the permaculture vision of broadacre integrated land uses has been slow to develop. Although I see rural communities as having great potential to realise this vision, several major impediments remain.

- Food remains cheap relative to wages, and returns from agricultural production are low. Thus the incentive for residents to use community land for developing land-based livelihood is low.
- Continued economic growth and affluence provide better opportunities to obtain income (from social welfare to consultancy).
- There are perceived conflicts between residential amenity and agricultural and forestry uses.
- There is little knowledge of suitable legal mechanisms for rights to land use, or of structures for integration and conflict resolution with other land-users.

Although the potential of these visionary attempts at integrating communities and land uses is yet to be fully realised, the combined effects of rural resettlement and rural tourism have stimulated many examples of integrated rural land use, often with a permaculture inspiration. People who come to the country as tourists to experience the environment are attracted by many of the same values and interests as people who move to the country to live. It is no accident that many new rural tourism ventures have been started by new rural settlers or by members of farming families disillusioned by the general trends in farming. One of the defining characteristics of the rural tourism boom is that visitors are seeking an integrated experience. They want to stay in accommodation made from local natural materials, eat local food produced on the farm, and experience a variety of activities in a diverse and harmonious landscape. Farms that have maintained or developed this integration for ecological and other reasons have the highest degree of integrity and value for developing rural tourism.

It is ironic that tourism is subsidising organic and other innovative sustainable agriculture in Australia, just as tourism is providing an economic base for traditional, sustainable culture and land uses around the world. Despite the contradictions and problems of this type of development, it is providing an economic incentive for integrated land use when most other economic forces and policies continue to accelerate the disintegration of rural landscapes. More importantly, it shows that in the modern world, a shortage of integrity — in the broadest sense — has created an economic value from what were once common and everyday aspects of integrated systems.

Patterns of Competition and Co-operation in Nature and Society

The emphasis in this principle is to increase the co-operative relationships as an essential aspect of integrated systems, but we must understand these relationships in the context of other possible relationships between living organisms.

Effective design for co-operation depends on a broad understanding of the functions of co-operation and competition. Simply deciding that competition and conflict are bad will not help us in this task. To the contrary, we need a balanced view of the contribution of all relationship dynamics to natural and human systems

In nature, systems that are immature and growing rapidly, in a situation of surplus free energy, tend to be dominated by competitive relationships; mature ecosystems, in which there is little free or surplus energy, show a high degree of mutualistic and symbiotic relationships. For example, a rainforest clearing with abundant sunlight and fertility will grow a profusion of competing annual weeds, vines and trees; in the mature forest, adjacent plants are more likely to have mutual or symbiotic relationships, such as trees that exchange nutrients via mycorrhizal fungi.

Pioneer communities

In rapidly growing human communities that are tapping new resources (pioneering and colonising), competition tends to prevail; in stable traditional societies where all resources are fully allocated, defined roles, mutual obligations, gifting, taxation and other social mechanisms prevail over competitive ones. In some hunter-gatherer societies that survived in very harsh environments, competition between group members was regarded as evil. For example, some Inuit groups in the Canadian arctic are reputed to have killed missionaries who attempted to introduce competitive team sports to their communities.[17]

In any rapidly developing area of knowledge or work, people are often surprised and dismayed that competition between like-minded people can be common. When the field is new and open, competition is a natural response. This competition is especially confusing when the field (such as permaculture) has a strong focus on co-operation.

Co-operation in capitalism

Over the last few hundred years, continuous growth and change based on the tapping of new and larger resources has created a global culture in which economic competition (capitalism) and personal competition (individualism) have become the dominant forces. This system has broken down many of the structures of mutual obligation of pre-industrial society. Early social Darwinists used evidence from nature to justify these radical changes. Although reference to nature as a justification for competition has reduced since the development of ecological science, the assumption that economic competition is an unbridled good has reached new heights at the end of the 20th century.

But even within capitalism, co-operation and trust are still essential elements. For example, small manufacturing businesses routinely accept substantial orders for products and contract work based on verbal communication without any legal contract with customers. Without some trust and co-operation, most small businesses could not exist. Highly sophisticated networking and co-operation between small manufacturing businesses is reputed to be a key factor in employment, wealth generation and economic efficiency in the Emilia-Romagna region of northern Italy.[18]

Niche markets

The ecological term "niche" is used by economists to describe small market opportunities with specific characteristics and needs. Small, specialised businesses have generally been able to service these markets more effectively than large corporations that focus on mass markets. This non- competitive behaviour by small businesses has allowed them to survive the growing power and domination of the corporations.

Nature determines society

Tim Flannery[19] has argued that Australian ecosystems and people have a strong disposition towards co-operative rather than competitive relationships because of the harsh and infertile environment. This seems true of indigenous ecosystems and culture. His extension of the hypothesis to post-settlement culture is harder to accept, because fossil-fuel-based capitalism and individualism masked the development of co-operative rural culture almost before it could become rooted in the soil late in the 19th century. Nevertheless, if we compare the pioneer history of Australia to that of the United States, there is a huge difference in the degree to which competition, conflict and outright warfare shape those histories. America's endowment of natural wealth was, and remains, much greater than Australia's, and its history is far more violent. The Middle East, as the world's richest region, has been far from peaceful in the 20th century while energy descent threatens even worse conflict.

Internal co-operation, external competition

Another aspect of the balance between competition and co-operation is the way self-organising systems, especially ecosystems, tend to internalise co-operation between member species and externalise competition with other ecosystems. For example, grasslands and forests compete for territory. Ecologists are familiar with the various mechanisms (fire, grazing, bird-distributed seed, shading, etc.) by which this happens.

Further, to quote Bill Mollison, "mature ecosystems exploit immature ecosystems".[20] Immature ecosystems tend to be less efficient at catching and storing energy, and they are more subject to leakage and loss of energy (as water, nutrients or biomass). These losses are often soaked up by more mature systems with larger, long-lived biomass storages and deep, well-structured soils. For example:

- the red gum floodplain forests of the Murray River system have, for thousands of years, caught most of the nutrients in sediment eroding from steeper, fire-prone forest and woodland catchments
- in northern Australia small areas of rainforest gain nutrients in smoke and ash from the seasonal fires in adjacent infertile savannah woodlands.

In farm landscapes, similar processes can involve relatively larger and faster transfers of energy. For example, wind eroded soil and fertiliser drift from cropland catch in well-established shelterbelts and natural vegetation on roadsides; these same tree belts also gain nutrients from perching and roosting birds that feed on the crops.

Permaculture examples of catching and storing energy often involve taking advantage of the losses of unsustainable land use. For example, nutrients in catchment run-off are intercepted by dams, soil and vegetation; and spoiled hay, collected from farms that have no use for rotted organic matter, is used to establish gardens.

These and other examples raise many ethical issues about exploiting the misfortune of others. We are more likely to be able to decide the ethical course of action in any particular situation if we understand this tendency to internalise co-operation and externalise competition.

Tribal Conflict

Tribal hunter-gatherer societies once had high degrees of internal co-operation, while competition between tribes resulted in occasional warfare. It could be argued that this occasional conflict on a contained battlefield[21] acted as a way to apply selection pressure on surplus young males without the conflict having too much adverse impact on the tribe as a whole.

The tribal pattern of internal co-operation and external competition also occurs between families, communities, businesses and institutions. Given that this pattern is so wide-spread in nature and culture, it would be naive to ignore it.

It is often suggested that we are fated by nature to use violence to resolve conflict. Tribal conflict and aggression is a pattern of behaviour that had more benefits than problems for most of the 100,000 years or so of human culture. Like all successful evolutionary strategies, it has become deeply embedded in our social nature and recurs in new forms time and again. However, the successful spread and proliferation of people in the last 10,000 years and the escalating power and scope of warfare have made tribal conflict a dysfunctional pattern.[22]

The lack of new territory or resources to exploit, and the degree of global interconnection and integration, have dramatically shrunk the opportunities for functional competition and conflict in human systems. A more inclusive and global understanding of interdependence also adjusts our interpretation of ethical principles. This global perspective requires us to consider all people and other lifeforms as part of the same living earth as ourselves, and therefore as participants in mutual relationships.

The appalling evidence of current global conflicts appears to suggest that violence and warfare are permanent and immutable aspects of human behaviour. I see these horrors as a condensation of the ancient "war between the tribes", analogous to a boil on the skin which condenses and excretes a deep systemic disorder that was previously unrecognised. In the last 2000 years, and especially the last century, humanity has been developing new patterns of collective identity and new ways to sublimate warfare into more acceptable forms of conflict, including diplomacy, commerce and sport. Violence and warfare are like a worn-out pair of shoes that we can discard as we fashion new shoes to wear.

Those new shoes include better conflict resolution mechanisms as well as new forms of collective identity. Inevitably some of those new solutions will end up recreating the problem in new forms. This classic problem in both evolution and design requires many iterative cycles before we expunge the dysfunctional behaviour. The most obviously dysfunctional examples of tribal conflict in the late 20th century — Rwanda, Bosnia, Palestine — are recognisable as boils on the skin of humanity. However, the forces driving these expressions of primitive behaviour are less obvious. Besides the pressure of more people on fewer natural resources, there is evidence that the very solutions we have created to avoid even larger-scale warfare are major contributors, directly or indirectly, to these outbreaks of unrestrained but localised violence.

Corporate Culture and Ecology

The rise of corporations from simple instruments of human commerce to super-organisms driving human evolution is the greatest example of the solution that turns into a new form of the old problem. Currently, most people are dependent for their needs on a global economy dominated by multinational corporations. It can be argued that these corporations, along with governments, large public institutions (increasingly seen as corporations) and international organisations (the World Bank, the International Monetary Fund, the World Trade Organisation) form a new global "ecosystem". This novel ecosystem has come into being based on the consumption of both the biological and fossil resources of the planet and the human resources of traditional cultures and values.

While the consumption of natural resources is well understood, the consumption of human and cultural resources is not so well recognised. Some traditional personal, community and cultural values — frugality, household self-reliance, the work ethic, the rule of law, free markets, universal education, concepts of national interest — have been important catalysts for the growth and development of global capitalism. However, that growth effectively undermines those values, just as the fertility of prime agricultural soils has been undermined by chemical monocultures.

The shift from segregated local and national economies to a highly integrated global economy reflects the maturing of capitalism; it is another sign that the flows of energy that spawned capitalism are slowing. Although this maturation process brings increased efficiencies and internal benefits, it tends to more effectively lock out nature and communities as competing systems to be consumed. The role of the nation state in providing captive markets and subsidies for the continued growth of the corporations gives the impression that these relatively recent and short-lived commercial entities are invincible organisms.

Although this new global ecosystem is clearly born from human organisation and culture, a strong case can be made that it is anti-people, anti-nature, and set on a course of self-destruction. When I describe the corporations and their global ecosystem as "alien", I do not mean that the people who work in and depend on those structures are in any way alien. Instead, I use this term to emphasise the emergent potential of institutions and structures that become self-organising without the constraint of human values and ethics. Serious researchers in artificial intelligence talk optimistically about continuing growth in technology that will allow human informational and organisational networks to generate intelligence that is a kind of hive mind.[23] The successful transformation of the large banks from service organisations employing large numbers of people into informational networks with little need for people, either as workers or real customers, baffles many people. Potential future developments that hybridise technology with biology, through genetic engineering and a limited number of "well-connected" and highly rewarded human employees, could see new autonomous symbiotic lifeforms emerge with remarkable speed.

Some of the latest developments in evolutionary theory suggest that this type of symbiotic leap in evolution is not unprecedented (see **Principle 12**: *Creatively Use and Respond to Change*).

The anti-globalisation movement has refocused our defensive natures from other people and communities to the predatory economic organisms of the global economy. Mainstream commentators constantly deride this as unrealistic and naive. If the pattern of internal co-operation and external competition that we see in nature and society is a given of self-organising systems, then this particular case of demonising the alien has a lot more to offer, with fewer risks, than the more traditional forms of xenophobia that are periodically recycled by powerful elites at times of emerging crisis.

The recognition of corporate globalism as a new enemy can stimulate the more self-reliant but locally interdependent ways of living that we call permaculture. Fear and loathing of an enemy can provide enormous motivation for action, but that does not of itself create alternatives to alienated dependence on the global economy. This requires the bottom-up evolution of greater degrees of internal co-operation and effective conflict resolution in new communities and culture. These models are desperately needed. They will be our best hope in future scenarios of either a gradual reformation of the social mainstream or a rebuilding of society after a more dramatic and precipitous collapse of global capitalism.

Materialism and Spirituality

While the corporate capital version of an emergent hive mind is in a race against the limits to growth in use of energy and materials, the real Achilles heel of this vision may be the denial of the spiritual side of humanity.

Earlier in this chapter I mentioned the separation of church and state as one of the productive uses of segregation in the history of western civilization but figure 5 in Ethical Principles shows the polar opposites of materialism and spirituality being drawn together. A creative and optimistic process is mirrored by a destructive integration of corporate materialism and religious fundamentalism. Almost inevitably, as a result of the modern history of the Middle East and Central Asia, the nexus between fossil fuels and religious fundamentalism is providing a focal point to draw these opposite worlds together in an emergent destructive union. Perhaps the empowering lesson in this graphical depiction of segregation and integration is that we can, to some extent, choose the form of integration, but not the inevitable process of integration.

Rather than constantly struggling to keep the spiritual and the material in separate domains, we need to recognise that the gravitational pull toward integrated systems in nature and human affairs is always at work. If we don't creatively design an ethical integration, then larger systemic forces will create one that is less sensitive to human values. The Yin-Yang symbol of Eastern spirituality, in which the polar opposites form an integrated whole and each contains the seeds of the other, is a brilliant graphical representation of systemic thinking.

Figure 28: Yin-Yang symbol of Eastern spirituality

Rebuilding Community

My speculative extension of this principle to describe the dynamics of integration in the global economy and the spiritual realm may be too airy-fairy for some permaculturalists, but almost everyone active in the permaculture movement would agree that stronger development of co-operative relationships between people, families and communities outside the large institutional structures is the perfect complement to personal and household self-reliance. Without this alternative, political strategies for taming the global institutions are like King Canute telling the sea to retreat.

Characteristics of a sustainable community

In **Principle** 2: *Catch and Store Energy* I described human organisation and cultural information as a store of energy, which needed to be transformed and rebuilt in ways suitable for an energy descent future. This is analogous to the ways we need to adapt and rebuild physical storages of energy. I expect that emergent sustainable cultures and forms of organisation will have the following characteristics:

- local and bioregional political and economic structures
- cross-fertilisation — biogenetic, racial, cultural and intellectual — giving natural hybrid vigour
- accessibility and low dependence on expensive and centralised technology
- capable of being developed by incremental steps with feedback and refinement.

While it is difficult, if not impossible, to predict sustainable culture, I think there are good systemic reasons for accepting these characteristics as useful indicators. We will look at them in turn.

Bioregional political and economic structures

As energetic descent demands small-scale, local use of natural energy and resources, structures of governance will need to be more localised. The bioregionalism movement which is closely associated with both permaculture and indigenous cultural resurgence, has raised awareness of the need to identify geographic governance boundaries that reflect natural systems, especially river catchments.

Cross-fertilisation — biogenetic, racial, cultural and intellectual

Although these new local economies and communities will have some of the characteristics of traditional and indigenous ones from the past, they will also be radically different in that they will be distilled from hybrid multicultural and migrant populations with genetics, rituals and ideas from around the world. This will produce new hybrid vigour, analogous to the hybrid ecosystems of exotic and indigenous plants and animals that will provide the resources for these new local economies. Thus although these cultures will be local in their action, they will be informed by global understandings and values, at least in the early stages.

Accessibility and low dependence on expensive and centralised technology.

In possibly chaotic and uncertain conditions, and without the global economies of scale, complex, centrally controlled technologies are likely to be unreliable, although it may be possible to maintain some elements of current communication and information technology for some generations.

Capable of being developed by incremental steps

Because the design of sustainable culture is beyond the capability of any mortal, the process must be organic and iterative. Each small step and stage should be immediately useful and workable and should provide feedback for refinement, and even changes, of direction.

Alternative culture

These criteria are reflected in many of the elements of the alternative or counter cultural movement of the last thirty years such as:

- home birth, homoeopathy, herbal and traditional medicines, self-healing and personal growth
- home schooling, Waldorf (Steiner) and other alternative schools, and a revitalised role for elders
- community gardens, city farms and subscription farming
- LETSystems[24] and ethical investment
- use of body corporate, co-operatives and other legal structures for the community ownership and management of land and other assets
- bioregionalism, spirituality of place and indigenous cultural resurgence.

Most permaculturalists would recognise at least some of these elements and concepts as being integral parts of a permaculture toolkit for building a more sustainable world, and complementary to the primary focus on sustainable land use. Some of the others may elicit support from permaculturalists for various reasons; ranging from expressions of permaculture principles to gut feelings and social exposure to such ideas.

Many of these examples could be used to illustrate several permaculture principles, especially **Principle 2**: *Catch and Store Energy*, **Principle 4**: *Apply Self-regulation and Feedback*, and **Principle 9**: *Use Small and Slow Solutions*. In the context of this principle, (8) they can all be seen as parts of an integrated alternative cultural system, which amplify the value and power of the primary permaculture agenda of sustainable land use as well as each other. The degree to which these examples reflect the design criteria listed above is a prediction of their likely persistence in a low-energy future.

Often when people attempt to move forward in one area of sustainable alternatives, the mismatch with mainstream culture reduces the potential value of the change. An integrated approach is essential if we are to succeed in creating a powerful alternative to dependence on the alien and energetically dysfunctional ecosystem being created by the corporations. Although much can be done at the personal and household level, many of the diverse aspects of design illustrated in the permaculture flower (Figure 1) can be best applied at the local community level.

See "Counterculture" in **Principle 11**: *Use Edges and Value the Marginal* for further exploration of the systemic importance of the cultural margins.

Designing community

Many of these strategies, such as LETSystems, subscription farming and other sustainable social and economic alternatives, can be developed within existing local communities with a critical mass of like-minded people. However, the most integrated application of permaculture principles and strategies is possible in establishing and evolving "intentional communities". These are communities that have been deliberately planned or designed by their participants, rather than unconsciously evolved by social and economic processes.

For more than a century intentional communities have been established to create alternatives to the dominant industrial society. Many have been spiritually based; most would be regarded as utopian to varying degrees in believing that human nature is capable of creating a much better society but that the path forward is by the establishment of small models rather than incremental mainstream change.

The most successful of these movements, the Zionist kibbutzim, succeeded in creating the nation state of Israel in 1948. While the existence of a threatening Arab majority in Palestine obviously contributed to cohesion and success of the kibbutzim, their cultural and religious unity is often overstated as a reason for their success. In fact, in could be argued that the hybrid community vigour generated by Jews coming together from many different countries and cultural backgrounds was as important as the unifying values of Judaism and Zionism. Few people realise that many of the early kibbutzniks were not only socialist ex-urbanites (many from Eastern Europe) but were also anti-religious if not atheist.[25]

The decline of the kibbutz movement since the 1970s is partly a product of its own success in creating a strong, more secure and more affluent nation, which has now followed the path of virtually all developed nations towards increasing individualism. We should not underestimate the contribution to this change made by the United States — both the government and the powerful American Jewish community — in aid, culture and politics.

The success and subsequent decline of the kibbutz movement provides an instructive case study of the conditions in which current attempts to develop intentional communities are likely to flourish.[26]

Co-housing

"Rugged individualism" is often cited as an innate characteristic that prevents Australians from choosing community living. This parochial view of national character ignores how the development of individualism and a socially autonomous existence as the default lifestyle wherever conditions of industrial affluence prevail. Interestingly, this universal trend is less apparent in affluent Scandinavia. The Danish co-housing movement[27] might not be the way of life for the majority in Denmark, but neither is it regarded as some weird alternative lifestyle, which could only be sustained by belief in some New Age guru. Co-housing generally involves a group of people jointly developing medium-density private apartments or houses, integrated with some degree of commonly owned and used facilities. The successful Danish model is being widely used by people in affluent countries wanting better ways to house themselves. Most co-housing projects are in urban areas, and many are mainstream in their architectural style and cultural demeanour. The common ownership structure and integrated design aspects of co-housing have made incorporation of ecological technologies a strong feature of recent projects. More important is the lesson that people can live together and effectively self-govern their community, a powerful alternative to the general view that co-operating with the neighbours is generally not possible or even desirable.

Intentional communities in the origin of permaculture

Intentional communities, including co-housing, provide the "invisible structures"[28] of land ownership, economic relationships, social services and decision-making processes that are necessary for a full and integrated development of the diverse aspects of permaculture such as land use, alternative technology and building. In the original conception of permaculture and its early spread, intentional communities were a source of ideas on ethics, as well as technical and social alternatives. The Farm[29] in the United States, a long-surviving intentional community of the late 1960s, had an influence on the design of the Tagari permaculture community which Bill Mollison and others established in Stanley, Tasmania, in 1979.[30]

At one stage Bill Mollison suggested that all the academic researchers of alternative communities get together and form a community instead of constantly knocking on the doors of communities that were trying to get on with life. Nevertheless, these studies have contributed to what I believe is an increasing survival rate of attempts at community; they have provided better design guidelines for those planning communities, as well as better information to the authorities that must remove the impediments to progressive housing and community models.[31] The development of the Global Eco-village Network[32] has been closely associated with the permaculture movement. However, for most people involved in intentional communities, permaculture remains an environmentally friendly method of growing food, rather than the design and philosophical basis of the community itself.

Use of ecology to describe community

Many permaculture designers have drawn on skills in facilitating groups, consensus decision-making, conflict resolution and other relevant tools that are used in eco-village networks to enhance the toolkit approach to permaculture. Others have found the language of permaculture design useful in describing, strengthening and rebuilding local communities. Some, such as Ian Mason,[33] have taken this further, using ecological terminology to directly and literally describe social functions and community design. Some of Mason's literal applications of ecological concepts are apt because ecological terms ("pioneer community", for example) were originally coined by ecologists to reflect common understandings from the social realm. Others, like "niche", have already been appropriated by economists to describe markets.

The usual criticism of this approach is that natural models are never adequate to describe human social complexity. My view is that, throughout human history and culture, we have used the astonishing diversity of nature as an encyclopaedia to understand and discuss human possibilities. The use of nature as a model for human behaviour can be good or bad; but to attempt to understand and design human systems without any reference to nature is arrogant, and may prove more dangerous than the risks of simplification.

The second criticism is by ecologists, who say that the classical ecological theory that informs permaculture and other social applications of ecological science has been rendered obsolete or overturned by a greater complexity of ideas and understandings.

Part of the role of this book is to use recent understandings of biological systems to help to make better designs for new land uses and communities. For example:

- the discussion of differing types of ecological relationships and their relevance reflects a more sophisticated ecological understanding than simply regarding co-operation as good and competition as bad
- although shared beliefs and co-operative behaviour are fundamental to the success of intentional communities, too much similarity in skills, ages, needs and personalities encourages competitive rather than co-operative relationships. See **Principle 10**: *Use and Value Diversity* for a discussion of the importance of social diversity within intentional communities

The development of more co-operative social patterns of behaviour and resource use is fundamental in a world of diminishing resources. Many people can see the enormous efficiency and savings of resources in shared ownership of land, infrastructure and facilities. But as long as individual and nuclear family autonomy remain possible, more co-operative living arrangements seem a remote dream (or nightmare) for most people.

The hope, that in a sustainable future we can continue to live isolated from each other is the social side of the illusion of a sustainable future in the technosphere. The belief that human nature demands that we live segregated and uncooperative lives is arguably a greater impediment to a sustainable future than the belief that technology and human brilliance can solve environmental problems.

Intentional communities are tackling the hard but rewarding job of evolving ways of more integrated community life; they are at the vanguard of the social solutions for declining energy. That task requires patience and a whole-hearted acceptance and embrace of complexity, rather than any illusion that forced simplicities can work.

Fryers Forest

Some of the lessons learnt from previous successful and not-so-successful communities have informed my own contribution in the design and development of the Fryers Forest community.[34] In the future I may be able to make a more substantial contribution to the distillation of design guidelines for communities. For the moment, the subtle possibilities of my own place in the emerging social complexity of our small community are more important than the detachment and perspective necessary to speak clearly about the process and its wider lessons.

Collective commons or feudalism

Most pre-industrial societies had broadacre commons, and all community members had some, if varying, rights to use them. The Enclosure Acts in England began the global process of privatising the commons, which is still proceeding in poorer countries. Reform of land tenure is one of the central sustainability and justice issues in the Third World. However, First World governments and corporations have resisted any positive initiatives which might provide livelihoods and justice, let alone new models of collective management.

In rich countries, our remaining commons in state forests, national parks and so on are an unlikely source of innovative and creative models of community management. Their fate seems to be determined by bureaucratic and increasingly corporatised management structures and combative public policy compromises; this leads to carve-ups of territory and functions, and in many cases to privatisation.[35]

Models for the management of public land are more likely to emerge from innovations in common land management within intentional communities. Perhaps more fundamental to the future sustainable society will be the tenure and management of our more fertile agricultural land. Without models of common ownership, redevelopment and management for the broadacre farmland commons, the default model for a low-energy future will be some form of feudalism. Although this word conjures up all sorts of negative emotions, ownership of vast tracts of land by a single family or company does have the potential to institute some sort of "baronial sustainability" where land is worked by non-owning farm labourers. The aggregation of most of our better farmland into very large holdings makes this feudal future most likely. At present most large farms tend to be industrial monocultures; in energy descent, more diverse and integrated uses of farm land will develop, which will be much more labour-intensive. Large farms will again become communities of some sort. Within this structure it is possible to imagine highly integrated and ecologically sustainable land uses, and even benevolent owners who look after the interests of their workers.

History shows that relying on the benevolence of elites is hardly desirable. How we manage and distribute the abundance from the broadacre commons is perhaps the greatest design issue for a society adapting to energy descent.

1 B. Mollison, *Permaculture: A Designers' Manual.*

2 Charles Darwin's emphasis on competitive and predatory relationships in driving evolution was based on some excellent observations of wild nature, but he was also influenced by his observations of the society around him. Early industrial England was a rapidly changing society, tapping new energy sources. Predatory and competitive economic relationships were overturning previous social norms and conventions. The social Darwinists used Darwin's work to explain and justify industrial capitalism and the free market. Peter Kropotkin was one of the first of a long line of critics of the social Darwinists. He provided extensive evidence from both nature and human history that co-operative and symbiotic relationships were at least as important as competition and predation. Kropotkin's work had a strong influence on my early thinking in developing the permaculture concept. See P. Kropotkin, *Mutual Aid.*

3 "Health Report" ABC Radio National, 20 May 2002.

4 Bill Mollison, pers. comm.

5 The alga, like all green plants, is a photosynthesising "producer" of carbohydrate energy from sunlight, water and carbon dioxide; the fungus is a "consumer", which in return for food energy provides habitat and assists in the provision of mineral nutrients and water to the algae.

6 In video *Bill Mollison The Permaculture Concept: In Grave Danger of Falling Food.*

7 An ill-fated project proposed in the 1990s by Japanese investors for Australia and captured by the South Australian government after much competition among several states.

8 A term coined by Indian eco-feminist Vandana Shiva to describe the reductionist thinking behind agricultural monocultures. See V. Shiva, *Monocultures of the Mind: Perspective on Biodiversity and Biotechnology* Zed Books 1993.

9 See B. Mollison, *Permaculture: A Designers' Manual* chapter 3.

10 See Holmgren Design Services website: www.holmgren.com.au

11 See D. Holmgren, *Trees on the Treeless Plains: Revegetation Manual for the Volcanic Landscapes of Central Victoria* for tree and shrub fodder design systems for integration into existing farming systems in local landscapes.

12 For example, roof run-off water, the sheltered microclimate against buildings for gardening, and animal manures from sheds and stables are all more difficult to use if everything is widely separated and designed around vehicle access. This approach causes people to drive between different parts of the homestead site.

13 See Rocky Mountain Institute website http://www.rmi.org/

14 Mollison defines a guild as a "harmonious assembly of species clustered around a central element (plant or animal)" See B. Mollison, *Permaculture: A Designers' Manual* chapter 3.

15 D. Holmgren, "Prospects for Rural Development" *Permaculture Journal* no.17, 1984.

16 See Article 5 "A Review of Rural Land Use 1991" in *David Holmgren Collected Writings* 1978-2000.

17 Bill Mollison, pers. comm.

18 Robert Putnam, *Making Democracy Work: Civic Traditions in Modern Italy* Princeton University Press 1993.

19 T. Flannery, *The Future Eaters* Reed Books 1994.

20 Pers. comm.

21 As occurred in the New Guinea highlands at the time of European contact.

22 See Article 26 "Tribal Conflict: Proven Pattern, Dysfunctional Inheritance" in *David Holmgren: Collected Writings* 1978-2000.

23 Kevin Kelly, *Out of Control: The Rise of Neo-Biological Civilization* Addison-Wesley 1994.

24 LETS stands for Local Exchange and Trading System. It is an interest-free currency and information system, developed in Canada and popularised through the permaculture movement.

25 Venie Holmgren, pers. comm.

26 For a brief history of Australian intentional communities and the personal perspectives of some of the pioneers of the recent history of communities, see Bill Metcalf, *From Utopian Dreaming to Communal Reality: Co-operative Lifestyles in Australia* University of New South Wales Press 1995.

27 For an overview of both Danish co-housing and the application of the concept in America, see K. McCamant and C. Durrett, *Co-Housing: A Contemporary Approach to Housing Ourselves* Ten Speed Press 1994.

28 "Invisible structures" is a term widely used in permaculture teaching to describe organisational elements in permaculture design, which must be considered along with the land, plants and animals, and physical structures.

29 In recent years permaculture has become a major element of The Farm's teaching and extension work. It is best known for its work in promotion and support of home birth. See website www.thefarm.org

30 Although Tagari as a close-knit intentional community with a "shared purse" lasted only a few years, some of its strategies were recognised as useful in the design of other communities—for example, the buying of cheap houses in an existing and relatively isolated small country town or village, with farmland owned by the community within walking or bicycling distance.

31 See David Kanaley, *Eco-villages, a Sustainable Lifestyle: European Comparisons for Application in Byron Shire and Australia* Environmental Planning Services Byron Shire Council, June 2000.

32 See their website: http://www.gaia.org/

33 See I. Mason, *The Living System: Modelling Human Communities on Plant and Animal Communities* a self-published discussion book from www.bluepin.net.au/sdn/ian_mason

34 See Fryers Forest page on Holmgren Design Services website www.holmgren.com.au and Article 21 "Starting a Community: Some Early Lessons from Fryers Forest" in *David Holmgren: Collected Writings* 1978-2000.

35 For my perspective on the tragedy of the commons in our public forests, see Article 15 "Wombat Forest Submission (to Victorian Government Minister Coleman, 9 January 1995)" in *David Holmgren: Collected Writings* 1978-2000.

Use Small and Slow Solutions

The bigger they are, the harder they fall
Slow and steady wins the race

Systems should be designed to perform functions at the smallest scale that is practical and energy-efficient for that function.

Human scale and capacity should be the yardstick for a humane, democratic and sustainable society. Whenever we do anything of a self-reliant nature — growing food, fixing a broken appliance, maintaining our health — we are making very powerful and effective use of this principle. Whenever we purchase from small, local businesses or contribute to local community and environmental issues, we are also applying this principle.

The speed of movement of materials and people (and other living things) between systems should be minimised. A reduction in speed is a reduction in total movement, increasing the energy available for the system's self-reliance and autonomy.

Speed, especially of personal movement, generates high levels of stimulation that drown out the subtle and the quiet. For example, when we drive somewhere new, we are stimulated by what we see in the landscape. When we travel the same route regularly, we may notice small changes but generally we lose interest and become bored. However, if we ride a bicycle or walk over the same route, our eyes, ears, skin and noses are opened to a new world of subtle stimulation that the enclosure and speed of the car keeps from us.

The spiral house of the snail is small enough to be carried on its back and yet capable of incremental growth. With its lubricated foot, the snail easily and deliberately traverses any terrain. Although it is the bane of gardeners, the snail is an appropriate icon for small scale and slow speed.

The proverb "the bigger they are, the harder they fall" is a reminder of one of the disadvantages of size and excessive growth. The proverb, "slow and steady wins the race", is one of many that encourage patience while reflecting a common truth in nature and society.

Energetic Limits

As with the other principles, there are good energetic reasons why this principle represents "enlightened self-interest". To understand this, we need to look again at natural systems for models of the effects of scale and speed on energetic viability.

Most people are familiar with the great plant-eating dinosaurs, the largest terrestrial animals; they know that these animals were slow-moving, and perhaps slow-thinking. One hypothesis suggests that smaller, faster and perhaps smarter dinosaurs may have replaced these huge creatures through competition. Whether this hypothesis is true or not, it explains a common-sense view about size and speed: that it is possible to have one or the other, but not both.

In simple energetic terms, a given energy supply can support a large mass moving slowly, or a small mass moving fast, but not both. If energy availability rises, systems can grow in size and increase in speed of movement. If energy availability diminishes, systems must shrink, or slow down, or do both.

A simple non-biological example is the difference between a speedboat — small, fast, manoeuvrable — and an ocean liner — large, slow, unwieldy. An ocean liner designed to have the speed and manoeuvrability of a speedboat would require energy out of all proportion to the advantage. Current materials technology and engineering would also find it difficult to make a ship strong enough to withstand the forces involved.

In the animal world, an ant can lift and carry many times its own weight, while that symbol of strength, the elephant, cannot. This reflects these energetic and material limits to scale.

Pursuing our example of an ocean liner and a speedboat, let's look at their engines. The liner has a very large, slow-turning engine, which also has a very long service life; the speedboat has a high-revving, two-stroke engine, which needs regular tuning and has a limited service life. Similarly, in the natural world, algae, bacteria and other micro-organisms come and go quickly, while large ones like forest trees and elephants are long-lived and reproduce slowly or infrequently. We can see that this also makes sense in the design of buildings and other material assets. Large, expensive public buildings should be designed to last a long time, while small shelters may be seen as temporary and built accordingly.

This common-sense understanding of the way the world works is a further example of the patterns of scale in space and time that we explored in **Principle 7**: *Design from Patterns to Details*.

Cellular Design

Cells, as we saw in **Principle 7**: *Design from Patterns to Details*, provide one of the most fundamental patterns in natural and sustainable design. Small cells, usually with a concentrated nucleus, replicate to create large systems. Most of the basic functions of living organisms operate at the cellular level, where relatively simple and very reliable processes with a long evolutionary history operate at the smallest possible scale. Each cell in the organism is as autonomous as possible within the constraints of the larger organ and organism. Cells have their own optimum size beyond which they do not grow. Where growth occurs to meet the requirements and potential of the larger system (organ or organism), division into two similar-sized cells is the result.

Cellular design in nature suggests that functions are best dealt with at the smallest workable scale, and that replication and diversification are the mechanisms for growth to support larger-scale functions. Although excessive growth in individual cells is generally impossible, uncontrolled overgrowth of cells by replication does occur; it is generally a sign of large-scale systemic disorder, which in medicine is called cancer.

Despite the almost universal nature of this pattern of growth and replication in nature, in human society we have come to regard individual growth and replication as symbols of what is good and successful.

Permaculture Scale

Permaculture has been closely associated, in fact synonymous in many people's minds, with food gardening. I prefer to use the term garden agriculture (to which permaculture principles are readily applied) and have argued for it to be taken seriously as a form of agriculture in Australia.[1] Food gardens are the smallest-scale and potentially most intensive form of agriculture; they represent the nuclei in the clustered cellular pattern of productive land use described in **Principle** 7: *Design from Patterns to Details*.

Households and gardens are, almost by definition, on a human scale: the tasks and the available yields fit the capacities and needs of people. These are normal scales for providing not only much of people's food needs, but also health care, education, and entertainment. Some other examples of small-scale, minimal-movement alternatives associated with permaculture include:

- stacking of plants to make full use of soil, water and sunlight in small areas
- multi-purpose buildings and integrated land uses that pack more functions onto less land
- production of perishable foods from gardens adjacent to housing
- low- and medium-density village and hamlet housing patterns
- local economic systems such as LETS
- bicycle transport.

We have come to regard such activities as small-scale only because almost all functions in industrial society reflect "economies of scale"; these dwarf what was considered sensible and efficient in pre-industrial times.

Slow is Sane

The imperative to reduce the physical scale of systems is better understood than the need to slow down.

In **Principle** 1: *Observe and Interact*, I mentioned the need to slow the hectic, over-stimulated pace of modern life in order to observe and understand natural processes that are slow-moving and subtle.

Working to produce anything of value can be a painstaking experience when we are used to seeing things apparently appear from nowhere. Houses (made from prefabricated components) are knocked together in weeks or months, while owner-built houses using more labour-intensive methods typically take years. The idea of building something once to last tends to occur to builders and other practical people later in life, after they have reconstructed a few things they thought were good enough at the time.

Melliodora

At Melliodora, the following aspects of our property, lifestyle and business all illustrate the small and slow approach.

- We use timber cut by small local sawmills and portable mills, which process logs at slower rates and thus get the best out of each unique tree.
- Home-produced food combined with infrequent bulk purchase of goods dramatically reduces "food miles" and speed. (Perishable food brought from great distances demands fast transport.)
- The use of site and local energies (passive solar and wood) illustrates small scale relative to the centralised energies of gas and electricity.

- Home birth and home education deal with life's processes without the need for large-scale, centralised medical and educational institutions.
- Working from home, with a commitment to the local and the regional over national and global, reduces the need to travel great distances and at high speed.
- Long-distance and overseas travel is for long-stay, multipurpose visits.
- Specialised publications in small print runs provide information on permaculture, which large publishers pursuing mass markets do not service.
- Slowly accumulated savings are used to fund gradual development, rather than using borrowed money for rapid growth.

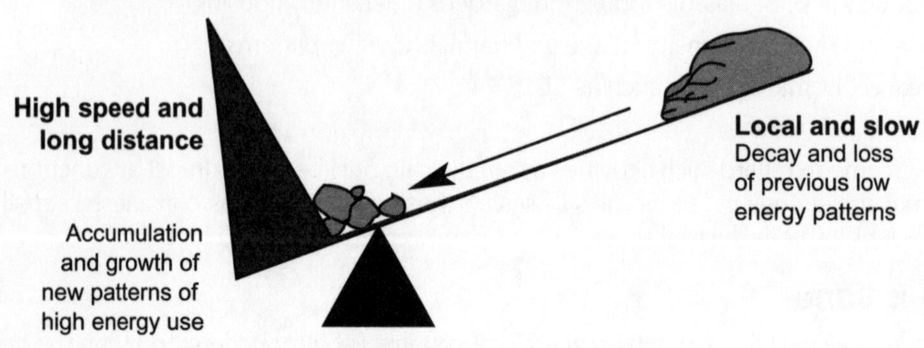

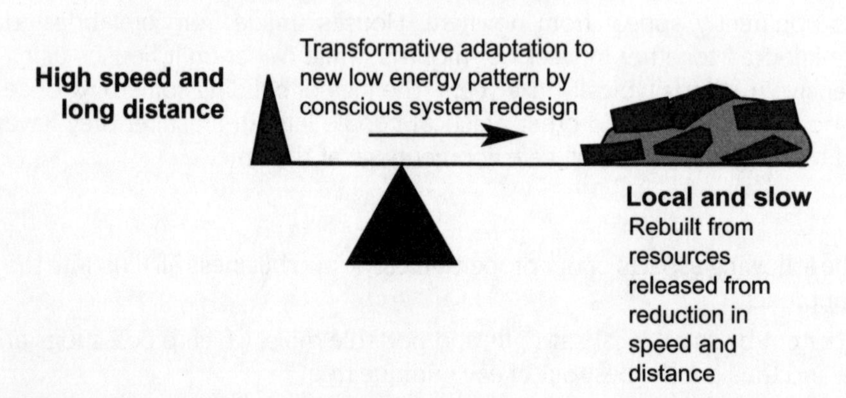

Figure 29: Speed and movement in industrial and post-industrial societies

Optimum Scale and Speed

It could well be argued that the principle is one of optimum scale; neither too small, nor too big; neither too slow, nor too fast. However, we must emphasise small physical scale and slow speed because of the built-in cultural bias in modern society towards growth in scale and speed as indications of what is good, effective and powerful. Beyond the need to hear this message and change habitual design and behaviour, the sustainable balance between these polarities of small/ large and slow/fast is not achieved when we have equal amounts of both.

Figure 29 shows how we should balance these qualities in a world of contracting energy availability. The beam balance, which has balance arms of different lengths to compare the weight of goods to calibrated small weights, is a more useful analogy than the seesaw, which has equal arms. The beam balance and the seesaw are both more useful analogies than the pendulum, because they illustrated the dynamic nature of balance rather than the final equilibrium (dead state) of the symmetrically balanced pendulum.

In traditional sustainable societies, speed and long distance travels (of people and goods) were potent factors that were balanced by the prevalence of local, slow movement (of people and goods). In the modern world, the growth of speed and mobility has tipped the balance, with consequent fracturing and decay of slow and sedentary systems. For example long-distance, high-speed transport of fresh food has displaced local production due to cheap energy and subsidised road transport.

To deconstruct fast systems and move up the steep slope to the other side of the balance requires high initial energy and commitment. The balance model emphasises the design and effort needed to consciously rebuild slow systems; it reinforces the idea that descent from an energy peak is not an easy coast down.

In rebuilding slow and small systems, overshoot is inevitable and necessary. This is analogous to rebalancing a set of scales or a seesaw by weighting the other end until it tips. The enthusiasm for home-grown food (among a small proportion of the population) has increased awareness of, and support for, local commercial production on a modest scale through systems such as subscription farming. This is a modest example of overshoot and rebalancing in action.

The asymmetrical balance model helps us to visualise the appropriate balance between large and small scale, as well as the other polarities that were mentioned in the Introduction as characterising the differences between industrial and sustainable culture:

- linear or cyclical flows of materials
- non-renewable or renewable resources
- positive or negative feedbacks
- centralised or distributed network organisation
- central or edge focus
- consumption or storage of assets
- reductionist or wholistic thinking
- episodic or rhythmic activity

Ethical Constraints on Size

The permaculture ethic of recognising and acknowledging limits provides a clear foundation for the "small and slow"' principle. In effect, "big is better" is a form of greed. The slogan "live simply so others may simply live" sums up this idea.

There are boundless examples of ethics and taboos that constrain growth and development in traditional societies. Many constrain simple immediate urges to catch another fish (for example); some control personal ego (the desire for the biggest house in the village); others warn of evil consequences of barely imaginable worlds.

Colin Turnbull provided a dramatic example of this[2] in relating a history/myth of the Congo Pygmies. This story told that their ancestors had lived in cities. This way of life failed due to technological hubris that conflicted with nature's laws. Our own cultural history provides us with a similar story; the Atlantis myth about the all-powerful civilisation brought down by natural forces. What is extraordinary about the Pygmy story is that, at the time it was told to Turnbull, the largest settlements these forest-dwelling people had ever seen were Negro agricultural villages in forest clearings in their region. The Pygmies were thought to have lived as forest-dwellers with the barest of possessions in a continuous lineage from human ancestors until the arrival of Negro settlements about 300 years ago, and yet they have their own "Atlantis" story.

We can take this as another fragmentary piece of evidence that perhaps the 6000 years of "historical" civilisation has been preceded by others;[3] or we can comprehend it as part of humanity's "collective unconscious" (in the Jungian sense) that includes some understanding of all human possibilities.

Since the Enlightenment, optimism about human (especially European) power and material expansion has gathered strength, so that the old ethical constraints on growth and expansion have had less and less influence in human affairs. Perhaps there was some truth in the belief of European colonists around the world that only they had to drive and "intelligence" to better themselves, while the natives had an "indolent" satisfaction with their situation. Lessons of the folly of excessive growth and size (such as "natural" disasters and economic depressions) have not been strong enough to counter the gathering momentum of the "bigger is better" culture. At the beginning of the new millennium, it seems we need a few more large disasters before we eventually learn what all traditional cultures understood.

After the Big Fish are Gone

It is natural for a fisher to focus on big fish when there are plenty to catch. As the supply declines, smaller fish become more interesting, but if you only have large hooks and tackle you won't catch anything. Thus as we move from fossil fuel scale to natural, human-scale systems and solutions, we are constantly searching for the big answers, but they don't work. For example, new opportunities for large-scale hydro-electric power are extremely limited: most prospective rivers have already been dammed, and the adverse environmental impacts of huge dams preclude new projects. On the other hand, there is substantial potential for small-scale hydro-electric development, both mini (for a small town) and micro (for a house or small business). Spillways of existing dams supplying irrigation water and very small mountain streams along the Australian east coast and Tasmania are being developed — partially because of the need for power companies to meet greenhouse gas targets.

Agricultural research and small-scale opportunities

The failure of agricultural research to tackle, or even recognise, the myriad of small-scale and situation-specific opportunities for innovation illustrates the difficulties of this change from macro to micro scale. In agricultural research and development, the issues and opportunities that affect whole industries receive the majority of funding. Because most sustainable agriculture solutions are small-scale, they tend to fall through the net and are ignored.

For example, in the 1980s and 1990s there was considerable interest and trialling of tagasaste (tree lucerne) for animal fodder on grazing properties across southern Australia. Many of the trials were at sites not suited to tagasaste; today it is only in parts of Western Australia where you are likely to see large stands as part of conventional farming systems. The current estimates of agricultural regions where tagasaste will grow well and be economically competitive with conventional pastures, has been scaled back from an optimistic 3 million hectares to 0.85 million hectares in Western Australia and 1.285 million hectares for Australia as a whole.[4] In this context, my own region of central Victoria does not rate a mention.

My own research on the potential for tagasaste in this region suggests that it may be economically advantageous on stony, well-drained ridges, which form a mosaic of sites covering perhaps 4000 hectares through volcanic, granitic and sedimentary farm landscapes. Although this amount is chickenfeed in terms of national industry development, tagasaste has the potential to address urgent salinity recharge issues while increasing farm productivity. This scale of benefit is ignored by our large-scale systems. A myriad of innovations that are local and specific to a particular site and situation are relevant to sustainable land use, but large-scale systems of research and development will never identify or deal with them. The implications for the diversity of sustainable agricultural research solutions is further explored in "Action Research in Agriculture" in **Principle 10**: *Use and Value Diversity*.

Self-Regulation

If we have a backpack to carry our possessions when travelling, we will fill it; if we have a car, we will fill that. This tendency to use, occupy and consume whatever is available is natural and can be widely observed in human affairs as well as nature. Nature is said to abhor a vacuum. If there is spare sunlight, water or space, some plant will colonise the available resource (see **Principle 3**: *Obtain a Yield*).

External environmental factors — food availability, disease, predators — ultimately constrain the excessive growth of any species. These external controls can be thought of as mechanisms by which the large-scale high-order ecosystem keeps its constituent parts in proportion. However, as explained in **Principle 4**: *Apply Self-regulation and Accept Feedback*, successful species often develop internal constraints to excessive growth. For example, behaviour patterns that limit breeding and reproduction typically interact with external factors to maintain growth and population at an appropriate scale and prevent booms and busts.

In human culture, ethics and taboos provided the internal constraints to endless expansion and growth. Because of the growth and development culture of the last 300 years, we tend to think that people have always exploited every opportunity for growth in numbers and material wealth and that it was only environmental and external constraints that contained humanity. However, there is ample evidence of traditional and indigenous people living in the midst of abundance that they did not seek to exploit.

Pre-industrial Land Use

Before fossil fuels, people depended on "solar" economies of hunter-gathering or agriculture where, allowing for variations due to rainfall and fertility, energy was dilute and distributed across the whole landscape. Consequently, human populations and development were decentralised and small in scale. Following the clustering patterns of nature, nodes of high-energy availability along edges (see **Principle 11**: *Use Edges and Value the Marginal*) between water and fertile land and between mountains and plains supported the development of towns and cities. The physical capacity of people was the unit of energy that determined the scale of technology and every aspect of society. The ability to co-opt the work of other humans (through organised warfare and slavery) was an important element in creating the wealth that we associate with ancient civilisations right through to the early industrial era.

Most importantly, travelling (by people) and transport of goods was minimal. People worked within walking distance of where they lived, and gained the bulk of their food from within a day's travel by bullock transport. The sun determined the patterns of daily life until the advent of the town clocks in the 14th century in towns in renaissance Italy.[5] Minutes, and even hours, were an unnecessary measure of speed.

The durability and beauty of an item was measured by the quality and the value of production by trade and craftspeople rather than how fast it was produced.

Only a few people ever travelled internationally, and for them it was generally a once-in-a-lifetime experience. The biological, cultural and economic consequences of travel by individuals were often very great — Marco Polo springs to mind. International trade was generally restricted to high-value, low-weight durable commodities for use by the elites, such as spices, silk and precious metals.

The shift from bullocks to horses as the basis of transport and farming in the lead-up to the Industrial Revolution (see **Principle 5**: *Use and Value Renewable Resources and Services*) represented a significant increase in speed and therefore power.

Industrial Scale and Speed

Fossil fuels from holes in the ground represented such a vast increase in the scale and concentration of available energy that human systems became reorganised around this new factor. Growth in total population and concentration (cities), greater complexity and scale in economic and governmental organisation, education, health and every other element of modern society has come to reflect the "economies of (fossil-fuel) scale" rather than human scale.

The disempowerment of people by this shift in scale has been the subject of endless debate and literature. Architects, urban and social planners are constantly searching for ways to provide human-scale environments and organisation because it has been recognised that people cannot be fully human without this.

Fossil-fuel scale also has the effect of overriding the fine-grained character of natural and landscape resources. The abandonment of small fields or their amalgamation by destruction of hedgerows and drains in the traditional British and European landscapes has been responsible for major degradation in wildlife and landscape amenity values.

Industrial speed

Even more dramatic than the increase in physical scale of the industrial world is the increase in movement and speed. Transport of goods of great bulk and low value completely reconfigured the economic geography of the world. Fast, long-distance refrigerated transport of perishable food has (in my lifetime) destroyed one of the last links to the patterns of the seasons in the life of ordinary people. *Commuting has become entrenched as a completely natural way of living, which begins at an early age with the daily trip to the childcare centre. The need to get in the car and go somewhere each day has become a fundamental one for an increasing proportion of the population.*

The concept of fast food, prepared as quickly as possible and disposed of almost as quickly, is bizarre — and yet normal.

Transport is an important factor in the urban sustainability debate. There is wide recognition and ample evidence that, the more car-dependent a city is, the more unsustainable (and unliveable) it tends to be.[6] A greater role for efficient and fast public transport on the European model is commendable for reducing energy waste and improving the equity and the quality of urban life, but these changes will not be adequate for the fundamental redesign requirements of declining energy. Radical commentators[7] have suggested that low movement and slow speed are fundamental sustainability criteria for cities. The current European situation, where people commute 100 to 200 kilometres on high-speed trains (often in neighbouring countries), has put in place a pattern that is inherently unsustainable during energy descent.

Techno-optimism

Although it may seem that travel and transport continue to get faster, I contrast these developments with the technological optimism I remember as a child in the 1960s. Besides predictions of space colonies on the moon, and even Mars, by the end of the century, the Boeing and Lockheed Corporations produced plans for large supersonic swing-wing passenger planes. The more technologically modest and smaller Anglo-French Concorde and its Russian competitor were the only supersonic passenger planes to fly regular services, and they have been commercial failures. More fuel-efficient, slower, larger-capacity planes, the 747 and the Airbus, dominate air transport both for passengers and cargo. Current plans to develop slower and much more efficient modern airships are a sign of a trend. Significantly higher oil prices could make this technology the only economically viable air transport.

As an Environmental Design student in Hobart in the mid-1970s, I remember seeing the 1962 Transport Strategy Plan; it showed Hobart ringed by freeways, which even today have not been fully built.

Similar techno-optimism and energy-optimism affected planning for new power stations across the industrialised world. In the early 1980s I remember seeing plans put forward by the State Electricity Commission of Victoria for a dozen massive coal-fired power stations (2000 megawatts each) in the Latrobe Valley, which would make Victoria one of the most energy-consuming regions of the world. These plans were based on bullish projections of increasing domestic and industrial demand for power. Back in the 1940s Sir Mark Oliphant and other advisers to the Australian government on nuclear energy were advocating nuclear-powered desalination of seawater and predicting that nuclear-generated electricity would be so cheap by 1980 that it would be "too cheap to meter".

The idea that faster is better in agricultural and industrial production, transport, communication and travel, eating, and almost every aspect of life is deeply entrenched as a cultural norm. The information technology revolution has resulted in an enormous boost to the idea that speed is good, at the same time that the material and energetic limits to giantism are beginning to be realised.

Small is Beautiful

One of the most eloquent and famous critics of "giantism" in economics and development was E. F. Schumacher, the English economist and author of Small is Beautiful[8] and other well-known books. Schumacher's research showed that modern technologies and organisation were ineffective, counterproductive and destructive when applied to Third World development problems. The inappropriate nature of modern technology is due to its large scale, its centralised and technically complex nature, and its inflexibility when applied in differing environments and cultural contexts.

Schumacher's Intermediate Technology Development Group,[9] and many other non-government groups working in developing countries, have promoted technology and development methods that:

- are small scale
- are simple to apply and maintain
- are labour-intensive rather than capital- or energy-intensive
- use local resources
- support local markets.

These intermediate technologies have been far more effective in achieving economic, social and environmental benefits than conventional development technology pushed by corporations and most government aid.

Other criticism of giantism has come from architects and social and urban planners who recognise the psychological need for a human (small) scale to our physical and organisational environment.

In the ecological housing movement in Europe there was early optimism about the possibility of houses with no adverse environmental impacts, which capture all of their own energy from the sun.[10] This optimism has been tempered by an acceptance that, in general, a small house is an ecological house.[11]

Apart from the obvious environmental costs — more materials are required for a larger house — the following factors become important:

- higher debt, which in turn leads to unsustainable lifestyles
- more space, which encourages purchase of more consumer goods
- more space also encourages an indoor lifestyle with less time spent in nature and the community
- higher heating, cleaning and maintenance costs.

Corporate Growth and Short-lived Solutions

Despite the evidence for the climaxing of energy, corporations and banks are getting bigger faster in order to pursue the remaining economies of scale in giantism. Often we hear that

Australia (with nearly 20 million affluent consumers) is not a large enough market to support the economies of scale required for modern manufacturing. The degree to which this globalisation is obtaining any remaining real economies of scale in giantism is hard to say. Paul Hellyer[12] and other critics of globalisation point out that ordinary people everywhere are the losers in this recent acceleration in the rush to giantism led by the big banks and corporations. He suggests that the collapse of capitalism is inevitable unless there is fundamental reform of the monetary system that allows banks to create money. Capitalism without the regulation of sound public policy is a system out of control, like an engine without a governor winding itself up to destruction.

The sustained pressure for deregulation and subsequent collapse of national economic sovereignty over the last two decades of the 20th century may itself reflect the fact that real energy and resource availability (if correctly measured) has been declining. The hoped-for efficiency of unfettered capital can be understood as the last top-down self-organisational attempt of "the system" to get more out of a shrinking cake by sacrificing its regulation mechanisms and apparently minor constituent parts — ordinary people!

Corporate alternatives

Some of the most progressive of the world's manufacturing companies[13] have realised the human limits to giantism. They have sacrificed engineering economies of scale for the sake of the human productivity and creativity gains of small working groups, where workers have control over a complete manufacturing task.

Gigantism creates the open niche

Overall, we do not see many signs of a swing back to modest scales. More commonly, we see new economic activities beginning at the micro scale and growing into the vacuum left by the rush to giantism. Boutique breweries and bakeries have emerged at the same time that mainstream brewers and bakers have become national and multinational. As Monsanto and other chemical giants have bought out all the established seed companies, new companies have sprouted to provide non-hybrid and non-GMO seeds to small farmers and gardeners. The do-it-yourself industry has expanded greatly in recent years, boosted by the availability of relatively cheap hand-held and bench power tools, even if these tools are themselves the creation of vast global production systems to perform tasks previously done by large, expensive factory machines. Desktop and Internet publishing have cut a hole in the market of traditional large book publishers (which have been also subject to takeover by integrated multimedia giants.) The explosive growth of the micro or personal computer and the global Internet — despite the plans of the corporations for centralised super computer dominance of the information economy — is perhaps the most potent symbol of "small is powerful".

To some extent, the advantage of small size in technology and organisation derives from the overall speed of change. Small size allows reallocation of available resources to flexibility, something widely recognised in relation to small business.

Corporate dematerialisation

The continued economic dominance of the global corporations can itself be attributed partly to their strategies to shed huge material and fixed assets (which reduce their flexibility) and move into controlling the capital and information flows that direct the production of goods and services. The survival and rapid growth of the global corporations since the first energy crisis in the 1970s suggest that, when embodied energy in infrastructure and information has accumulated to high levels, and flows of raw energy remain high and stable, then big and fast is possible. However, the chances of those global corporations persisting for long are very low.

Corporate lifespan

Many people are surprised, even shocked, to learn that most corporations have a shorter lifespan than humans. The cause of this surprise is the common-sense understanding that big, powerful systems should be slow-changing and have a long lifespan. *World affairs are now dominated by enormous government and corporate institutions, operating at scales that affect global climate, but driven by an intelligence and planning horizon which is shorter than individual humans are capable of. This mismatch of scale and lifespan is close to the heart of the unsustainability of industrial culture.*

Long-term Thinking

The fable of the Tortoise and the Hare and the associated proverb "slow and steady wins the race" reminds us of the value of persistence and taking a long-term perspective. It is natural for young people to focus on the immediate and ignore the long term. As a middle-aged person trying to explain why slow and steady are fundamental values in energy descent, I am relieved that *Permaculture One* is evidence that I had the same ideas when I was in my early twenties, and my preference is not simply the effects of ageing.

In **Principle** 4: *Apply Self-regulation and Accept Feedback* I mentioned the "adolescent" nature of our society, where people want to have without consequence. This same culture of adolescence, which dominates our personal, economic and political affairs, leads to the valuing of the fast, flashy and novel over durability, permanence and the evolved. This should not be taken as a criticism of young people; rather, it is a collective failure of older people to grow up.

Most of the debates about the economy versus the environment revolve around the issue of short-term versus long-term benefits. The degree to which short-term thinking dominates politics and business today is shocking to people who remember the behaviour of leaders 40 years ago as the norm; historical evidence suggests that 100 years ago thinking was even more long-term.

For example, today's corporations and investors struggle with the long-term investment required for timber plantations with rotations of 25 years. In 18th-century Britain, replanting oaks to provide timber for shipbuilding required a 200-year plan. Because steel ships made those oak forests redundant (for shipbuilding), Britain still has ancient oak forests today. This gives a clue as to why the planning horizon keeps contracting. In a situation of rapid and unpredictable change, long-term investment strategies often go astray because of novel factors. Faith that technology and human brilliance will get us out of any fix in the future prevails, despite the mounting evidence to the contrary.

Building stone walls

Stone, the most durable and timeless of materials, remains the best solution for many built structures because of its resistance to rot, termites and fire, combined with its low embodied energy and absence of toxicity.

When I was building dry stone retaining walls from stone excavated from our house site in 1986, I wondered why I was building garden walls when I did not even have a house. That questioning led me to rush the job. Within ten years, I wished I had paid more attention to the quality of the stone, and over the last few years I have progressively rebuilt the walls.

If I had found the wall was in the wrong place, I would certainly have regretted putting so much effort into something that I would later remove. Because of the innovation and experimental nature of much of what people do with permaculture, the solid, long-term approach can be counterproductive. By investing too much in current ideas, we may miss new opportunities as they develop. Although this might be seen as an argument for temporary and contingent (fast) solutions, it can also be interpreted as a reason to slow down and think carefully before we act. Once we decide to act, we should not be easily diverted or rushed in our task. (See **Principle 12**: *Creatively Use and Respond to Change* for discussion of the appropriate use of temporary solutions.)

Slow-Growth Strategies in Agriculture and Forestry

The benefits of slowness in biological growth and development are less evident to us than those of speed. In nature, fast growth is certainly a competitive advantage in disturbed areas. Pioneer plants, which are adapted to these conditions, are the source of most of our crops, as well as most weeds. Fast growth must be one of the most ancient and persistent selection criteria for agricultural crop plants. However, in the absence of disturbance, slower-growing but longer-lived plants tend to take over from the pioneers (this is the story of the tortoise and the hare).

Perennial crops

While it would be hard to ignore the advantages of fast-growing, short-lived, predominantly annual plants in providing for human needs, the benefits of slow-growing, long-lived, perennial species in creating sustainable systems was a key concept expounded in *Permaculture One*. This strategy of making greater use of perennial plants in providing for human needs is perhaps the most fundamental demonstration of the principle that "slow is sane".

In gardening, perennial vegetables have the advantage that they do not require replanting every year. Many of them are older types of our annual vegetables. Although they may not yield quite as much as highly selected annuals, the savings in energy and soil disturbance in annual cultivation and planting is a contribution to more sustainable systems.

New Zealand flax is a perennial rush; it produces high-quality fabric that can substitute for several annual fibre crops, including cotton. Cotton is the most environmentally damaging of all broadacre crops because of its very high demand for nutrients and water and its susceptibility to pests. This substitution[14] would result in huge savings by preventing soil degradation, conserving water, and reducing chemical and fertiliser use.

In the long term, the development of perennial grain crops from promising native grasses has the potential to substitute for at least some of the current dependence on wheat and other annual grains.[15]

Vegetable production

Modern vegetable production systems aim to harvest a crop as quickly as possible. Soluble fertilisers, maximum irrigation, and even hydroponic techniques have all been employed to this end.

Some of the adverse effects of this approach include:

- minerals are generally at a low level and unbalanced, leading to poor flavour, keeping qualities and nutritional value
- the very high water content of produce further dilutes flavour and nutritional qualities
- the luxury uptake of nutrients, especially nitrogen and potassium, leads to unmetabolised nitrates in the product that have been shown to be carcinogenic.

With slightly slower organic production systems, plants are predominantly fed nutrients as needed by symbiotic microbes, which draw on insoluble organic and mineral sources. If the market were able to recognise nutrients and other hidden qualities, organically produced fruit and vegetables would gain a price premium independent of any certification.[16]

Well-hardened nursery plants

It is acknowledged in the tree nursery business that good-quality tube stock for farm plantations and shelterbelts should be hardened off before it is planted out. The perfect growing conditions of the nursery produce fast, lush growth, which cannot endure exposure to the elements of wind, sun and frost. Consequently, any nursery producing quality stock hardens the plants by gradual exposure to environmental stresses, which slows the growth rate and literally hardens the soft lush foliage.

Long-rotation forestry

Forestry provides strong evidence for the importance of growing trees slowly.

Before industrial times, forests were the measure of the wealth of nations—as important as agriculture. Fast-growing European trees, such as pine, poplar, birch, alder and hazel, all had a myriad of uses, but it was the strong, durable, slow-growing timbers, primarily oak and beech, which were the measure of sovereign wealth.

In the last 100 years, fast-growing pioneer trees — conifers, poplars, eucalypts, acacias — have come to dominate plantation forestry. Short rotations of these species for pulp, sawlogs and composite wood products have been planted to replace the more diverse, slower-growing native forests across the world.

As woodworkers know, in the world of wood you never get something for nothing. Fast-grown trees tend to produce timber that is:

- non-durable in exterior situations or in the ground
- poor in strength and toughness
- small in cross-section
- prone to high shrinkage, and therefore deformation and cracking
- plain and featureless in aesthetics.

Three important factors have allowed foresters to focus almost exclusively on fast-growing trees producing poor-quality timber:

- continued harvesting of old-growth native forests to provide high-quality wood
- industrial substitution with high embodied energy materials, such as steel, aluminium, concrete and plastics
- industrial (high-energy) processing of wood to improve performance (laminating and finger-jointing, reconstituted fibre boards).

The low environmental values of fast-growing, short-rotation, monocultural plantations are well recognised. The vast plantations of blue gum and other fast-growing eucalypts around the world have given eucalypts a bad name as timber that splits like a celery stick, and is brittle and non-durable, only suitable for pulp or firewood. As firewood these fast-growing species are fast burning. Few people outside Australia are aware of the wonderful high-strength, high-durability, beautiful wood from our Boxes, Ironbarks and other slow-growing eucalypts.

The same situation exists with respect to exotic trees grown in Australia. The vast majority of plantation timber in Australia is *Pinus radiata*, a conifer from the Monterey Peninsula in California. Its very fast growth in early arboreta and trial plantations of conifers made it the focus of research and successive generations of plantations, even though it produced awful timber from a tree of poor form with large branches. Nearly a century of natural and human selection and product development has resulted in respectable softwood that supplies the bulk of the house framing and other major timber markets in Australia.

A huge range of more valuable exotic timbers has been ignored because of apparently slow growth in early arboreta and trial plantations.

These early arboreta and trial plantations around Australia (mostly planted in the 1930s "work for the dole" schemes) have been sites of pilgrimage for me over the last 20 years for many reasons. One of the lessons from those plantings takes us back to the story of the tortoise and the hare. While many of the species planted have not produced wood in volumes to justify wider planting, others have grown to trees of great size after 60 years. The question that troubled me for many years was why foresters believed these trees were too slow growing to justify wider planting. Many of the best timber trees grow slowly when young and may require sheltered conditions, but after a decade or two, their growth rate increases.

The Californian Redwood story

Californian redwood (*Sequoia sempervirens*) provides a dramatic example. Despite the less-than-ideal climate, these trees are frequently the largest in parks and arboreta in southern Australia. The wood is excellent stable softwood, similar to western red cedar. In 1983, the catastrophic Ash Wednesday bushfires in Victoria provided the opportunity for my colleague, Jason Alexandra, to salvage logs from a diverse range of exotic and native trees from the charred gardens of Mount Macedon. Among the thousands of tons of dead and fire-damaged trees that he salvaged was one particular redwood, which we suspected was the largest exotic tree in Victoria (2.1 metres diameter). It came from a garden designed by that giant of early Australian economic botany, ecology, and landscape design, Baron von Mueller. Our house at Hepburn is a celebration of the beautiful native and exotic timbers from Mount Macedon, much of it from trees 40 to 80 years old, and all species that were regarded at the time as too slow growing for commercial plantations.

My fascination with Californian redwood (among other timber trees) intensified during the mid-1980s when I became aware of the redwood plantations belonging to the Ballarat Water Board. The board's foresters were thinning and taking the time to measure growth rates.[17] Although the 60-year-old trees that they were harvesting were just large enough for sawlogs, the plantations were very dense. The annual growth rate of the whole plantation over the 60 years had averaged 20 cubic metres per hectare, and the plantations were currently growing at an annual rate of 44 cubic metres per hectare. For most people outside the forestry profession, these numbers mean nothing. For a forester, they stretched credibility.[18]

After 20 years of making my own (unqualified) judgments and asking many foresters about the value of a wider selection of plantation timber species (including Californian redwood), my reaction to this information was frustration and anger at the forestry profession's collective blindness to this natural abundance.

The trees at Ballarat are not growing in the ideal climate for redwood, unlike a small stand of trees in the Aire Valley of the Otway Ranges. Those trees are a similar age, and before they reach 100 years they will probably set a record for the largest volume of wood standing on any hectare of land in Victoria. The Water Board forester's eyeball estimate in 1988 was 2000 cubic metres on one hectare of land; this approaches the standing volume of our largest old-growth forests of mountain ash (*Eucalyptus regnans*), which are more than 250 years old![19]

Californian redwood is not unique. Many of our Australian rainforest timber species also demonstrate this growth pattern of slow at first, fast later.

Today a new generation of qualified foresters and farmers is including slow-growing timber trees in plantations in order to provide high-quality wood, despite the continued chanting of the forest economists' mantra that interest rates will reduce the current net value of such plantations.

Animal husbandry

In the field of animal husbandry, intensive feedlot production systems are designed to maximise the speed of growth up to the age at which animals are slaughtered. Apart from the outrageous waste of energy and resources, the pollution and the ethical issues directly and indirectly deriving from these systems, there are issues about the quality of meat produced. For example:

- hormones and other additives that are used to achieve these growth rates sometimes persist in the meat
- the water and fat content can be very high, while nutrients and flavour are reduced
- many animals raised in these systems would die of degenerative diseases if they were not slaughtered within a year or two of their allotted time. A similar situation exists with dairy cows, which are brought into production at a young age and dispensed with after a few years.

This outcome of a system designed to maximise milk output per kilo of fodder can be contrasted to the slow, steady approach required on dairy farms run by Hare Krishna communities, where every calf born, female and male, must be cared for over its natural life. To produce dairy products under this tradition (without unsustainable growth in animal numbers) it is necessary to maximise the milk production per calf by continuing to milk a cow for many years before getting her in calf again. With careful management, it is possible to have about seven milking cows in a total stable population of 80–90 cows and bullocks.[20]

Of course, the other half of this curious equation is to find a job for the bullocks. In India the bullocks were used for pulling carts and ploughs. They were more valuable than cows, which yielded only small amounts of milk because of a shortage of year-round good fodder.

To slow down to bullock pace is a great challenge in today's world, even for devotees of Lord Krishna. What appears to be a mad and unsustainable taboo, is — or at least was, in its cultural context — a lesson in the limits to good things (dairy products) and the need for a slow and steady approach.

Childhood Development

There is also increasing evidence relating to humans that growing up fast physically and mentally may not be such a good thing. The emerging key value is balance rather than speed. The familiar notion of learning to crawl before you walk has a much wider currency. For example, Waldorf schools, based on the educational concepts of Rudolf Steiner, pursue the idea that teaching children to read and write at a very young age creates imbalance and stunts the development of other faculties.

In human nutrition, the practice of feeding children concentrated food, especially processed foods that are rapidly absorbed into the body, has many negative effects. There is now plenty of evidence about the health problems created as a result of children drinking too much pure fruit juice (which is a concentrate rather than a whole food).[21] It is possible that over-nutrition is causing children to mature earlier and grow taller than previous generations. Degenerative diseases (as with the feedlot pigs) may be more common as a result.

Slow Food

One of the signposts to a turnaround in the valuing of slow processes is the emergence of the Slow Food movement. It celebrates the loving preparation and consumption of food as an alternative to the aesthetic (as well as physiological and ecological) obscenities of fast food.[22] The movement is reputed to have 60,000 members world wide.

The Information Economy

The acceleration in the material world has been small in comparison to the acceleration in information, media and culture, where there appears to be no limits to speed. Most of the discussion of this phenomenon focuses on the wondrous increase in speed and capacity of technology but ignores the value of what is communicated, and the side-effects of speed.

The throughput of authors and titles by publishers, which began to accelerate in the 1970s, astonished and frustrated my parents who ran a specialised technical bookshop in Perth, WA. Today, new information is generated and becomes redundant almost as soon as it is produced. The contraction in the number of minutes, and now seconds, during which the general viewing public can be expected to focus on any TV news item is a cynical and shocking measure of increasing speed and collapsing attention span. The mobile phone, in a period of only 10 years, has severely damaged people's ability to make simple travel plans and inform others of them without the aid of continuous contact; it has redefined what it means to be alone.

As computer and other systems of artificial memory grow exponentially, the breathtaking lies of politicians and corporate PR representatives, and lack of interest in history, reflect a general loss of memory and an inability to understand what older people are talking about when they remember the past.

The continuing acceleration in computer speed, and therefore capacity, has become an ever-changing yardstick for measuring everything. The reports, plans and presentations of consultants in every field follow the escalation in technology for getting the message across, while the focus on content has shrunk to a point where content-less presentations at high-level conferences receive earnest attention and praise. For media technology sceptics in the 1990s, the Microsoft Powerpoint computer slide presentation became a symbol of this world of "all show, no substance".

The information aspects of the service economy have been called the weightless economy because what is being bought and sold often involves no materials or direct energy despite the long chains of embodied energy (EMERGY) involved in their production. It is the miniaturisation, and consequent speed, of information technology that has made possible the decoupling of economic activity from materials and energy consumption.

Although information technology does allow us to do more with less, there is ample evidence that consumption of materials and energy continues to rise in affluent countries (even without increases in population). The much-heralded "paperless office" has not eventuated. Instead, the increasing capacity and reduced capital cost of creating paper copies of electronic documents, along with a whole range of habits and expectations, have resulted in even more paper in the bin. Apparently, office paper use has quadrupled over the last 10 years.

E-commerce, which was touted as a way to reduce travel time and cost in shopping, has generated an explosion in the individual door-to-door delivery of goods. In the United States the private logistics corporations that are replacing the postal service are some of the fastest-growing companies, stimulated by e-commerce and subsidised by cheap oil and a publicly funded highway network.

Many high-level government and corporate planners appear unaware of the linkage between e-commerce and increase transport costs.[23] Perhaps the dot-com crash and increased oil prices have brought some greater sense of reality.

Permaculture and information technology: dancing with the devil?

The value or otherwise of this information economy is a contentious issue for people involved in permaculture. Many see it as simply another tool, which can be used for good or ill; they argue that we need to use it to understand, communicate about, and manage natural and human systems. To reject it is like refusing to use two arms when everyone else is. Some enthusiasts see the self-organising, networking and evolving nature of information technology as reflecting natural (permaculture) principles, and believe the promise of reduced materials and energy use.

Others mistrust it for:

- reducing our connection to, and ability to comprehend, the slow-moving processes of the natural world
- accelerating the consumption of materials and energy, despite the promises of reducing waste (the myth of the paperless office)
- increasing the centralisation of power, despite the promises of devolution of information and power (for instance, electronic democracy).
- increasing dependency on complex technology that cannot be locally produced.

As a consultant and author, I have used a computer for 14 years; I know that the technology has shaped my work, for better and for worse. I am aware that my ability to treat computer technology as a tool in my hand is very different from the shaping effect that it is having on younger generations, who could see it as an integral part of their body and senses. This may increase efficiency in the use of these tools, but it will probably follow the pattern of previous innovations in media technology that atrophy other human capacities.[24]

I use the following guidelines in planning my relationship with computers.

- Stay abreast of the technology if you need to in your work, but never operate at the cutting edge of new hardware and software[25] because you become part of a free debugging service for the technology developers.
- The costs and depreciation rates are very high, which in turn require full-time commitment to using the technology (you become a slave to it).
- Don't reward the corporations for producing more of the new stuff rather than refining what already exists.
- Don't let the gee-whiz factor get in the way of the task at hand.

Instead:

- update with second-hand hardware and software that is well proven and does not carry high costs and depreciation rates
- accept that the Holy Grail of fully integrated and compatible systems will never eventuate while rapid growth in speed and capacity continue
- assess all new and old functions to gauge whether it is worth the effort of computerising them; include realistic estimates of the time it will take to learn to use the system and the inevitable downtime (crashes, malfunctions, printer jams)
- value the local over the global and use access to global information networks to strengthen your use of local resources
- always maintain non-computerised back-up systems for all essential functions
- balance time at the screen with time spent with people and with nature
- try to minimise the exposure to computers of children below the age of 12
- remember the map is not the territory; the virtual world is never reality.

The Paradox of the Radical and the Conservative

As energy availability peaks, it is slowing the rate of change in many ways. The change of direction from growth to contraction, from materialistic to more spiritual values, is so fundamental that it will turn the world on its head. Our experience of this shift is likely to be one of incomprehensible and chaotic change — that is, an acceleration of what we are experiencing now! Think of being in a speedboat as it accelerates; it lifts and bounces on the waves. After a while, we get used to the speed and the exhilaration, but making a tight 180-degree turn intensifies the thrill, even if we slow down in the process of the turn.

This presents a structural paradox about the change from material growth to contraction, which is inherent in permaculture, and more broadly, in the counterculture and radical movements.

Radical and revolutionary ideas suggest the need for change from some stable norm, a break with tradition and the established ways. However, when the norm is itself one of continuous and radical change and the new idea relates to durability, permanence, persistence and sustainability, we have a contradiction in terms and ways of being. The conservative is the radical, and the radical advocates a new conservatism. This paradox underlies much of the confusion of current environmental and political debates. The term "conservationist" epitomises this paradox.

Fundamentalism and Reactionary Values

As well as the radical critics of the status quo of accelerating change, there are also reactionary critics who want to return to what they see as traditional values. Within the spiritual domain, the religious fundamentalists who preach precise adherence to specific versions of the "word of God" are gaining influence worldwide in response to the impacts of modernity. Although the critiques of the mainstream by the radical and the reactionary have much in common, the reactionary seeks to rebuild the past, piece by piece, while the radical aspires to utopian possibilities that must be constructed or evolved. For those in the current mainstream, the radical and reactionary critiques are indistinguishable; they are unable to imagine anything other than the present instant and a constantly reconstructed history, which gives the present its illusion of lineage and permanence.

Conclusion

The evidence that "small is beautiful" and "slow is sane" is all around us. The more we come to terms with the implications of energy peak and descent, the more we can recognise giant systems as dinosaurs of the era of fossil-fuel abundance. As we accept our own fallibility and mortality and tune in to nature's patterns, we see that slow and steady does win the race.

When an adolescent sense of immortality and values of speed, novelty and endless growth define a whole civilisation, I think we are close to its demise and the birth of a new cultural paradigm. Watch it slowly unfold.

1 See Article 7 "Gardening as Agriculture" in *David Holmgren: Collected Writings 1978-2000*.

2 C. Turnbull, *The Forest People: A Study of the Pygmies of the Congo* Simon and Schuster 1962.

3 For a serious discussion and research related to this hypothesis, see Institute of Meta History website: http://www.imh.ru/main_e.htm

4 T. Lefroy, J. Cook and B. Peake, *"Tagasaste in Australia" in Tagasaste: Proceedings of Workshop Review of Tagasaste Research in Western Australia* Centre for Legumes in Mediterranean Agriculture Occasional Publication no. 19, 1996.

5 Louis Mumford, *The City in History* Penguin 1961.

6 P. Newman and J. Kenworth, *Sustainability and Cities: Overcoming Automobile Dependence* Island Press 1999.

7 Including Professor Allen Rodger, pers. comm.

8 E. F. Schumacher, *Small Is Beautiful: A study of economics as if people mattered* Blond & Briggs 1973.

9 Founded in 1965. See ITDG website: http://www.itdgpublishing.org.uk/

10 In north-western Europe the proportion of heating that can be obtained from the sun is 30% (Declan Kennedy, pers. comm.), while in southern Australia it is close to 100%.

11 Proposed new energy efficient housing regulations in Victoria make no reference to size as the most important environmental impact factor. The regulations will require all new houses to gain a 5 star rating using the "First Rate" software which predicts heating and cooling energy use per square meter of occupied building space but does not consider total or per person energy use.

12 "Background Briefing" ABC Radio National, May 1999.

13 See Ricardo Semler, *Maverick* 1995, about the revolutionary management changes at Semco, one of Brazil's biggest manufacturing companies.

14 Grown in Tasmania during World War II (Bill Mollison, pers. comm.) and widely adaptable to seasonally wet bottomland in high rainfall southern Australian.

15 See website of the Land Institute in Kansas http://www.landinstitute.org

16 One recent study by the Organic Retailers and Growers Association of Victoria of vegetables from a single Victorian organic grower showed levels of key food nutrients 10 times higher than those in a random sample of conventional market produce.

17 For more details on this and other examples of forgotten and rediscovered timber species, see D. Holmgren, *Trees on the Treeless Plains: Revegetation Manual for the Volcanic Landscapes of Central Victoria*.

18 The average annual growth rate of our good *Pinus radiata* plantations over a 25-year rotation is 20 cubic metres per hectare. The figure of 44 cubic metres per hectare is as high as some of the record measurements for sub-tropical pulpwood eucalypts over a 15-year rotation.

19 Vern Howell, pers. comm.

20 Gokula Dasa, pers. comm.

21 American Academy of Paediatrics, *The Use and Misuse of Fruit Juice in Paediatrics* (RE0047) See http://www.aap.org/policy/re0047.html

22 See C. Petrini, (ed.), *Slow Food: Collected Thoughts on Taste, Tradition and the Honest Pleasures of Food* Chelsea Green Pub.

23 Steven Bright, management consultant, pers. comm.

24 For a discussion of how media technologies have damaged cultural and human capacity, see Article 24 "Do Media Technologies Scramble Young Minds" in *David Holmgren: Collected Writings 1978-2000*. For arguments for a realistic appraisal of the value of the Internet, see Jerry Mander, "Internet: The Illusion of Empowerment" in *30th Anniversary Whole Earth Catalogue* 1998.

25 The "leading edge" is now known as the "bleeding edge".

Use and **V**alue **D**iversity

Don't put all your eggs in one basket

The spinebill and the humming bird both have long beaks and the capacity to hover, perfect for sipping nectar from long, narrow flowers. This remarkable co-evolutionary adaptation symbolises the specialisation of form and function in nature.

The great diversity of forms, functions and interactions in nature and humanity are the source for evolved systemic complexity. The role and value of diversity in nature, culture and permaculture is itself complex, dynamic, and at times apparently contradictory. Diversity needs to be seen as a result of the balance and tension in nature between variety and possibility on the one hand, and productivity and power on the other.

It is now widely recognised that monoculture is a major cause of vulnerability to pests and diseases, and therefore of the widespread use of toxic chemicals and energy to control these. Polyculture[1] is one of the most important and widely recognised applications of the use of diversity, but is by no means the only one.

Diversity between cultivated systems reflects the unique nature of site, situation and cultural context. Diversity of structures, both living and built, is an important aspect of this principle, as is the diversity within species and populations, including human communities.

The proverb "don't put all your eggs in one basket" embodies the common sense understanding that diversity provides insurance against the vagaries of nature and everyday life.

Biodiversity Conservation

General awareness about loss of species biodiversity in wild nature has emerged as an important environmental issue after 200 years of fascination with, and classification of, biodiversity by botanists and zoologists. Public recognition of the rapid loss of biodiversity in cultivated varieties of plants and breeds of livestock has been slower to emerge, even though it may have adverse effects on humankind than losses in wild nature.

Most of the responses to these problems are essentially reactive attempts to minimise loss of wild and cultivated biodiversity. Some examples are arboreta, botanic gardens and zoos, national parks and nature reserves, wildlife protection laws, endangered species breeding and restocking programs, seed and germ plasma banks.

Diversity of nature is a constant theme of biological science. With the gathering pace of biodiversity loss due to human impacts, it has become common to think of environmental issues as always involving a conflict between nature's drive for diversity and human demand for productivity. But this view of environmental issues is very limited. While permaculture incorporates strategies to conserve biodiversity, it also seeks a more fundamental redesign of all we do, so that biodiversity becomes a valued and functional part of our world. By extension, it can also provide the support of "natural principles" for the continuing social and political struggle worldwide to value human cultural and individual diversity.

Balancing productivity and diversity

The discussion of the Maximum Power Law in **Principle** 3: *Obtain a Yield* shows that the systems that most effectively use available energy (more correctly EMERGY) for survival tend to prevail over those which do not. So survival of the fittest operates at a system level, as well as an individual level. *Nature is equally concerned with diversity and with power and productivity. Teaching of environmental science and popular environmental culture tends to ignore this aspect of nature in an effort to counter the obsession with the prevailing economic measures of productivity and power.*

Diversity of elements and functions has already been discussed as one of the key characteristics of integrated systems (see **Principle** 8: *Integrate Rather than Segregate*). Proverbs such as "don't put all your eggs in one basket" and "variety is the spice of life", remind us of the value of diversity. The well-known quote from pioneer American conservationist Aldo Leopold, "the key to intelligent tinkering is the save all the pieces", reminds us of the problems of biodiversity loss from human intervention in nature. But we need a much richer contextual sense of diversity because it is such an important balancing value to the drives for functionality in permaculture expressed through the **Principles** 2: *Catch and Store Energy*, 3: *Obtain a Yield* and 4: *Apply Self-regulation and Accept Feedback*.

Those principles focus on broadening our understanding and evaluation of what is "functional", "productive", and even "good", from the narrow, rigid econometric values that are the distilled wisdom of fossil-fuel-based capitalism. This will no doubt be interpreted by some as ruthless functionalism, where anything that does not prove to be "useful" or productive should be dispensed with. This is ecological rationalism, which cannot accept the wonder and abundance of life without measuring its value against fixed criteria.

For a deeper, wholistic understanding of the use of diversity, we need to see it in dynamic balance and complementary tension, in all systems, at all levels. With this, we should be in a better position to work with, and design for diversity at all levels, instead of worshipping particular classes of diversity while ignoring or ruthlessly destroying others.

Specialisation in Nature

One broad ecological classification of plants and animals is as generalist or specialist species. Generalists are the jacks-of-all-trades, able to thrive in a range of habitats and with a range of nutritional sources. Generalist plants make good pioneers, moderating and improving the environment for more sensitive specialists. Specialists are more like the master crafts-person, efficient in a particular habitat but inevitably not very flexible or adaptable.

Landscape patterns of diversity and specialisation

At a landscape scale, the diversity of ecosystems is mapped onto the underlying biophysical variation created by climate, landform and bedrock minerals. Different species and ecosystems are most efficient and powerful in using the available resources in different places. These "different places" range in scale from bioregions to micro-climatic niches occupied by single organisms. Diversity in terrestrial plants is especially great, simply because the potential growing sites vary so greatly in available resources of sunlight, heat, water and nutrients. For herbivorous animals (able to move around and select what they consume), the extremities of environment are moderated by the storages and variety of plant fodder available, and by habitat diversity. Higher up the food chain, omnivorous and carnivorous animals are supplied with a more stable (if hard-to-catch) food supply, which further dampens the seasonal variations that affect plant foods.

Consequently, animal diversity, while still great, is constrained, following fewer successful patterns which repeat themselves across many ecosystems. This is especially true for many of the higher animals, which are generalists adapted to a range of environments. *Homo sapiens* and our associated pests, such as rodents, are the ultimate generalists.

The dynamics of competition and co-operation were explored in **Principle 8**: *Integrate Rather than Segregate*. The observation that diversity tends to reduce competitive relationships and foster co-operative and symbiotic ones was demonstrated with examples from nature.

Where available resources are little differentiated into niches, a single generalist species tends to be most efficient at using the resources; it therefore comes to dominate, resulting in low diversity. Large numbers of individual organisms of the same species fully occupying those resources inevitably compete with one another, because all individuals have the same needs. Where resources are highly differentiated into niches, different species tend to prevail in different niches. This typically occurs as a mosaic spatial pattern, where different species are in close proximity but do not compete for exactly the same resources. Temporal differentiation can also occur, when different species or individuals occupy the same space at different times of the day or season. Beyond the absence of competition, mutual and even symbiotic relationships can develop because of the different needs and outputs of those species.

Diversity Creating Stability?

Total biodiversity, as measured by the number of species, either at a site or within whole ecosystems, can vary greatly. There was a substantial debate between ecologists over several decades as to whether greater diversity was a major, or even a contributing, factor in the stability and of ecosystems and their resilience to stress.

Diversity provides alternative pathways for essential ecosystem functions in the face of changing conditions. This makes sense from our understanding of the traditional organisation of human systems, where diversity of crops and resources provides insurance against failure in one or another function.

Tropical rainforests, which are some of the most stable ecosystems in the world, have high biodiversity. Large-scale climatic disturbance cycles (ice ages) periodically eliminate much of the biodiversity from temperate latitudes, especially in the northern hemisphere; in the tropics, evolving diversity can continue to accumulate.

On the other hand, some apparently simple ecosystems appear to be very stable. For example, many forests go through successional stages, from pioneer stages involving a great number of initial species, to a very stable climax dominated by one species of slow-growing, long-lived trees. Examples are the yew (*Taxus*) forests in western Europe and myrtle beech (*Nothofagus*) forests in Tasmania.

Diversity of structure

The structural diversity in the complex matrix of roots, litter and compost, trunk buttresses, faults and hollows, as well as a complex canopy structure, are features of yew and myrtle beech forests, which in turn support substantial insect and microbiological diversity. In addition to this structural diversity, the potential lifespan of these trees (many hundreds of years in the case of myrtle beech and thousands in the case of yew) is the key factor giving stability in the face of seasonal fluctuations.

Diversity of age

Capacity for regular regeneration, and therefore diversity in the age structure, of forest trees may also be a factor in ecosystem stability. In the case of tall wet sclerophyll[2] forests of mountain ash (*Eucalyptus regnans*) and other species that occur in south-east Australia and Tasmania, it is generally thought that these forests regenerate only after catastrophic wildfire, and that all the trees are a single age. Without fire, it is assumed rainforest species invade wet sclerophyll forest and become dominant after the eucalypts die. This suggests that single age structure is a weakness in ecosystem persistence, unless drought and resulting wildfires hot enough to burn wet sclerophyll forest are regular features of the climate. The great ribbons of bark shed in the hottest months may be an adaptation to encourage a regenerating fire.

The common view of catastrophes such as fire and insect plague in forests used to be that they were a dysfunctional disturbance. The idea that they may be a co-evolved aspect of a system which pulses periodically for maximum power and health is explored further in **Principle 12**: *Creatively Use and Respond to Change.*

Most foresters and ecologists have assumed that trees in wet sclerophyll forests can live for 250 to 300 years. However, limited evidence from dating of old eucalypts, combined with historical evidence on the size and form of these trees, suggest that ages of 500 to 1000 years may have been possible on favourable sites. On such sites, especially in Tasmania, these trees grow in what are described as "mixed forests", because they always include rainforest trees. It is possible that these mixed forests are not simply a succession phase between wet sclerophyll and rainforest, but a third ecosystem. This system is able to maintain itself in the absence of fire by developing an uneven age structure as a result of the occasional regeneration of eucalypts following wind throw of seed from large mature trees.[3] Given that it is hard to tell the age of very old eucalypts, these mixed forests may well be uneven in age, even if no obviously young trees (less than 100 years) are present. Whether or not my hypothesis about our most spectacular and fought-over forests is correct, it illustrates how diversity of age could increase ecosystem stability.

Genetic diversity

Another factor that may be missed by simply focusing on species diversity is the role of diversity within dominant species. Many widespread and dominant plants show great varietal diversity. For example, in the case of widespread and dominant eucalypts, such as manna gum (E. viminalis) and red gum (E. camaldulensis) in southern Australia, the provenances[4] and even subspecies show great variation from one another. This reflects adaptation to specific local conditions, and sometimes isolation from other populations of the same species. These natural variations between local and regional types are of critical importance as a foundation for the evolution of new species and selection and the breeding of plants and animals for human use.

Within any local population of common and dominant plants, great individual variety also is common.[5] This genetic diversity allows adaptation to changing conditions as the more vigorous individuals reproduce more successfully. It is perhaps surprising that this process of survival of the fittest does not lead to a monoculture of identical individuals. Plant and animal breeders understand how hard it is to completely breed out so-called undesirable traits, and how quickly they reappear once selection pressure is withdrawn. Because environmental conditions rarely remain the same for long, positive feedback to entrench a particular characteristic may be periodically reversed. The generation of new and novel diversity by mutation and other "creative" forces may also contribute to maintaining diverse populations.

Whatever the mechanisms, diversity is a proven survival strategy for plant and animal populations and species. Nature rarely seems to place all her eggs in one basket.

On balance, I believe it is reasonable to say that diversity of individuals, populations, species, ages and structures in natural systems are contributing factors to stability, even if the concept of stability itself requires further clarification (see **Principle 12**: *Creatively Use and Respond to Change*).

Pre-Industrial Cultivated Diversity

In pre-industrial agriculture, polyculture was the norm; where monoculture did exist, it was on a small scale.[6] The reasons for polyculture in traditional pre-industrial agriculture are themselves diverse.

Self-reliance demands diversity

Where agriculture provides for household (subsistence) needs rather than serving markets, diversity of crops is essential to provide nutrition, variety and, to the extent possible, regular supply. In many regions of India, for example, legume crops such as lentils provide the protein that is nutritionally complementary to grains (as well as fixing nitrogen and providing a disease break for the staple grain crop, rice). Thus self-reliance demands diversity.

On the other hand, when central markets redistribute food, they provide for people's diverse needs. This makes possible the specialisation and marketable yield increases of monoculture as part of an evolving fossil-fuel-based industrial ecosystem, but it damages the ability of the agricultural ecosystem to provide for people's needs in the long term.

Security from diversity

A diversity of crop varieties and species provides some degree of security or insurance against seasonal failures and pest or disease attack. This is one reason peasant farmers persist in the cultivation of some low-yielding or otherwise inferior varieties alongside higher-yielding or superior varieties. In plant selection there is often a trade-off between high-yield, easy-to-harvest characteristics on the one hand, and drought tolerance, resistance to pests and diseases, and general hardiness on the other.

Diversity as cultural maintenance

The third reason for crop diversity is that people often grow varieties for aesthetic, sentimental, cultural and spiritual reasons. These desires and obligations may celebrate a lineage of kinfolk (growing Grandpa's special bean, for example). Sometimes this commitment to diversity is like homage or a tithe to the abundance of nature, which symbolically acknowledges the importance of things other than our own immediate needs.

Improvements in crop breeding or shifting the goal posts?

Some may argue that modern agricultural selection has provided fundamental improvements in plant varieties and animal breeds. This view reflects a progressive paradigm: that modernity creates constant improvement in all areas of human endeavour. The so-called improvements of modern varieties and breeds over traditional ones were achieved by selecting for a novel set of conditions provided by non-renewable resources and technology.

Many scientists and others involved in the development and introduction of the "improved" strains of rice and wheat that constituted the Green Revolution appeared to be unaware of the trade-off between monocultural productivity and resilience.[7] Of course the "deeper wisdom" behind the Green Revolution was that inputs of fertiliser, pesticides and water would compensate for the loss of resilience in "improved" strains. The inputs would be purchased from multinational corporations[8] (or in the case of water supply, which was not profitable, from large government dams and irrigation projects funded by international aid and constructed by corporations).

In Australia, the early history of scientific crop improvement by the Commonwealth Scientific and Industrial Research Organisation (CSIRO) did give priority to hardiness and natural disease resistance in crops. This was a key factor in establishing more natural, low-input farming as the norm in Australia long before current environmental concerns arose. The diminishing returns from traditional breeding of traditional crops, combined with a shift from public to corporate and industry funding, is driving CSIRO's work toward genetic engineering and corporate objectives.

Breeding of vegetable varieties has been more counterproductive than traditional field crops. The infamous machine-harvestable tomato that has no flavour is a good example of a change in the selection criteria in crop breeding resulting in the loss of other values. Even with optimum (organic) fertilisation and watering, some trials[9] have shown that "improved" market hybrids consistently fail to yield as much as traditional varieties.

Thus, the concept of "improving" already highly bred, ancient, traditional crops through plant breeding is often erroneous. What passes for improvement is generally an unconscious change in the selection criteria, and a blindness to lost attributes.

Balancing Productivity and Diversity

The growing of varieties that are not the highest-yielding, and other examples of diversity maintenance, can be seen as reflecting a much broader tension and balance in all cultivated systems between productivity and resilience. At a landscape scale, it is between the cultivated systems and wild ecosystems. The cultivated system provides high yields, but is dependent on intensive management. On the other hand, the wild ecosystem provides low yields with little or no management other than harvesting (see **Principle 3**: *Obtain a Yield* and **Principle 11**: *Use of Edges and Value the Marginal*).

Animal Breeding and Diversity

Maintenance of rare breeds of domestic livestock is an important part of agricultural biodiversity conservation. Like old varieties of vegetables, the rare breeds of poultry, cattle, horses, dogs, and so on represented careful selection over many generations for characteristics adapted to specific uses and environments. The characteristics are frequently appropriate to permaculture designed systems. For example, most dog breeds once had specific hunting, retrieving, herding or other functions. (See **Principle 5** : *Use and Value Renewable Resources and Services*) The use of conformation[10] to select good breeding stock was always balanced by functional criteria, which included performance in the chosen task as well as fecundity, growth, resistance to disease, and general hardiness.

Industrial animal husbandry

In animal breeding today there has been a separation and intensification of selection. In production livestock industries, elaborate performance criteria for factory farming of poultry, pigs and beef and dairy cattle are turning animals into machines barely able to exist outside intensive rearing facilities. Over bred and weak European and North American cows still get shipped to Third World aid projects even though the unsustainable and inappropriate nature of these modern breeds is well documented.[11] This degeneration from over breeding results from what Vandana Shiva has incisively described as "the monoculture of the mind".[12] The most limited yardsticks measure what is productive, functional and good, without reference to the wider context or adverse side-effects.

Breed fanciers

Not only have animal breeds been degraded to drive agricultural economic ideology. Even worse, breeders of poultry and pet animals, especially dogs, have ruined many of the breeds they sought to conserve. Breeding by conformation without the more basic performance and hardiness criteria is the source of the problems. These latter criteria prevailed when the animals were useful rather than ornamental. For example, a long narrow nose might have been a conformation sign for Collie dogs, but a dog with a long nose that was also small-brained and skittish would have most likely been done in, and certainly not used for breeding. In poultry breeding, even common and widespread breeds like Black Australorp have been largely maintained for decades by fanciers who are not necessarily interested in the bird's ability to forage without being caught by foxes or even in their egg-laying capacity in free-range systems.

It is easy to restore the wild hardiness in poultry and other livestock, but in the process lose the high productivity and other characteristics of the breed. For example, I've seen people recreate the wild jungle fowl of South-East Asia in only a few generations of free-range breeding from domestic birds. The result is a very impressive and hardy bird—but try finding the eggs or containing the birds in a run, and it's like being a hunter-gatherer!

Traditional animal shows are partly to blame for rewarding breeders only on looks and purity of breeding. In-breeding, the curse of over-selection, is the result. Where animals are still predominantly bred for a useful function in a natural environment, such as cattle and sheep dogs, looks are less important than performance. The popularity as pets of the Blue Heeler, Kelpie and other working breeds threatens to destroy the intelligence and energy that people love in these animals.

Breeding selection by aesthetic criteria did have merit in the traditional context in which the breed emerged. In that environment a particular conformation and set of patterns and colours represented an aesthetic summary of the breed, a visual shorthand for a complex mix of invisible traits. But, without the appropriate environmental context, the aesthetic rules are as meaningless as a European christmas celebrated in an Australian summer.

Diversity in this context is a paradox. The diversity of traditional breeds is maintained first by purity of breeding to exclude wild diversity and hybridisation, which would destroy the breed. But without an appropriate environmental context, the breed is destroyed by maintenance of purity. This paradox is analogous to the debates over functionalism and aesthetics in horticulture (see **Principle 3**: *Obtain a Yield*) and in architecture (see **Principle 7**: *Design from Patterns to Details*).

Clearly the balance and integration of productivity with aesthetics, purity of line with hybrid vigour, and fixed specialised form with diversity, is delicate. It needs to be constantly reconsidered in every situation and context. Ultimately, we have to ask ourselves whether what we are doing represents care of the earth and its lifeforms, care of people, distribution of surplus, and acceptance of limits.

Geographic and Cultural Diversity

So far I have focused on diversity within cultivated and managed ecosystems. What is even more evident is the great diversity between the cultivated ecologies of different places documented by ethnobotanists and more generally anthropologists. Every bioregion of the world had its own characteristic ways of cultivating plants, husbanding animals, and organising the landscape, and often its own language to describe reality, which all reflected local conditions and a degree of isolation from neighbouring regions.

Food fermentation cultures

A subtle, invisible example of lost cultural food diversity and its close links to biodiversity is microbiological diversity in food fermentation cultures. The continuously maintained sourdough culture of a bakery in any European village was a co-evolved ecosystem of many different yeasts and other micro-organisms; it reflected the varieties of grain grown in the village fields, the methods of the baker, and even the microbial environment of the (often ancient) bakery. It is not surprising that the bread of each village was unique.

In the 1920s, Japanese food scientists began "purifying" the culture for miso production in the interests of efficiency and productivity. Today, virtually all the world's fermented foods are produced with pure (monocultural) starter cultures. We have no idea what we have lost or how long it takes to recreate such complex fermentation cultures.

While travelling in Europe in 1994, my partner Su Dennett was offered a sourdough culture reputed to be 500 years old. Apart from the impossibility of taking this back to Australia because of quarantine controls, she found the idea of responsibility for such an old culture too great. To anyone who has kept a sourdough or yoghurt culture going for years, the idea of a domestic and community economy that are stable enough to maintain one for 500 years is almost beyond comprehension. As in nature, specialised co-evolved diversity requires great stability in the environmental conditions.

Language diversity and loss

The diversity of human cultures, which astonished early travellers and remains a source of wonder and conflict in the modern world, is itself nested within the bioregional diversity that gave rise to them. Even in Europe, where the power of kings, empires and nation states has for centuries tended to create homogeneity and punish cultural diversity, variety redevelops as soon as the opportunity arises. The diversity of dialects and cultures from one region to another reflected the constantly renewed power of the land as a formative force in human culture. Nevertheless, Europe is the home of less than 4% of the world's languages;[13] this may reflect past losses as much as an inherited lower cultural diversity.

According to the World Watch Institute, about 90% of Australia's 250 Aboriginal languages are near extinction; only seven have more than 1000 speakers; and only two or three are likely to survive the next 50 years. The predicted loss of most of the world's languages in the next 100 years rivals the loss of biodiversity as a devastating consequence of industrial culture. It represents a direct loss of indigenous knowledge and local sustainable design, most of which has not been documented or passed on.

Cultural Globalisation and Renewal of Cultures of Place

Responses to globalisation

Most people on the planet today find themselves with one foot in a tradition of place and one in modernity, which they frequently perceive to be the culture of some other (dominant) people. For Aboriginal Australians, it is white culture; for Israeli arabs, it is Jewish Zionism; for Corsicans, it is metropolitan French culture. For many people worldwide it is American culture, increasingly delivered by media and the internet.

Elements of local domination of one people and culture by another persist as a major element of politics and society. This often disguises the fact that "modernity" as delivered by the media and corporations is the greatest cultural imperialism the world has ever seen. Although predominantly Anglo-American in origin, with English as its lingua franca, modern global capitalism (or whatever we want to call it), driven by fossil-fuel wealth, is consuming human cultural, agricultural and natural diversity and replacing it with a global monoculture at all levels.

Almost everywhere in the world there is anxiety, anger and conflict about loss of local culture and meaning in the face of globalisation. Most indigenous and traditional people recognise that, if their culture is removed from nature, land, food and other practical expressions, its most valued aspects — language, kinship, spiritual beliefs — become a disconnected and eventually dead tradition.

This global culture is accepted as normality, and one's own tradition becomes some anachronistic leftover of a parochial past. Whether people view modernity positively or negatively, they generally understand it and their own personal sense of tradition, in terms of each other. Few people have much understanding of the commonalities between all traditions of place, or on the other hand, of the unique nature of modernity as a global culture of no-place.

New cultures of place

Permaculture uses the patterns that are common to traditional cultures for design principles and models. The diversity of design solutions, strategies, techniques and species are a toolkit towards new cultures of place. Wherever we live, we must become new indigenes.

Ironically, some people may see permaculture in this context as "cultural appropriation", and therefore to be resisted. I would see it as contributing to the recognition of the value of traditional culture, knowledge and biological diversity, which is a pre-requisite to "fair cultural trade". So long as sources are clearly acknowledged and respected, and the use of the knowledge is not critical to generating substantial private or corporate wealth, this use of indigenous knowledge is an important aspect, even a foundation, of permaculture design in many parts of the world.

We are beginning to recognise that the sharing of experiences between traditional peoples can provide an understanding of diversity within basic patterns that are widespread or even universal. The commonalities between traditions of place provide a powerful alternative to the usual comparison between the local and the global, which assumes that the global culture of no-place is normal.

I assume that globalisation will run its course to the extent allowed by non-renewable energy and resources, and that new cultural and natural biodiversity will need to be constructed from the remnants of many different systems, both natural and cultural. If the engine of the global economy stopped tomorrow, the aftermath of gathering change in both nature and culture would ensure that reconstruction of pre-industrial culture (and nature) is impossible. Everywhere some pieces of old traditions, relatively intact, will be combined with diversity from other places to create new local cultures with hybrid vigour.

The image of us collecting pieces for a new jigsaw puzzle (see **Principle 1**: *Observe and Interact*) provides a way of responding to natural and cultural diversity. It is a search to understand and value what nature and other people have to contribute and at the same time a determination to create new indigenous cultures of place from the best of all relevant diversity.

Self-Reliance and Diversity of Yield

The application of self-reliance to the food supply (see **Principle 4**: *Apply Self-regulation and Accept Feedback*) has enormous implications for diversity within production systems. In current high-energy systems the diversity of the marketplace provides for peoples' needs and wants, even though the production systems that feed those markets are almost all monocultures. By extending the lines of supply across bioregional boundaries, and even continents, marketplace diversity has increased dramatically in recent decades.

In growing our own food, we are immediately confronted with the simple reality that people have diverse needs, so the systems that supply these needs must be diverse in their products.

Affluent countries

I have argued that we need to regard gardening as a serious form of agriculture, and that we need a major increase in the number of skilled food-growers who can feed their families and neighbours from food gardens.[14] If this garden agriculture is to begin replacing central market systems in the supply of fresh produce, before energy and food costs force these changes, the diversity of yields that are possible from those systems will be a key factor in their success. One of the sobering realities for the garden food gourmet is that not even the Garden of Eden could provide the diversity of ingredients that many people have grown accustomed to in the fridge and supermarket culture.[15]

Therefore, gardening strategies that maximise diversity of yield will generally be more successful in alluring jaded consumers away from their dependencies on high-energy systems. This issue may account for the popularity of "permaculture gardens" over more traditional food gardens. Much of what is called traditional gardens is small versions of market gardens; they produce a handful of varieties in great abundance, but with little variation and/or surprise.

Similarly, the rapidly increasing popularity of heritage vegetable varieties can be attributed to politics and the search for novelty as much as practical need, but it is having the effect of maintaining diversity abandoned by agribusiness.

Poor countries

In poor countries today, especially for the urban and landless poor, gardens provide the only nutritional balance to a monocultural diet of the traditional staple. They can obtain a moderate diversity of fresh food (including animal protein from poultry and other small livestock) from gardens to supplement the staple, which may be a root crop (cassava) or grain (millet, corn or rice). In these situations, garden diversity can mean the difference between bare survival and well-being.

In rich countries, the same issues may emerge in the future. The remarkable growth of urban and organic agriculture in Cuba since the collapse of the Soviet Union and the US economic blockade provides one of the best models for redevelopment of diverse fresh food production to meet local needs. Although relatively poor, Cuba is quite urbanised and has a long history of modern market-based agriculture.[16]

In any incremental development of less energy-dependent systems, a diversity of sources of supply is important. For many people, a home garden, a local market, or subscription farm or barter with friends, plus some purchases of organic produce from the central market system, can satisfy needs and wants and provide greater fresh food security.

Melliodora

Some of the strategies that we have applied at Melliodora to maintain diversity of food production include:

- physical design — raised beds, keyhole layouts — which is suited to sequential sowings of small batches of seed
- taking advantage of micro-climatic variation, including that created by the built environment, to grow crops that are early or climatically marginal
- use of varieties suited to early and late yields
- use of varieties suited to storage and preservation
- inclusion of semi-wild and self-sown varieties and species that can provide a yield when more favoured ones fail.

The Diversity Debate in Permaculture

Some people see the emphasis on diversity in permaculture as meaning that a random mix of species makes a system stable. In response, Bill Mollison suggests that it is the number of functional connections between species, rather than the number of species, which makes for stability. We can think of zoos or botanic gardens as very diverse but not stable, because they lack the functional connections that contribute to a self-regulating system. (See **Principle 8**: *Integrate Rather than Segregate* and **Principle 12**: *Creatively Use and Respond to Change* for more about stability and change in ecosystems.)

There is good reason to doubt that planting a chaotic (or even well-planned) diversity of plant species will necessarily contribute to a stable and self-regulating ecosystem directly, although it may contribute indirectly.

The dynamics of diversity: proliferation and culling

Although permaculturalists can become obsessed with growing the widest possible variety of plants, for most experienced food gardeners, the fascination with diversity is balanced by the need to concentrate on what produces the greatest weight and nutritional value of produce for the least space, water, fertiliser and work. At Melliodora, after 12 years of not purchasing vegetables, we are now a little more inclined to concentrate on growing what does best and is easiest to harvest, and maybe replacing some trees that are not doing so well. Some might see this as a slackening of the commitment to permaculture ideas, or alternatively as pragmatic realism. I see it as harvesting some of the fruits of diversity to create a more refined, more functional system.

Many years ago, one of Victoria's pioneers of commercial organic berry production, who started out inspired by the permaculture concepts,[17] told me that the diversity strategy didn't work. He cited his recent culling of all his apricots and most other stone fruit after years of failure to fruit well. Further, he had simplified the production system to rows of berries with mown clover paths, and apples interplanted with chestnuts at very wide spacings. What he had seen as a failure of principle, I saw as natural progression: proliferation, followed by culling to produce a more refined and functional system. In fact, I believe that this is a perfect example of the success of the diversity principle.

The key to designing more self-reliant systems is that we always proceed without knowledge about what will work and how it will work. There are several reasons for this:

- the lack, or limited relevance, of local sustainable traditions as models
- the inherent complexity and individuality of integrated systems designed for energy descent
- novel factors such as the availability or presence of new species, knowledge or technology (for instance, the role of trace elements, the use of drip irrigation)
- natural co-evolutionary forces, which operate once systems are established.

This inherent uncertainty provides a strong incentive to experiment, to try lots of species and different strategies. It is inevitable that much of what we try will fail, but frequently we find novel solutions that we could have never known from the beginning. It is natural that we should see the culling process as what finally makes things productive; but without the diversity, there is literally nothing to cull.

Melliodora

At home, I have resisted the pressure to ruthlessly cull what has not worked for many reasons:

- our aims are to provide for our own needs, not commercial production. Aesthetic variety and interest are non-material needs that our place provides for us, thus reducing our demand for more costly and less sustainable forms of entertainment and stimulation
- as a permaculture research and demonstration centre, experimentation is an essential, continuing, long-term function
- realisation of the persisting mineral imbalances in our soils some years ago led to corrective measures, followed by a reappraisal of varieties of vegetables and fruits that had been either unproductive or poor in quality. We are finding some varieties of vegetables that we had given up on are now successful, and fruit trees we had considered culling or regrafting are now producing more and better-quality fruit.

Forestry models of diversity and culling

In **Principle 5**: *Use and Value Renewable Resources and Services* I explained the rationale for thinning regrowth forests. This focus on culling to increase productivity relies on reducing diversity of form in the timber trees although it may contribute to forest diversity in other ways, such as allowing understorey species to thrive. The prolific regenerative capacities of most dominant forest tree species following disturbance (especially in the temperate climate zones) make this culling a sustainable and appropriate process.

Similarly, plantations established by direct seeding generally produce far healthier and better timber than planted trees because hundreds of thousands of seedlings provide a much greater diversity to thin out, leaving only the very best. Thus, it is fine to focus on culling for productivity as long as natural processes maintain diversity. We must never forget the value of the diversity, and remember that the gains from culling are made possible by nature's diversity.

Natural proliferation

Nature appears profligate in the production of seed, insects, young plants, and even animals, which are culled not only to provide sustenance for those higher up the food chain but also to ensure the "survival of the fittest". One limitation of the Darwinian view of evolution has been the over-emphasis on natural selection that leads to survival of the fittest, and the lack of attention to the processes that generate diversity in the first place. In other words, Darwin provided an explanation for nature's culling or editing process, but not much about her creative process. The orthodoxy of neo-Darwinian evolutionary theory is now challenged by a plethora of ideas that focus more closely on nature's creative process (see **Principle 12**: *Creatively Use and Respond to Change*).[18]

This *proliferation of diversity followed by culling is a fundamental pattern of nature. It operates in all systems, at all scales, from the whole earth over geological time through to growth and development of business organisations. It is like the wild enthusiasm and experimentation of youth, which is replaced in middle age by a more sober focus on our role in society (whatever that may be). Without experimentation, we do not know what the possibilities are.*

Much of the evolution of modern agriculture and modern economics can be seen as a ruthless culling of apparently inefficient traditional diversity in order to produce more powerful centralised and productive systems. In this context, permaculture is about rebuilding diversity, but the major value of diversity in cultivated systems will emerge in the future, as we cull what doesn't work.

A deep respect for both natural and human diversity seems part of the wisdom in most spiritual traditions; diversity was often protected by taboos as a restraint on over-zealous culling. A reference to this principle, and ultimately to the Care of the Earth ethic that requires us to consider and care for all nature's diversity, can help us find the right balance.

Rebuilding Diversity

The use of current wealth to rebuild diversity is an excellent investment strategy for dealing with an uncertain future. We need to focus our efforts in rebuilding diversity on what we believe is most likely to be adapted and useful. Because of the inherent uncertainty and lack of knowledge that prevail, chaotic proliferation of almost random biodiversity is a valuable strategy, despite the high failure rate.

For example, seedling fruit trees mostly yield fruit that we consider inferior to that of selected cultivars; but if we have thousands of wild trees, we are certain to find a few that are valuable. At Melliodora we have found that in the case of nectarines, as few as 10 seedlings have generated what we think are superior types.

Revegetation

Over the last 30 years of the 20th century a wide variety of Australian native trees and shrubs were planted in rural landscapes for many reasons. The knowledge that informed those plantings ranged from brilliant to ignorant; the successes have become prevalent by their physical dominance in the landscape, by natural regeneration, and by follow-up plantings. What is surprising from my own research[19] is the degree to which random diversity, and even ignorance, has generated surprising successes. Some of the most useful planted native trees are successful well outside their natural climatic range. If we planted only species from similar climates, we would never have discovered these ready adaptations.

For example, spotted gum (*Eucalyptus maculata*) comes from the east coast, which is dominated by summer rainfall. It is one of the most vigorous and valuable farm timber trees in parts of southern Australia dominated by winter rainfall.

Further, we find examples of species that are little used for revegetation in their natural environment due to disease or insect predation, which thrive in areas outside their climatic limits. Tuart (E. *gomphocephala*),which is indigenous to the Perth coastal plain in Western Australia, is severely affected by beetle attack across its rainfall range, from 450 millimetres in the north to 1000 millimetres in the south. In contrast, tuarts planted along the moist to semi-arid southern coast of Western Australia and South Australia are thriving. In coming decades, they will dominate some of these coastal landscapes from follow-up plantings and natural regeneration.

Perhaps the most extraordinary tree in exposed Australian coastal environments is the ubiquitous Norfolk Island pine (*Araucaria heterophylla*). It comes from a high-rainfall, subtropical climate with relatively mild onshore wind conditions. This tree has been planted and has slowly grown to massive mature size in some of the harshest onshore coastal environments in Australia, with winter-dominated rainfall as low as 450 millimetres, where the tallest indigenous tree is about 3 metres.

Because of this constant element of surprise in plant adaptation and vigour, the strategy of "try everything", (maximum unselected diversity), or even simple ignorance, will continue to be a source of new adaptive diversity in the future. (See **Principle** 12: *Use and Respond to Change* for a discussion of chaotic proliferation of new and unregulated diversity as a mechanism for accelerating evolution.)

Indigenous restoration

While exotic diversity is often a source of species that thrive, indigenous species that are rare or have died out locally are not necessarily maladapted to prevailing conditions. In many ecological restoration projects, the results of reintroduction of both plant and animal species can be surprising. Even when species appear doomed to extinction, a change in one factor may allow them to thrive. The recent success (especially in South Australia) of wildlife refuges with predator-proof fences in allowing the rapid proliferation and breeding of many threatened small marsupial species is well known. At the Food

Forest[20] permaculture demonstration farm at Gawler, SA, a predator-proof fence has allowed bettongs to thrive (and become a potentially harvestable animal) while controlling weed growth.[21]

Economic and Social Diversity

The patchwork quilt of diverse systems

One of the great challenges in energy descent is to replace mass solutions and systems with a great diversity of systems and solutions to suit the particular nature of sites, situations and cultural contexts. Much of what is described as "design from patterns to details", or a shift from "segregated simplicities to integrated complexities", or from "big to small scale", is by implication, a shift from a mass monocultural solution to a diversity of solutions. From the bottom-up perspective, a particular, local, integrated, multi-faceted solution replaces a universal and simple recipe. The diversity of these local solutions shows up when we take a more global, top-down perspective and see them like a patchwork quilt of many colours and patterns.

Action Research for diverse agricultural solutions

The evolution of modern agriculture provides a clear example of the challenge to think systemically and accept diversity. The success of modern scientific agriculture can be attributed in part to the agricultural education system which educated future farmers in modern methods, and the extension systems of government, agricultural colleges and chemical companies that provided information to practising farmers. Over time, a series of generally applicable and standardised farming systems developed replacing the previous diversity created by the unique interaction between land, culture, family history and personal character in pre-industrial times. This transformation of agriculture from something embedded in nature to something embedded in human-controlled industrial systems has been much slower than was the case with the manufacturing and construction sectors of the economy because the diversity and complexity of nature are less amenable to mass solutions than the more controlled, non-biological fields of human activity.

Although it is natural for me and other critics of modern agriculture to focus on its failings, agricultural research and development has been extraordinarily successful. It has created methods which, when followed by a great diversity of farmers on very variable land, produce more or less reliable results. Much of the doubt among farmers about modern methods is due not so much to the failure of these forced simplicities, but to the relentless downward pressure on commodity prices. As prices fall, even farmers who follow the recipe perfectly and get high yields of good-quality produce receive barely enough to cover their costs.

As the pressure to find more economically viable and ecologically sustainable farming methods has increased, governments and agricultural researchers have responded by attempting to find better methods. Progress in this new direction has been slow for many reasons. A greater political will and financial support for sustainable agricultural research would be useful but there are fundamental impediments to this top-down approach.

Many researchers and others fail to recognise that the unsustainability of modern methods is due, in part, to mass solutions applied to diverse conditions. More sustainable systems will be characterised by site-specific and situation-specific solutions. Further, instead of the solutions arising within established fields of research — animal health, pasture agronomy, silviculture — they will be found between and across research fields. In fact, the solutions will be part of integrated systems which will include the unique nature of the land and the farmer. In this context, conventional approaches to research are as good as useless.

A more useful approach to research, emerging partly by design and partly ad hoc, is the recognition of innovators who appear to be "doing interesting things". A collaborative and wholistic systems approach could then be used to measure and document the aspects of the system which the farmer and the researchers think might be useful for ongoing development, or that are interesting in terms of some more widely applicable strategy or technique. The results of that research will never be conclusive or certain; it will rarely produce models that are very widely applicable as a whole, but it can provide information that others might find useful in developing their own unique sustainable systems.

This type of action research has the effect of recognising the innovative practitioner as the creative source, with the researcher performing a secondary role like that of technical assistant, record-keeper and librarian. Many interesting innovations in sustainable agriculture in Australia, including organics and the use of tree and shrub fodder, reflect this model, although formal research support for these innovations is rare. The much lower cost of this type of research, compared with the conventional controlled trials in standardised conditions, is a major advantage.

For many within the research community, these progressive approaches represent a hopeless loss of scientific (reductionist) rigour. They do not provide large-scale influence with solutions that are applicable to millions of hectares and millions of dollars worth of produce (and therefore of funding). Unfortunately, unless the research community faces up to this issue it will continue to generate more problems than solutions by attempting to simplify complex integrated systems into a few controllable factors. The call for the development of a new field of sustainability science is a direct recognition of the failure of current scientific methods and disciplines to deal with the diverse, integrated and complex nature of interactions between humans and their environment.[22]

In the permaculture movement, the need for a myriad of diverse and local demonstration sites has been recognised as essentially complementary to the teaching of general principles and design methods.[23] I have emphasised the value of local demonstration sites and case study documentation because these allow us to experience what integrated complexity looks and feels like, even if they provide no blueprint for what to do somewhere else. All the detailed complexity of our science and large-scale systems blind us to the diverse and integrated complexity all around us.

Mass and niche markets

When the world of business adopted the ecological term "niche" to describe emerging markets, it signalled a recognition that the mass markets for products and services built up in the early phases of the industrial economy were fragmenting into a greater diversity of smaller, more specific and specialised markets.

There were early signs that this break-up of mass markets would result in a flowering of small business and challenge the continued increase in the power of the global corporations.[24] To the extent to which this has happened, it has been more than balanced by the collapse of small businesses in many areas and the adaptation of the corporations to the demands of niche markets. The computer revolution has been the main factor that has allowed the corporations to hijack niche markets and use technology to provide an apparent substitute for individual service. The banks, with their endless stream of new products tailored to every imaginable consumer group, are the grand masters in this wholesale replacement of people with technology and junk diversity.

The explosion of diversity in the service economy illustrates the point that diversity is not of itself necessarily useful. Rather, we need to identify the diversity of functional connections and work out which systems are nourished by those connections.

Community commonality and diversity

One of the burning issues for intentional communities is the bond of common belief, values, experience, ownership and so on which hold successful communities together in the face of the individualistic atomising forces which prevail in the modern world. While it is clear that many people in mainstream society yearn for the missing sense of community, the energetic realities that continue to allow people to pursue more segregated and individual lives undermine any commitments to collective endeavour. It is unfortunate that rapid growth in intentional communities is unlikely to occur until economic and social options in the wider society contract.

On first analysis, individualism appears to be a case of too much diversity, which needs to be constrained or culled for the common good. In practice, it may be the opposite. In response to the perception that difference is the source of conflict, many people have made great efforts to form communities with like-minded people. This has often resulted in communities of people of similar age, social background and wealth getting together, only to find that their expectations of commonality with others were rarely fulfilled. If they were all raising young children and building their own houses, sharing brought some benefits; but over time the need for social and economic relationships with people outside the community (distant parents, the mechanic in town) were often stronger than those within the community.

Experience has shown that intentional communities need diversity in age, interests, livelihoods — and possibly even wealth — in order to:

- provide a basis for economic exchange and interdependence
- allow for the emergence of natural authority[25] in relation to different issues and fields of community activity
- foster recognition that commonality and difference are in a dynamic tension.

Multiculturalism

The wider community has come to value a measure of diversity in multiculturalism — although for some people, this amounts to not much more than increased shopping opportunities. *Multiculturalism itself contains the same paradox as the permacultural use of biological diversity, where the process of valuing and making use of nature's diversity contributes to changing it. Acknowledgement of the value of differing traditions goes hand in hand with a promiscuous hybridisation to create new local cultures of place.*

This contributes to the eventual dissolution of those original ethnically diverse cultures — which was the expressed aim of the now discredited assimilationist values and policies.

Perhaps the significant difference is that the change and transformation of the cultural mainstream is an implicit part of multiculturalism, whereas this was strongly resisted in assimilationist Australia.

Modern multiculturalism is very different from the carefully maintained balance between ethnic groups that used to be the norm in many cities of the pre-industrial world. Ancient cities were like beehives, highly dense but on a human scale; people of different ethnic groups often mixed freely in their public lives but fastidiously maintained their own culture for generations through separate language, education, religion, and even professions. These diverse human ecosystems mostly worked well, despite the periods of violence that punctuate the historical record.

Modern multiculturalism can only represent a transitory stage in the evolution of new cultures of place. This is analogous to the permacultural combination of biotic diversity in pioneering combinations, rather than the ancient stability of pristine biodiversity — the icon of the conservation movement.

We do not know whether multiculturalism can have any impact on stemming the mass extinction of most of the world's remaining 6800 languages over the next century to allow some more gentle hybridisation and integration of knowledge and culture. A rapid and wholesale collapse of industrial civilisation would probably save much more of this vast cultural diversity than the combined effects of multicultural tolerance and pro-active linguistic conservation. Which loss would be greater for the long-term survival of humanity — thousands of distinct languages and cultures of place maladapted to a world changed forever by industrialisation, or the vast internal diversity of industrial culture maladapted to declining energy?

These impossibly tragic outcomes, about which we can do little, may inspire gloom and paralysis. I take them simply as evidence for getting on with the practical business of creating local cultures of place, which will eventually create new global biological and cultural diversity. This can only be done when we become rooted to place and community, rather than the global culture of no-place, which both accelerates the losses and leaves people mourning the direct or indirect consequences of past losses.

Human Biological Diversity

The biological diversity of human kind at the racial, ethnic and individual levels is extraordinary, and has been used throughout human history to include and exclude people from social groups of all descriptions. This diversity led many scientists in the late 19th century to suggest that racial and even ethnic differences reflected genetically fixed groups that had followed separate evolutionary development. However, modern genetic evidence shows that even strong features such as skin colour are relatively superficial responses to environment and social selection pressures, which operate on a common human genetic pool that is relatively fixed and long-standing. It is even possible that some of our so-called ancestors thought to be other species, such as the Neanderthals, were no more than racial selections.

In this view of Homo sapiens, the diversity we see is akin to the diversity in breeds of domesticated animals. We have applied selection pressure through a diverse range of cultural preferences and taboos, which interacted with natural selection pressures to produce the physical and other characteristics of tribes and races. We probably began this process before we did the same to livestock. Thus humans were the first domesticated animals. These selection pressures continue to influence human characteristics through aesthetic preferences. Cultural change is so fast that a selection pressure for, say, blue eyes may not be maintained for enough generations to create a new global culture dominated by blue eyes. The important point about human self-selection is that it operates through cultural co-evolutionary processes not controlled by any person or power structure.

In the 1920s and 1930s eugenics, the science of "breeding better people", was a respectable field of research; its many enthusiastic proponents influenced public policy and laws at least in the United States and Germany. The Nazi experiments in breeding a "super-race" and, more importantly, the systematic elimination of "inferior'" peoples gave eugenics a bad reputation. Since the Holocaust, it has been widely recognised that it may not be possible to meddle with the diversity of humankind without running the risk of similar horrific and dysfunctional insanities. Because this is such a sensitive subject, it is difficult to discuss the most fundamental issues concerning human fitness and diversity.

Some of those questions include:

- has inter-breeding of previously separated races and ethnic groups injected hybrid vigour into Homo sapiens, especially over the last few hundred years?
- is the disappearance of ethnic and racial identity inevitable with inter-breeding?
- have medicine and other benefits of the modern world reduced human fitness in the way over-breeding has reduced fitness in animal breeds?
- is the re-emergence of eugenics inevitable if biotechnology continues to provide new ways of intervening in human reproduction and genetics?
- have technology and rapid change made conventional notions of fitness irrelevant?
- are new diseases of crowding, and other novel environmental stresses, introducing new selection pressures?
- what will be the effect of declining energy on old and new selection pressures?

Using the ecological systems perspective to consider the global situation over the last and next few thousand years, it seems the rapid growth in the human energy base has broken down old biological and cultural diversity, and simultaneously led to proliferation of human numbers and individual diversity. The climaxing and decline of energy will inevitably result in a reduction of human numbers, and possibly individual diversity, while stimulating the re-emergence of localised biological and cultural diversity.

The details of how this could take place range from the benign and enlightened to the horrific and genocidal. While we need to acknowledge that human self-selection through cultural forces will continue in some form, *the wisdom of the ages has been to value human diversity as a gift of nature or of God. While accepting this gift, we need to also acknowledge that we have neither the power nor the wisdom to conserve every aspect of human individual or ethnic diversity, any more than we have the power to conserve and maintain all the planet's lifeforms*. It seems that even conserving the diversity that has been built up over the last few thousand years in cultivated plants and domesticated animals may be beyond our capacity and wisdom.

In accepting this situation, we are back to the third ethic: recognising diversity as a form of abundance to be cherished, while accepting that forces larger than ourselves will set the limits to that abundance.

1 Polyculture is the cultivation of many plant and/or animal species and varieties within an integrated system.

2 Forests such as mountain ash, dominated by sclerophyll (hard-leaved) trees (eucalypts), but with an understorey of soft-leaved and moisture-dependent species including many species found in rainforests.

3 While living in these forests in Jackys Marsh in the late 1970s, I occasionally saw young eucalypts growing on the clay soil attached to the massive upturned root fan of a forest giant, giving individual replacement of canopy trees.

4 A forestry term to describe a naturally occurring local variety of forest tree

5 Unless the individuals result from clonal reproduction, such as that from a suckering stand of trees that are genetically identical.

6 Larger fields of staple crops were grown in some places, but the larger the crop area, the harder it was to control weeds. The weeds themselves represented a great biological diversity, which (along with reducing crop yield) would have helped prevent the spread of pests and diseases through the crop by a number of mechanisms: harbouring and feeding pest predators and parasites, interfering with the spread of pests and diseases from plant to plant.

7 Local people and the natural environment felt all the adverse effects. For example, the Green Revolution grains that required more fertilisers and pesticides also produced less straw for feeding animals and soil organisms. See Vandana Shiva, *Monocultures of the Mind* and her other books for detailed evidence of the effects of the Green Revolution, especially in India.

8 The current hard sell of genetically engineered crops shows this lesson has not been learnt because the Green Revolution was a great success for corporations and market capitalism.

9 Trials by Diggers' Seeds, Victoria. See C. Blazey, *The Australian Vegetable Garden Book* 2000.

10 In animal breeding, the form or outline that is used as a visual cue to the breeding purity.

11 Phil Larwill (local vet and aid project worker in Mexico), pers. comm.

12 Vandana Shiva *Monocultures of the Mind.*

13 P. Sampat, "Last Words" *World Watch Institute Report* June 2001.

14 See Article 7 "Gardening as Agriculture" in *David Holmgren: Collected Writings* 1978-2000.

15 If we analyse the typical modern diet rather than that of the health food gourmet, we find that a handful of plant and animal species provide 90% of the content of an amazing range of highly processed and refined unhealthy food products; the remaining food diversity provides little more than an occasional garnish.

16 For an extraordinary model of nationwide sustainable development under stress, see the Project Censored website: http://www.projectcensored.org/c2001stories/12.html

17 Phil Rowe, pers. comm.

18 See Kevin Kelly, *Deep Evolution: The Emergence of Postdarwinism* Addison-Wesley 1993.

19 See D. Holmgren, *Trees on the Treeless Plains: Revegetation Manual for the Volcanic Landscapes of Central Victoria.*

20 See the Food Forest website: http://www.users.bigpond.com/brookman

21 Because there is no legal market for bettongs (other than more wildlife refuges), geese remain a more important yield at the Food Forest.

22 Prof. Ian Lowe on "Ockham's Razor" ABC Radio National, 24 June 2001.

23 See D. Holmgren, "Permaculture Movement and Education: Searching for Ways Forward" *Permaculture and Landcarers* (later *Green Connections*) vol. 3, Spring 1995.

24 See P. Hawken, *The Next Economy* Henry Holt & Co. 1983

25 By "natural authority" I mean the recognition within a community of the particular ability of an individual in some field that justifies accepting their opinion as having greater weight (in that field). This natural authority is earned over time and is not necessarily dependent on formal position or qualification.

Use Edges and Value the Marginal

Don't think you are on the right track just because it is a well beaten path

The icon of the sun coming up over the horizon with a river in the foreground shows us a world composed of edges.

Within every terrestrial ecosystem the living soil — which may only be a few centimetres deep — is an edge or interface between non-living mineral earth and the atmosphere. For all terrestrial life, including humanity, this is the most important edge of all. Deep, well-drained and aerated soil is like a sponge, a great interface that supports productive and healthy plant life. Only a limited number of hardy species can thrive in shallow, compacted and poorly drained soil, that has insufficient edge.

Eastern spiritual traditions and martial arts regard peripheral vision as a critical sense that connects us to the world quite differently to focused vision. This principle reminds us to maintain awareness and make use of edges and margins at all scales in all systems. Whatever the object is of our attention, we need to remember that it is at the edge of any thing, system or medium that the most interesting events take place; design that sees edge as an opportunity rather than a problem is more likely to be successful and adaptable. In the process, we discard the negative connotations associated with the word "marginal" in order to see the value in elements that only peripherally contribute to a function or system.

"Don't think you are on the right track just because it is a well-beaten path", reminds us that the most common, obvious and popular is not necessarily the most significant or influential.

Landscape Edges

Coastlines

On a global scale, coastal ecosystems are diverse and ecologically productive interfaces between terrestrial and oceanic domains. While the deep oceans are ecological deserts, the shallow coastal waters fed by nutrients from the land support a huge diversity and abundance of fish and other lifeforms. Tidal estuaries are a complex interface between land and sea, which can be seen as a great ecological trade market between these two great domains of life. The shallow water allows penetration of sunlight for algae and plant growth, as well as providing forage areas for wading and other birds. The fresh water from catchment streams rides over the heavier saline water that pulses back and forth with the daily tides, redistributing nutrients and food for the teaming life.

Within terrestrial landscapes, water bodies such as rivers, lakes and wetlands support freshwater aquatic and semi-aquatic ecosystems that are also diverse and productive.

Vegetation immediately adjacent streams and waterways (riparian vegetation) is often more diverse in species and has greater density than vegetation further from the water.

In contrast, where the sea and land meet at the abrupt edge of a sandy beach or cliff, the interface is minimal; the extreme and erratic energies of the surf and wind make it difficult for more than a limited number of hardy species to colonise. Animal life living off the detritus delivered by those wild energies is predominant. In the tropics, the generally calm weather and moderate seas allow mangroves, the ultimate development of living interface between land and sea, to move out of the sheltered tidal estuaries and colonise parts of the open seashore, especially rocky headlands.

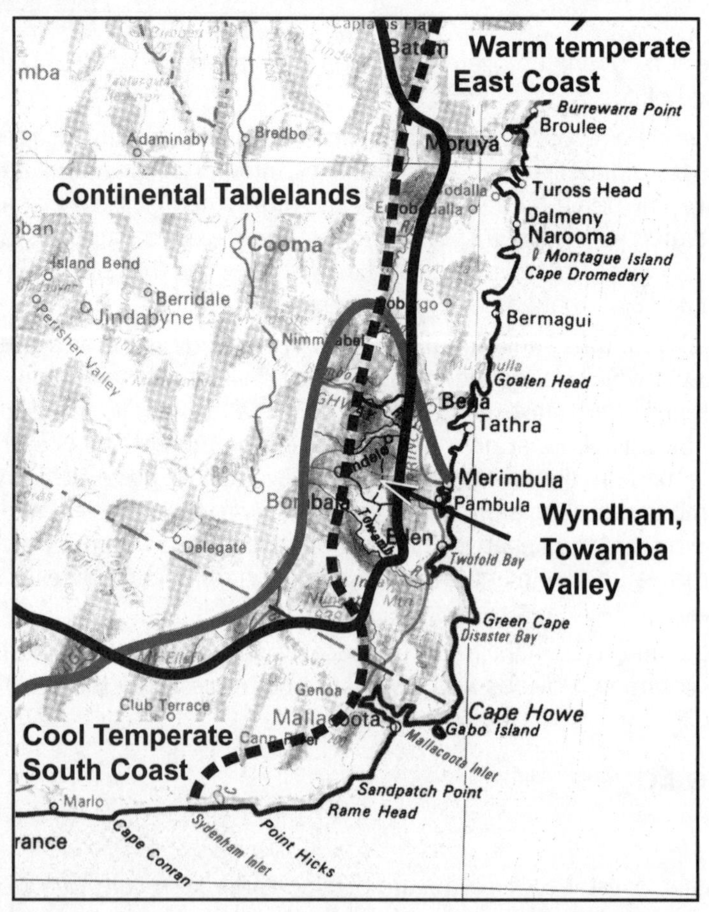

Figure 30: The ecotone overlap between major bioregions on the far south coast of New South Wales

Ecotones

In bio-geography, an ecotone is edge between two bioregions where the distribution of species from both regions overlaps, creating greater biodiversity than in either of the respective regions. Ecotones generated by altitude may be quite narrow (less than 1 kilometre); those generated by latitude or by distance from the sea (continental versus maritime) may be tens of kilometres in width.

The Towamba Valley on the far south coast of New South Wales is the centre of one of the most important ecotones on the Australian continent: a three-way edge between the warm temperate east coast, the cool temperate south coast, and the cold dry plateaus of the Monaro region (Figure 30).

While developing a bush property in the Towamba valley in 1980,[1] the presence of 13 species of eucalypts from all three of those bioregions on less than 75 hectares was a sign we were close to the heart of that major ecotone. Elsewhere on bush properties of a similar size in southern Australia I was used to seeing two eucalypt species, or six or even eight, but never 13. Apart from the complexities of wood value and management in such forests, these species were the sign of a local climate that produces some bizarre combinations of weather, of critical importance to what could be grown in gardens.

Geographers recognise that older cities and towns were usually located at the edge between at least two major resource regions or land systems. Typically, cities are built on harbours and sheltered bays along the coast, adjacent to great rivers and often at confluences of major rivers, or on the foot slopes below mountains and above adjacent plains.

Land systems

Changes in soil type, slope (break of slope) or aspect (ridges) can create rapid transition in vegetation types over distances as small as a metre. Bushwalking along these edges gives us a strong sensation of being in a great garden where some unseen hand is doing the weeding and the cutting. The most dramatic of these involve edges between forest and grass, heath or swampland. Such edges support a greater number of bird species than either vegetation system because the resources of both systems are available. They also often mark the boundaries between different land systems (see **Principle 8**: *Integrate Rather than Segregate*).

Particularly distinct edges in wilderness landscapes are spiritually uplifting, perhaps because it is there that we recognise the designer in nature/God. One of my favourite roads in Australia is the Chesapeake Road on the wild south coast of Western Australia. It runs through wildflower-filled heathlands, with a mosaic of stunted heathy eucalypt woodland and reed and rush swamps on peaty and sandy soils. These sandy landscapes were laid down over the last ten thousand years by the relentless onshore south westerlies that blow the tops out of the massive coastal dune fields 10 kilometres to the southwest. Rising out of this plain of sand, peat and tea-coloured water are islands of red granitic loams that support towering forests of karri with a soft-leafed and ferny understorey. The transition between heathy woodland 6 metres tall to dense karri forest 40 metres tall is complete in as little distance as the height of one mature karri. Over a 25 kilometre stretch from the Gardiner to Shannon Rivers, the Chesapeake Road passes straight through more than a dozen of these spectacular karri islands, which are now thankfully protected within the D'Entrecasteaux National Park. The driving experience is awe-inspiring.

Micro scale edge

Within the bodies of plants, animals and living soil, edges are where the action happens. The lungs are perhaps the ultimate development of a fractal pattern of tissue (a type of delicate skin) that allows maximum gaseous exchange between the atmosphere and red blood cells.

Plant roots have surprisingly little mass, but enormous surface area (edge) to allow for the osmotic absorption of soil water, gases and dissolved nutrients. The root surface and its immediate environment support (literally, with a flow of plant carbohydrates) myriads of micro-organisms, which make available and regulate the supply of mineral nutrients.

Clay and humus provide complex surfaces (indicated by soil cation exchange capacity), which hold and release nutrients for uptake by plants. Soil structure is composed of interlocking particles (peds) with spaces in which water and air flow. A huge part of what makes a soil productive is an open friable structure that amounts to maximum edge.

Edge as a Systemic Property

From these examples across many systems and scales, it can be seen that edges are dynamic and productive parts of all natural systems where exchange of materials and energy take place. They are the places where both co-operative and competitive relationships between system elements and whole systems are played out.

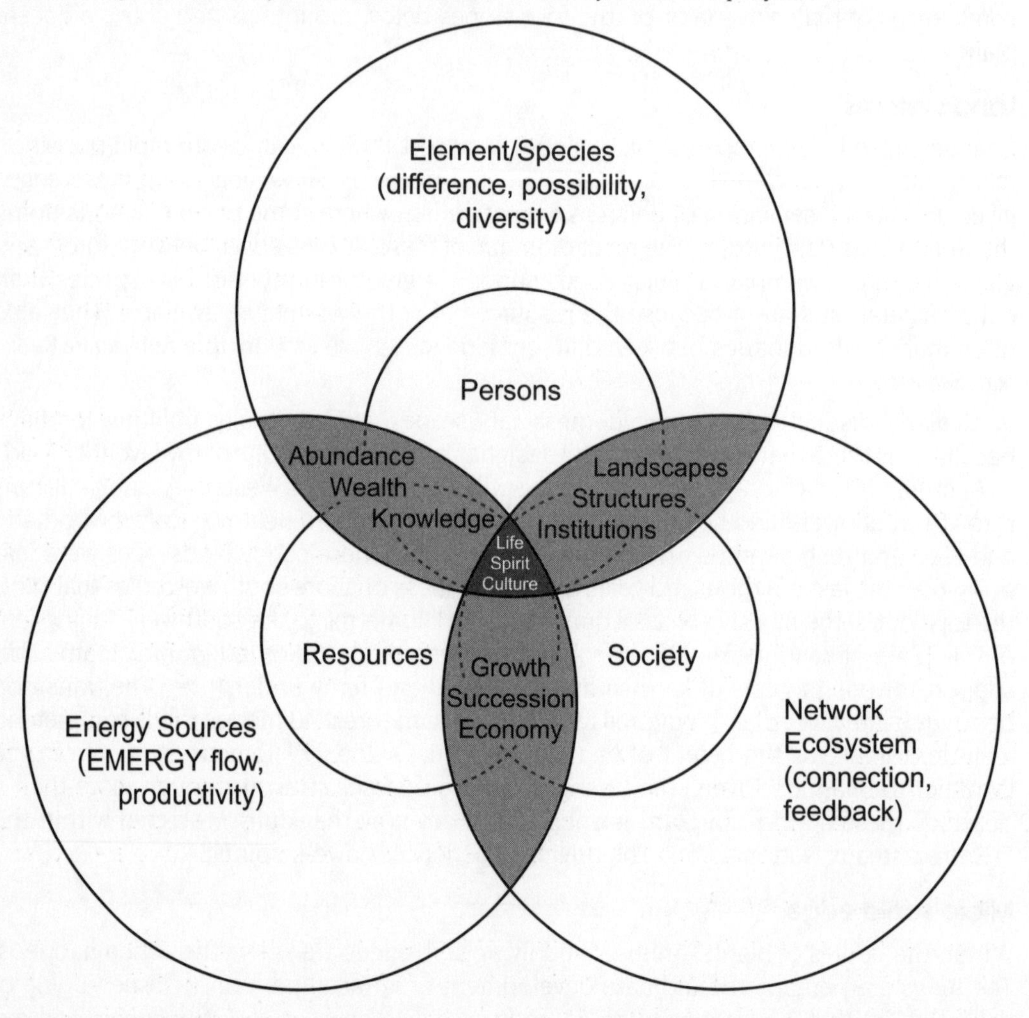

Figure 31: Conceptual musing on self-organising systems.

What constitutes an edge and what is a system in its own right is a matter of scale and perspective. At a given scale, edges represent only a small part of any system. However, increasing edge is one important way to increase system intensity and productivity. It is possible to see increasing complexity of edge as an alternative mode of growth and development to gross system expansion. In this way increasing use of edge feeds into the principle of Small Scale.

Conceptual edges as overlap

Venn diagrams, named after the English logician, John Venn, have been widely used to explain the overlap between sets of elements in mathematics. The ecotone map, in Figure 30, is like a Venn diagram in showing the ecotone as the overlap between different bioregions. Venn diagrams are also useful in helping us organise our thinking about complex conceptual overlaps and interactions. In Figure 31, overlapping small circles represent the human domain as a subset of larger conceptual overlaps between the energetic, elemental and network aspects of nature. Rather than being a rigid picture of reality, such tools help us to grasp fuzzy overlapping concepts and phenomena.

Edge in Cultivated Landscapes

Old England

The patchwork fabric of traditional English farm landscapes was made up of irregular fields bordered by hedgerows, with woods and small copses of trees, tightly clustered villages and a network of roads and lanes, streams, ponds and drains; in other words, it was full of carefully maintained edges. These diverse cultivated landscapes reconstructed some of the values that probably existed in a less intensive form in much older, partially cleared, Anglo-Saxon landscapes. A hedgerow or shelterbelt can be thought of as a double forest edge, intersecting crop and pasture land. Drainage and irrigation channels, ponds and dams all increase the land water interface.

The change from small-scale intensive systems to large-scale monocultures had the effect of eliminating much of the landscape complexity in edges. The removal of hedgerows from traditional British and European farm landscapes to create larger fields is a well-known example. By reducing the amount of edge in farm landscapes, costs of management and labour were reduced and marginal increases in cropping productivity were sometimes achieved. Simplification of the farm landscape goes hand in hand with high throughput of fossil fuel energy and higher outputs of monocultural yields. Unfortunately, these changes reduced environmental values such as wildlife habitat, and rural resources such as wild food, herbs and construction materials, as well as landscape amenity.

Today, the traditional English countryside, or what remains of it, is regarded as a national treasure for its aesthetic, environmental and heritage values. In fact, it is also an economic treasure, earning millions of pounds in foreign exchange from tourists who flock to see and experience "old England". More important to a sustainable future are the efforts to use these living museums as models for the design of modern sustainable farm landscapes.

Mediterranean terracing

From the 14th to 16th centuries, the terracing of stony hillsides degraded by pastoral farming in the Mediterranean region increased agricultural productivity by heroic construction and meticulous maintenance of landscape edge. This and other examples of terracing illustrate the general pattern in the development of cultivated landscapes from primary forests. Initial clearing for grazing and limited cropping increases landscape edge, ecological diversity and agricultural productivity. Further clearing reduces edge and leads to degradation and reduced productivity. Eventually, after substantial investment of resources in new designed edge, ecological values and agricultural productivity rise.

The amount of structural edge in a landscape can be seen as a "leading indicator"[2] of biological and resource diversity and, eventually, economic productivity.

Indigenous Australian landscapes

Active landscape management by the Aboriginal people of Australia was a stimulation and inspiration in the original conception of Permaculture, well before the Mabo High Court decision overturned the legal fiction of *terra nullius*.[3]

The early European settlers in Australia were confronted by a landscape that was alien and strange in all its elements; its diversity and edges were too subtle for many to consider it anything but monotonous. Certainly, the differences from the cultivated landscapes of England were immense, but those differences were not as great as is commonly imagined today. Large parts of the indigenous landscape were managed by people (using fire) to create open pastoral woodland with distinct patches of rainforest or thickets, with very sharp transitional edges. Today we often see either vast tracts of cleared farmland, or dense regrowth forests. Areas with a mosaic of cleared paddocks, woodland and forest are generally less degraded, more biologically productive and attractive for rural residential development) than either of the extremes of cleared farm or dense forest.

Aboriginal land management

As an inspiration for sustainable management of landscape in a low energy future, it is useful to examine Aboriginal systems of management in more detail.

Childhood musings on primary school history lessons in my home state of Western Australia were a trigger for my research of this subject, years later. I had found it very curious that the slight rises of the Darling Range remained a barrier to exploration and settlement for many years. Were the early explorers so incompetent that gentle hills contained them? Years later, I found that the answer to my question was to do with the nature of the vegetation and its management by aboriginal people. On the coastal plain, a complex landscape of heath-covered sand plains dotted with wetlands, limestone ridges with tuart woodland and rich colluvial foot slopes of the Darling Range in marri woodland, blended with the extensive estuary of the Swan River. In contrast, Dale, the explorer who made it over the Darling Range, described a dark and gloomy monotonous jarrah forest, which went on and on as a trackless wilderness, that their Aboriginal guides abandoned after two days. When he reached the granite soils beyond the jarrah forest he described "a landscape as fine as an English gentleman's country estate with trees so widely spaced one could scarcely begrudge the space they occupied and native camp fires to be seen everywhere".[4]

Aboriginal people made use of the most Australian landscapes, though to varying degrees some areas were heavily populated and managed while other areas, such as extensive rainforests, were virtual wildernesses. The most densely populated and used areas tended to have one or more of the following characteristics:

- high mineral fertility
- moderate rainfall

- edges along permanent streams, lakes, wetlands and coastlines, as well as between major land systems, supporting different food and other resources.
- vegetation structure tending to open forest or woodland, but with sharply defined edges of denser vegetation along gullies and other sheltered sites
- pockets of elevated mineral fertility and organic matter.

The first three characteristics may be seen as fixed aspects of the landscape, but even these changed dramatically over the long cultural memory that directed Aboriginal land management. Ability to adjust to these changes with no substantial cultural changes evident in the archaeological record is indicative of the great adaptability of the culture.

The last two characteristics were to a large extent a creation of land management practices over centuries. In fact, it is appropriate to regard the landscape at the time of white colonisation as a cultural landscape well suited to human habitation, rather than a wilderness. This is supported by historical and ecological evidence.

The seasonal movements of Aboriginal groups followed specific, culturally embedded patterns, which allowed for the harvesting of food and application of management practices to the particular land system. Harvesting and management were often achieved by the same activity done in (apparently) the most casual manner. Timing of these activities was determined by precise synchronous cues in the natural environment.

Fire was the major management tool used by Aborigines to shape whole landscapes, and make them habitable and comfortable places to live. Most importantly, it enhanced production of traditional foods. Prehistorian Rhys Jones coined the term "firestick farming" to describe this land management process. There is debate about the frequency and intensity of burning in different vegetation types and bioregions, but there is substantial evidence that the net effect was far from random or chaotic. The landscapes that Europeans found could not have evolved without great regularity and precision in the use of fire over thousands of years.

In my view, most of Australia's fire-prone land systems, such as heaths and grasslands, were burnt as often (each annual dry season) and as soon as they would burn. By this process only the driest sites will burn—generally northwest aspects and the ridges that have the shallowest, lowest-fertility soils. Gullies, southerly slopes and more fertile sites would act as firebreaks.

The effects of the fire regime were manifold:

- it created open accessible ground along all routes used to traverse the particular land system
- it shaped mature trees, creating nest hollows where burnt branches died back and base hollows which were used for shelter, including ancient revered birthing trees
- it stimulated the growth of lush green grass, high in protein, which attracted kangaroos and other herbivores, stimulated flowering in heath land plants (rich sources of nectar), and was frequently critical in stimulating germination, fruiting or tuber formation of a diverse range of food plants (orchids, yams and beans)
- it created, over time, an incremental decline in the mineral fertility and organic content of the soil, with an associated ecological drift to more fire-prone vegetation. Although this may appear to be a form of land degradation, the nutrients lost from the burnt areas, in smoke and water, were mostly absorbed by unburnt areas adjacent and often down slope. Over time, these areas became more fertile, with ecological

succession to less fire-prone vegetation (often rainforest). These "islands" provided different food sources and habitat for food animals that utilised the seasonal productivity of the burnt areas.

The culturally embedded land management practices produced a pattern of localised degradation and enrichment, which was self-reinforcing. It illustrates the fundamental pattern of energy distribution in Aboriginal landscapes described in **Principle 7**: *Design from Patterns to Details*. It is also a brilliant example of the use of edge, between burnt and unburnt areas, to increase total landscape diversity and productivity.

Urban Examples of the Use of Edge

The urban rural fringe

Edges are dynamic, diverse and productive sites in both natural and cultivated landscapes, and these characteristics can also be seen in modern industrial economies and landscapes. *The urban sprawl, lamented by generations of urban planners, is one of the defining characteristics of modern car-based settlements. But while the car and cheap energy make the sprawl possible, what pushes it is the constant search for that edge between town and country, between the human and natural worlds*. The edge of suburbia, with its open paddocks and abandoned spaces, has been a great place to grow up for generations of Australians, but it is inevitably destroyed as the suburban monoculture takes over from the rural resources. The sought-after edge moves further out as the city grows.

Suburban landscapes such as the North Shore of Sydney have always been attractive real estate because the constraints of topography have retained a network of escarpment bush and harbour inlets through the suburban fabric. Progressive suburban design maintains green space along watercourses for amenity and stormwater management, as well as attempting to create nodes of denser, pedestrian-dominated urban development,[5] which provide services within bicycling distance of the suburban fabric. These designs can all be seen as drawing on the amenity of natural edge or the creation of designed edge.

City shop front as edge

In the retail centres of cities, shopfront is a limited resource in great demand, which real estate agents sell and lease to the highest bidders. Shopfront is the interface between the public domain of the street and the private stores of goods for sale, the edge across which information about what is for sale is presented, and through which people move to participate in economic exchange. As city retailing intensified, the development of arcades and redevelopment of service lanes increased the length of edge where exchange could take place.

Today's shopping mall has captured and enclosed the edge so that the whole character of the previously public street domain is managed and manipulated to maximise the exchange of money for goods. This management and manipulation diffuses the shopfront membrane in a confusing fractal structure, in which the public naturally and unconsciously takes up goods and is relieved of their money before being ejected into the monoculture of the carpark!

Classic Permaculture Examples

If use of edge is everywhere, the important issue is the appropriate use and application of edge. Examples of the use of edge are given in many permaculture books.

- Keyhole garden beds increase the accessible edge for a given area of path and bed.
- Ponds and dams that are built with a sinuous edge of peninsulas, marshy flats and islands have greater total biological productivity than those with a simple shoreline; greater edge increases the yield of farmed fish in the dam by providing more habitat and feeding surface for aquatic organisms on which fish feed.

Revegetation

Shelterbelts and hedgerows are traditional farm landscape examples of edge, which was also a major feature of my own work on farm revegetation.[6] Although broadacre permaculture strategies often suggest returning whole landscapes to some sort of forest, in other areas — such as high-fertility arable soils with middling rainfall, and lower-rainfall pastoral farming areas — continuous forest may be unsuccessful or a poor use of resources.[7]

In these landscapes, the most effective and productive revegetation is often linear (not necessarily straight) belts of trees. If these belts are deep enough (five or more rows) to provide for forest edge functions and species, they can optimise the passive functions and yields from trees without the disadvantages of fence-to-fence reforestation of cropping or grazing land.

Agroforestry

Wide-spaced tree systems (agroforestry), where tree-growing is completely integrated with cropping or grazing to form a third land use, has received a lot of publicity in Australia and New Zealand. Although these agroforestry systems are good expressions of a number of permaculture principles, especially integrated systems, they are also subject to a number of problems:

- most trees, even the hardiest, prefer to get established in a self-sheltering stand of dense young trees where the canopy suppresses grass, the worst enemy of young trees
- establishing good timber form in wide-spaced trees is difficult and labour-intensive
- the spacing of trees creates inter-row spaces that dictate the type of cropping or haymaking equipment that can be used.

These problems have naturally led to the development of alley farming[8] and shelterbelt forestry[9] systems, where the trees are concentrated in belts but there is still strong beneficial interaction between the trees and the cropland.

These systems are good examples of the use of edge, without completely dissolving the independent nature of both the crop or pasture system and the forest. Playing with edge in the design of revegetation, increasing its length, and ensuring that it reinforces fundamental underlying patterns of landform and soil, are important ways in which we can get the benefits without the problems of complete integration.

Another example of use of edge in revegetation is the planting of shrubs along the edges of, rather than throughout, plantations. Although it seems less "natural", siting shrubs along edges has the following advantages:

- branching problems on the first row of timber trees is reduced
- the forest floor is sheltered against wind and fire

- healthy growth in the shrubs is maintained for longer, due to better access to sunlight and moisture, providing aesthetic variety with more prolonged and visible flowering.

This edge design of plantations actually mimics the development of natural forest edges once tree canopy reduces the light available for understorey shrubs.

Designed Edge as Appropriate Segregation

Principle 8: *Integrate Rather than Segregate* acknowledged segregation as a sometimes useful, if simplistic, strategy for dealing with the competition and conflict between system elements. The balance between integration and segregation of system elements and functions is a fine one. Often a design that allows the separate nature of the system elements and functions to be fully expressed will be more successful than complete integration.

The farm revegetation strategies referred to above involve adding elements to cleared farm landscapes, so they could hardly be called segregation. On the other hand, the efforts made by previous generations of farmers to maintain the integration of the classic pastoral woodlands by retaining shade and shelter trees when clearing land have not been very successful. While the natural pastoral woodlands worked well when they had many trees and low-intensity pasture management, intensive stocking and few trees has led to salinity, tree decline, and other problems. By way of contrast, the carefully designed edge created by retained timber belts proposed and demonstrated by P. A. Yeomans[10] and other progressive farmers in the 1950s was appropriate segregation of grazing animals and trees and shrubs; it has stood the test of time and become the model for most current farm revegetation.

Effects of Intensity of Use

The complexity of edge between different vegetation systems that it is possible to manage is dependent on scale and intensity of use. For example, if we mow or cut by hand an edge of grass along garden beds, that edge can be sinuous, scalloped or lobed. If we do it with a mower, the edge must have much fewer changes of direction and they need to be less acute. Edges controlled by tractor slasher or animals contained by a strained fence must have much less variation.

If we are applying minerals to balance a soil, we might treat garden beds or slight natural variations in soil type as different, with their own tests and treatment. In a field crop it may not be practical to treat small areas of different soil differently, so we are forced to ignore them in testing. The imperative of small-scale systems suggests we should always be considering the possibility of changing our designs and management to reflect small site variations. But for any scale of operation, there is a limit to what can be managed effectively: there will always be fine-grain variations that must be ignored. This scale difference is embodied in the permaculture zoning concept described in **Principle 7**: *Design from Patterns to Details*. The strategies and techniques for applying the use of edge in Zone 1 may be inappropriate in Zone 2 and certainly in Zones 3 and 4.

Our reference point for increasing the length and complexity of edge must be the default or typical management for that type of land use. Although permaculture-designed farms might look somewhat like gardens, they can never be gardens in the complexity of edge that is appropriate or practical.

Kakadu National Park and the Loss of Edge

Fire management in Kakadu National Park over the last two decades provides an example of the consequences of failure to consider the importance of edge effects. The burning of the savannah woodlands of Kakadu for ecological reasons has always been contentious, but park management burnt about half the park every year during the 1990s. Falls in small mammal numbers were noted in the 1993 surveys (compared with base surveys in 1986), and they accelerated dramatically in the 1999 surveys.[11] This is hardly a good result for our iconic national park with its special federal funding.

The idea of burning vast areas in one year might make sense for park managers looking at a map, but it creates a monoculture in nature that reflects application of concepts at too large a scale to maintain edges between burnt and unburnt savannah. Annual burning destroys habitat in hollow logs and reduces fruiting shrubs and trees as well as directly killing many animals. Although the appropriate burning regime remains contentious, it is now recognised that a patchwork pattern of burnt and unburnt country is probably better than past prescriptions.

If the grasslands were burnt only in the early dry season, then only the driest parts of the landscape would burn. In successive seasons, the same areas would tend to burn. Over time, the edges between burnt and unburnt areas would begin to become self-reinforcing as the processes of ecological drift to more and less fire-prone vegetation increased the differences between adjacent areas. This use of edge on a grand scale would mimic traditional Aboriginal management.

The Value of Marginal Systems

This section builds on the discussion in **Principle** 3: *Obtain a Yield* and **Principle** 6: *Produce No Waste*.

Wild foods as marginal systems

It is natural for us looking back to pre-industrial Europe to think that all food and other needs were supplied by the cultivation of grains, flax and other field crops, and animal pastures. But this is not the case. The patchwork fabric of the landscape, referred to previously, provided many of the foods and products that people consumed and used. The woods and hedgerows provided not only fuel and structural materials but also animal fodder (foliage, acorns, beechnuts), human food (berries, hazelnuts, rabbits, mushrooms), medicinal herbs (hawthorn, comfrey, elder). Aquatic systems provided fish and plant materials for baskets and other constructions.

This diversity among the productivity of field agriculture was critical in providing the needs of the poorest people; during times of famine, whole communities depended on it for survival.

The cultural bias that ignores or undervalues the wild is not restricted to modern affluent society. In many peasant societies, because wild diversity required so little work to maintain, it could to a substantial degree, be ignored.

Second, since the value of these resources was greatest for the poor or in times of crisis, it was often denigrated. The Russian author Alexander Solzhenitsyn, in *The Gulag Archipelago*, recounted that they were forced to eat nettles as an illustration of the extremity of deprivation. Nettles are actually highly nutritious and flavoursome in soup, but if your primary experience of any particular food is in dreadful conditions, it will naturally colour your attitude.

The well-recognised persistence of taste memory allows those negative attitudes to last for a lifetime. Many Australians who consumed rabbit during the hardship of the Great Depression have never been able to see it as flavoursome meat. At Positano on the Italian Amalfi Coast, my son and I collected delicious carob pods, which the locals ignore as human food. Older people told us that they remember eating carobs as children during the hard times of the war that ended when the Americans came with chocolate.[12]

In discussion with Italian environmentalist Giannozzo Pucci,[13] I asserted that the trouble with tradition, as a basis for sustainability, was that people within a tradition rarely understood the functional basis of traditional practices. Giannozzo responded strongly that what I was describing was merely the "shadow of tradition, and that in true tradition people understood exactly why they did things". He gave the example of a share farmer, a man in his seventies who ploughed with the rare white Chianina bullocks, gave the appearance of dancing while forking hay, and used complex seasonal patterns and signs to determine sowing times. This may be so, but today it is hard to find people like that share farmer, let alone learn what they know.

At the other end of the social and cultural spectrum, wild foods have gained a special status in affluent urban society. There are several reasons for this:

- the increasing rarity, and therefore status value, of many wild foods
- wild foods provide a sense of reconnection to nature, which is lacking in modern life
- concerns about the toxic and "unnatural" industrial agriculture
- the high cost of many wild foods, due either to rarity or to high labour costs in harvesting.

This valuing of wild food, although superficial in some ways, may be part of an antidote necessary to expunge the deep-seated disrespect and devaluing of wild and marginal systems.

Rundown neighbourhoods as marginal systems

Another example of the importance of marginal systems concerns the sources of both cultural and enterprise innovation and their relationship to urban renewal. Jane Jacobs[14] pointed out that one of the values of rundown urban neighbourhoods was that they provided cheap rents in old warehouses, shops and houses, where small start-up businesses and companies could establish themselves. She provided evidence that the elimination of these areas by urban renewal programs in the 1950s and 1960s was killing the economic life, as well as the artistic and cultural life, of American cities.

There is tension between the "tidy up" mentality that wants to recycle and make full use of everything, and the mindset that values leaving things be to see what emerges. The balance is a fine one, whether we are working in the garden or planning a city.

The city and the hinterland as centre and margin

Another more abstract, but universal, example of the importance of marginal systems is the conceptual divide between the city and the rural hinterland. For thousands of years, the city and its institutions have distinguished so-called civilisation from more simple cultures.[15] But over most of that time, the relationship between the city and its hinterland was that of the nucleus to the cell described in **Principle 7**: *Design from Patterns to Details*. Although exploitation of the hinterland by the city is a repeating theme in history, there were generally limits to that exploitation because dependence on the hinterland was never completely overcome.

Today the city has become — conceptually, if not physically — the whole; rural life, farming and nature are some sort of surrounding membrane or fringe. These marginal systems continue to wither as every aspect of nature becomes controlled and managed (often from the city) and commodity prices for rural economies are screwed down further and further. After providing the wealth on which the city is built, the countryside and its people are increasingly seen as dispensable. When it appears that cities can thrive on their global connections, rural hinterlands die.

On a national scale, small regional economies such as Tasmania or Nova Scotia in Canada suffer from being economically marginal; and all small towns are definitely non-viable, the economic gurus tell us. On a global scale, the relationship between rich and poor countries, and even whole continents,[16] has led to the hinterlands being seen as dispensable.

The environmental crisis is one for the whole of human civilisation, especially the cities. There are good reasons to believe that we will only succeed in dealing with the environmental crisis when we do so in the cities. However, I believe the inspiration, examples and wisdom for the solutions come not from the centre but from the margins, where people live at the edge between culture and nature, between modernity and the past. The idea that the hinterland provides a wellspring of human biological vigour, values and renewal for civilisation is an old one, but I believe the ways in which this is happening are diversifying and intensifying as we approach the end of the fossil-fuel era.

Origins of permaculture at the margin

People (generally from North America and Europe) sometimes ask me why I think permaculture emerged from somewhere like Tasmania. Tasmania is a place where modernity and nature are in close proximity. It is far enough away from the sources of the dominant paradigms of society, but it has had the benefits of modern education and relative affluence for as long as anywhere else in the world. It is a site where the inspiration and lessons of nature and rural culture can be infused back into urban and intellectual culture.

Hobart, capital of Australia's second-oldest and most decentralised state, is not set within some settled European landscape but clings to the foot of the wild slopes of Mount Wellington. For me it symbolises this interaction between civilisation and wilderness. From the property on the footslopes of Mount Wellington where the permaculture concept was born in the mid-1970s, it was possible to drive (or catch a bus) 5 kilometres in one direction to the city centre or the university. In the opposite direction, 5 kilometres on walking tracks put you on the alpine southwest face of the mountain, with nothing more manmade than a fire trail between you and the great wilderness of southwest Tasmania.[17]

Bill Mollison bridged these worlds. A fisher and bushman who left school at fourteen, he went on to become a wildlife researcher, university academic, environmental activist, co-originator and teacher of permaculture around the world.

Apart from permaculture, Tasmania had the world's first green political party, and it was the place in Australia where organic agriculture first grew from isolated individual farmers to a vibrant network.

Early in the 20th century, Australia was a nation where the benefits of education, political democracy and affluence were fused with exposure to the natural world and rural economy to generate an extraordinary range of scientific and technological innovation (when compared to its small population). New Zealand, with its even smaller population, has an even stronger record of innovation in science and technology.

I believe that another reason for radical innovations to emerge from the fringes of the affluent and democratic world is that, in places like Tasmania, New Zealand and Denmark, the small scale of political and social institutions gives those with radical ideas the feeling that they can make a difference. In the big cities of the great nations, the massive scale of establishment culture and institutions makes for apathy and acceptance that the world is a bad place and cannot be changed.

In 1994 I travelled to Europe to teach permaculture and study sustainable systems, both traditional and innovative. Although I saw many inspiring examples of innovation, it was only in Denmark where I had the sense that the vitality and relative scale of the various "sustainable alternatives" was comparable to or greater than what I was familiar with in Australia and New Zealand. In Britain and Germany, as in the United States, there are a great many interesting projects, groups and individuals, but they are few in number compared to their populations.

Ideas and science at the fringe

Taking further this idea of innovative renewal from the margins, the university as an institution is losing its grip on intellectual culture more than 2500 years after Pythagoras founded the first university. A combination of factors has dramatically weakened the hold of the universities on intellectual culture and allowed those at the physical and intellectual fringes of society to contribute more to progressive ideas and action.

These factors include:

- devaluation of undergraduate education by its increasingly universal consumption and the declining quality of, and rewards for, teaching
- failure to deal with the intellectual limitations of reductionist thinking within the scientific disciplines, to develop a culture of wholistic integration, or to reward cross-discipline endeavours
- replacement of the community of scholars with a corporate culture selling products, with associated loss of financial and political autonomy.
- new communications technology and media allow those at the fringes more equal access to information and ideas.

As a university student in the 1970s, I thought this process was already fairly advanced. Today I regard Australian universities as a tragic joke, although I respect many people who continue to work within them. I recognise my own relative isolation from university culture has had some disadvantages, but overall I have found it has given me the freedom to explore other sources of intellectual stimulation in nature and those who work with it.

University-educated participants in our Permaculture Design Courses often comment on the intellectual challenge and stimulation that our course, and permaculture generally, have provided, compared with the staid and discipline-bound thinking presented at university. Many years ago, a consultancy client of mine (with an M.B.A). left his merchant-banking career to take up full-time farming, and after a year described it as the most intellectually challenging thing he had done. A shame he was not able to make a living by farming.

James Lovelock, the independent atmospheric physicist who works from his cottage and laboratory in Cornwall, provides one model of the new scientist no longer tied to the bonds of the city, the university, the discipline, and the funding institutions. He is the co-originator of the Gaia hypothesis and, arguably, of the whole field of Earth System science.

Never in human history has there been such a great need and opportunity for cultural and intellectual renewal from the margin. The countercultural movement of the late 1960s and early 1970s is perhaps the most dynamic cultural example of the value of marginal systems in social transition. The counterculture, misunderstood and denigrated in the cultural mainstream as a failure characterised by naive and silly ideas, has been a major source of innovation. I see permaculture and the counterculture within a longer tradition of alternative movements within modernity, which have the potential to spark the transformation of civilisation necessary for inevitable energy descent.[18]

The American poet, novelist and organic farmer Wendell Berry has written eloquently about the values of margins, both physical and conceptual.[19] Within a critique of modern industrial agriculture and the associated destruction of rural life and community, he saw modern agriculture as controlled by orthodoxy, both ignorant and arrogant, and characterised by its progressive contraction around increasingly limited notions of what is productive and economic. Around this orthodoxy, a "widening margin of the divergent possibilities" exists to be explored. At the same time that these conceptual margins are expanding, he saw "the margins of geography and practice" as having narrowed drastically. By this he means that the places and people who know how to farm outside the orthodoxy are rare. As Berry was writing these words, permaculture was another divergent possibility being conceived in Tasmania. Since then there has been a rapid growth of organic agriculture and associated activities, despite the expanding hegemony of agribusiness over farming generally. Just maybe, Berry's margins of geography and practice are beginning to expand and that edge is where the action is.

1 Documented in D. Holmgren, *Permaculture in the Bush*.

2 "Leading indicator" is a term used by economists to describe an observable and often measurable parameter, such as interest rates, which tends to pre-empt general economic activity and can be used to predict events. The dynamic nature of this concept could be applied in sustainable land design and management to enhance the "biological indicator" concept, which aims to measure general ecosystem health through the presence or absence of a particular species.

3 *Terra nullius* was a legal doctrine that the colonial authorities used to get around the readily observed fact of Aboriginal management, which under English common law accepted the title rights of native peoples. The Mabo decision in 1992 famously overturned this doctrine in Australian law. This legal revolution has an ecological parallel. The orthodoxy of Australian ecosystems as essentially wild and natural is being replaced with a recognition of the deep-seated cultural aspect in virtually all Australian indigenous ecosystems.

4 For further exploration of indigenous land management, see Article 4 "Aboriginal Land Use" in *David Holmgren: Collected Writings* 1978-2000. The source of the quote from the explorer Dale's journal was Sylvia Hallem, *Fire and Hearth* Aboriginal Studies Institute Canberra. An excellent review of the historical evidence on the use of fire by aborigines in the S.W. of Western Australia.

5 See P. Newman and J. Kenworth, *Sustainability and Cities: Overcoming Automobile Dependence* Island Press 1999.

6 See D. Holmgren, *Trees on the Treeless Plains* (chapter on shelterbelt design). For the best exploration of the functions, design and management of shelterbelts and hedgerows, see D. Soltner, *L'arbre et la haie* (The Tree and the Hedge) 10th edn. Collection Sciences et Techniques Agricoles 1995 (in French).

7 The current plague of speculative blue gum pulpwood plantations replacing three generations of dairy and grazing land use from fence to fence, often destroying farm assets, remnant trees and houses, gives tree-growing a bad reputation.

8 The alley farming systems developed in the Victorian Mallee by permaculture designer Kim Kingdon and grain farmer Anthony Sheldon are an excellent illustration of this approach, which is designed around the natural sand ridge/loam flat pattern of the Mallee landscape. See Museum of Victoria "Future Harvest" Exhibition website: www.mov.vic.gov.au/FutureHarvest/fffuture.html

9 A shelterbelt of high-pruned *Pinus radiata* harvested from a New Zealand dairy farm in the 1983 with a royalty value of $30,000 per kilometre made a big impact in the farm forestry fraternity.

10 See P. A. Yeomans, *The Challenge of Landscape* Keyline Publishing 1958.

11 "Up in smoke: what is killing the wildlife of the Top End?" *The Age* (Melbourne) 30 September 2000.

12 American troops invading the Italian peninsula in 1943 were issued with chocolates and cigarettes to "pacify" the not very resistant locals.

13 Organic farmer, publisher (including *Permaculture One* in Italian) and green politician from Florence.

14 J. Jacobs, *Death and Life of Great American Cities* Vintage Press1961.

15 L. Mumford, *The City in History* Penguin 1961, provides the classic overview.

16 Africa is frequently referred to by those at the centre as an economic "basket case" which can fail without substantial impact on the global economy.

17 Now recognised as World Heritage and the focus of the environmental battle over the Franklin River in the early 1980s.

18 For further exploration of this important aspect of the principle, see "The Counterculture as Dynamic Margin" in *David Holmgren: Collected Writings* 1978-2000.

19 W. Berry, *The Unsettling of America: Culture and Agriculture*.

Creatively Use and Respond to Change

Vision is not seeing things as they are but as they will be

This principle has two threads: designing to make use of change in a deliberate and co-operative way, and creatively responding or adapting to large-scale system change that is beyond our control or influence. The acceleration of ecological succession within cultivated systems is the most common expression of this principle in permaculture literature and practice. These concepts have also been applied to understand how organisational and social change can be creatively encouraged. As well as using a broader range of ecological models to show how we might make use of succession, I now see this in the wider context of our use of, and response to, change.

In *Permaculture One* we stated that although stability was an important aspect of permaculture, evolutionary change was essential. Permaculture is about the durability of natural living systems and human culture, but this durability paradoxically depends in large measure on flexibility and change. Many stories and traditions have the theme that within the greatest stability lie the seeds of change. Science has shown us that the apparently solid and permanent is, at the cellular and atomic level, a seething mass of energy and change, similar to the descriptions in various spiritual traditions.

The butterfly, which is the transformation of a caterpillar, conveys the idea of adaptive change that is uplifting rather than threatening.

While it is important to integrate this understanding of impermanence and continuous change into our daily consciousness, the apparent illusion of stability, permanence and sustainability is resolved by recognising the scale-dependent nature of change discussed in **Principle 7**: *Design from Patterns to Details*. In any particular system, the small-scale, fast, short-lived changes of the elements actually contribute to higher-order system stability. We live and design in a historical context of turnover and change in systems at multiple larger scales, and this generates a new illusion of endless change with no possibility of stability or sustainability. A contextual and systemic sense of the dynamic balance between stability and change contributes to design that is evolutionary rather than random.

The proverb "vision is not seeing things as they are but as they will be", emphasises that understanding change is much more than the projection of statistical trend lines. It also makes a cyclical link between this last design principle about change and the first about observation.

Systemic Perspectives on Change

Top-down managerial change

The top-down and bottom-up perspectives explained in **Principle 1**: *Observe and Interact* and **Principle 4**: *Apply Self-regulation and Accept Feedback* are useful as a framework for considering change. When we intervene in systems over which we have some substantial design or management influence — gardens, farms, business, family — we can make use of change in ways that reflect our power and relationship to the system. This is top-down control.

The systems in which we participate are also changed by the effects of our co-participants, such as the plants and animals in the garden or farm, our partners in business or family members. Our hierarchical relationships with co-participants, and the degree to which we are engaged in an interactive exchange or dialogue with them, will determine the degree to which top-down control is possible or appropriate. We appear to exercise arbitrary control in the garden when we plant, shift or remove plants, but other forces may also be planting, shifting (by reproduction) or removing species from our garden — for instance, wild birds, insects and diseases. In any enterprise, decisions frequently depend on others to implement change processes. When we act in co-operation with other agents, our effective power to change systems is amplified. Even in trying to change our own behaviour we frequently fail because our whole being does not support our rational decisions.

This experience of apparent lack of power to bring about what we believe is appropriate change gives us an opportunity to think more systemically about the situation. There are vast literature and consultancy processes about change management in organisations, partly because efforts by decision-makers to implement change have been so fraught with difficulty, conflict and unexpected outcomes. Much of the change management during the 1980s and after has been driven by the ideology of economic rationality and so-called efficiency. John Ralston-Saul[1] and others have argued that this ideology has failed to reinvigorate corporations or public institutions.

Some management consultants have used novels and short stories[2] to try to introduce more accessible ways of thinking wholistically about organisational change. Perhaps the most dramatic and positive example of creative top-down change in business is the story of Semco,[3] the giant but failing Brazilian manufacturing company. It was turned around by the owner Ricardo Semler, who instituted management transparency, flattened the hierarchical structure, provided human-scale rather than industrial-scale workplaces, established worker share ownership and decision-making, and made other radical changes in an integrated approach to restructure the company for the benefit of all participants.

It is not surprising that creative and positive change in large-scale, complex systems is so fraught with difficulty when we have trouble instituting creative change at the smaller scale — managing small businesses, community groups, families, and perhaps most of all, ourselves as self-directed individuals. Again, at the personal level, there is a vast literature on personal growth and transformation of the self.

Permaculture design as top-down change

The permaculture approach to these issues is to start by dealing with the most concrete and mundane aspects of our material existence through the self-audit process described in **Principle** 4: *Apply Self- regulation and Accept Feedback*. By making small changes, we increase our confidence to tackle larger and more difficult changes. It is surprising how a rational audit process of material needs and processes can be a fast track to dealing with the most abstract questions of motivation, value and meaning.

The permaculture design process can be thought of as a top-down change management process. It might begin by focusing on external factors of a physical and biological nature, but it is increasingly drawn back to a personal change process. In both consultancy and Permaculture Design Courses I have found, for example, that as people begin to think more systemically about their house and garden design the issues of the way they live are drawn into the picture. As a result of this more wholistic approach, personal change is often shown to be an easier, faster and better process for improvement than the external environmental changes that are the initial focus. So often the maxim that "to change the world we need to change ourselves" is true. Despite this, the initial focus of permaculture is on the external systems that provide our basic material needs, especially food, by designing and managing home gardens. This can be useful in exploring the processes by which we bring about creative change.

In the garden we are free to explore and experiment with top-down change processes because we can exercise great power (relative to other system elements) if and when we choose. The changes we implement are not earth-shattering in either good or bad ways, but they provide an excellent opportunity to learn to think more systemically and act in more wholistic and less arbitrary ways.

The mysteries of soil and nature's complexity are an ever-present challenge to our hubristic control and an opportunity to learn; seasonal harvest and ease of management are rewards for appropriate action.

The general problem with top-down control over systems is that we do too much too quickly. A little change goes a long way if used carefully. Identifying where we can be most effective in using our limited resources and power to gain the most leverage is more important than rushing around trying to keep or make everything just right. Human motivation and energy are wonderful resources, but excessive intervention in natural systems is a mistake that we seem to make over and over again. Masanobu Fukuoka's eloquent articulation of "do-nothing" natural farming suggests that it is always a good strategy to observe first and think hard before we change anything, but especially when dealing with systems for the first time. *In attempting to fix any system, we may damage another that is working perfectly well.*

Fukuoka's "do-nothing farming" and Mollison's "self-maintaining permaculture Garden of Eden" should not be misinterpreted to mean simply an alternative way to be a passive and sedentary consumer of a "natural" drip feed. Some of us need constant reminders that it is often better to "let nature take its course"; others need the proverbial "boot in the backside" or the "fear of God" before we act to save ourselves. These two responses, over-energetic and slothful, come to the fore in most of us at different times. Permaculture involves the transition from dependent consumer to responsible producer, but it is just as important to acknowledge that it is our energetic creative side trying to control and manipulate nature that is the root of the environmental crisis.

Resilience to change from above

Sometimes we find ourselves responding to change that comes from above, driven by forces beyond our control or even comprehension. Gaia, the living earth, delivers change in many forms — the tectonic forces of earthquakes and volcanoes, weather events and climatic change, species invasions and disease. Governments and global corporations make decisions that we cannot significantly affect. How, and how much, these larger forces affect our lives and systems depends on our independence from, or resilience and adaptability to, specific or general change.

Natural systems at all scales must develop resilience to pulses of catastrophic change that arrive from large-scale forces. Resilience in ecosystems is the continuity of basic system functions and critical elements, despite fluctuations in the environmental conditions and even the balance of species populations. The ways in which species, ecosystems and whole landscapes develop resilience to these larger destructive forces is a central issue in ecology and, by conscious design, in permaculture.

Flexibility

When dealing with powerful external forces, flexibility becomes more important than resistance and rigidity. For example, the large, long-lived trees that occupy fertile alluvial flats in many ecosystems are replaced by pliable trees and shrubs along stream banks and on islands, where destructive floods make rigidity, strength and great age a disadvantage. So it must be with our own design, where there are always natural and human forces to which we must bend.

This flexibility produces a dynamic stability that is more like riding a bike than the inert stability of a concrete slab.[4]

In the face of extreme and unpredictable change, human behaviour is driven to more and more opportunistic responses. We can see this at many levels. The homeless person who lives by hustling and handouts takes advantage of opportunities as they arise, generally without any plan or even much expectation. In semi-arid range-land, the pastoralist who sows a crop on a lakebed or floodplain to use the fertility and moisture delivered by rare floods takes the — perhaps once-in-a-lifetime — opportunity to make more money than is yielded by sheep or cattle grazing in normal years (and this opportunism is perhaps no more damaging to the land than grazing).

On a larger scale, human society has had to maintain a high degree of flexibility in the face of both natural environmental change and social change. Large-scale warfare between established and permanent armies appears to be associated with the first cities in the Middle East about 6000 years ago, and it has remained the scourge of humanity ever since. It is possible that annual grain agriculture and grazing developed in this region partly because of the destruction of ancient managed "food forests" during the periodic wars. Annual crops and grazing animals could be more readily restored after the invading armies scorched the earth than fruit and nut trees hundreds of years old.[5]

Entropy as a constant source of change

Besides the stresses and opportunities created by precipitous change from above, there are also the predictable and inexorable changes due to the effects of entropy, or the tendency to disorder. These lead to ageing and gradual decay in all living systems, but they especially affect the built environment.

Renewability versus durability in the built environment

Learning to see change in the built environment as an organic process will be important as energy declines, because buildings designed for high energy have to be progressively adapted to new uses with limited resources.

Principle 9: *Use Small and Slow Solutions* emphasised the value of long-term, durable design and materials in the built environment. **Principle** 5: *Use and Value Renewable Resources and Services* considered the use of living structures that are largely self-maintaining.

Principle 6: *Produce No Waste* discussed the importance of traditional values and behaviours associated with regular maintenance. Neatness and tidiness may have provided a socially visible icon closely associated with functional maintenance, but it also has often persisted as obsessive make-work activity constantly battling nature's recycling processes. However, maintenance could be described as a reaction to inevitable change, a reaction that is repetitive and uncreative.

Rather than always building things to last, the opposite strategy of using a cheap and abundant renewable option has merit. This strategy avoids the waste of expensive or valuable materials when they are not needed, and adapts to the inexorable effects of entropy by going with the flow. The banana leaf was used as a plate for take-away food in South-East Asia prior to the spread of polystyrene, plastic and paper, an excellent example of a simple throw-away solution that is sustainable.

In tropical countries, several factors favour the temporary over the permanent in the built environment:

- rapid decay rates for most materials
- high level of natural disasters (cyclones, earthquakes, floods)
- availability of rapidly grown renewable materials (bamboo)
- availability of cheap and skilled labour
- largely outdoor lifestyle.

In some cases, it may be better to leave things to weather and decay rather than constantly trying to keep them in pristine condition.

In the garden we are always making use of small structures, trellises, animal hutches and so on. When we consider the high decay rates of wood and metal exposed to weather, compost-rich soil and animal wastes, we can see that it is best to take one of two courses: either build very durable structures, or build temporary ones with reused or easily renewable materials. Often half-hearted attempts at durability end up being costly, and even toxic, when a renewable solution would be much better. Examples include:

- it is often unnecessary to use galvanised or cadmium-plated nails and other fasteners in outdoor structures, because the timber will often rot well before the nails holding the garden structure together have rusted away
- it is unnecessary to treat naturally durable poles in the ground until the natural preservatives in the wood have broken down. The ability to add preservatives after some decades might be an important consideration in permanent buildings
- painting timber in an attempt to increase durability can be counterproductive. Unless the painting is renewed often, the life expectancy may actually be reduced by flaking paint that holds in moisture.

I remember discussing the merits of untreated and supposedly non-durable peppermint eucalypts for weatherboards with a bush sawmiller while standing outside his modest house. The rough-sawn weather boards were a beautiful silver-grey colour from 30 years of sun and rain. The miller thought they had another 20 years in them. In that region a peppermint would grow into a sawlog in less than 50 years. I thought about the generations of builders and homeowners who have fought to avoid the silver-grey of weathered timber with the aid of paints, oils and sandpaper, and the growth rate of peppermint; I had to agree that, from an environmental perspective, it was a good argument for natural weathering and replacement when necessary.

The Danish research on pig husbandry referred to in **Principle 6**: *Produce No Waste* made use of carefully designed but very simple seasonal shelters made from straw bales as part of a mobile free-range system. Each season's shelter was allowed to compost, and next year's house was built in a fresh location. This was a much simpler way to minimise disease risk than maintaining hygiene in expensive, climate-controlled, high-tech, permanent buildings.

Knowing when and how to allocate resources towards renewable structures and systems instead of maintaining durable ones can be a key issue in permaculture design. Conditions that suggest the relative merit of the renewable approach are where:

- a low-input renewable resource (such as a forest) can grow replacement materials in less time than decay or degradation of function demands replacement; for instance, saplings or bamboo for bean poles
- the process of replacement can be easily done without disruption of other elements of the system; for instance, replacing house cladding, compared with replacing the structural frame
- novel design conditions and high degrees of uncertainty reduce the possible value of durable solutions; for instance, urban squatter settlements without land tenure
- the solution is only required to address an ephemeral or occasional need; for instance, tarpaulin covers for storage of bumper grain harvests on farms.

In *How Buildings Learn*, Stewart Brand provides a key concept for working out how to combine durable and renewable approaches to making buildings both enduring and adaptable to changing needs. The Six S's of buildings,[6] from the most permanent to the most ephemeral, are:

- site
- structure
- skin
- services
- space plan
- stuff.

By not embedding the ephemeral aspects of buildings in their more permanent structure, we maintain flexibility for current and future users of the building to adapt it to their needs. This approach is well established in the design of modern commercial buildings because of the rapidity of change of occupants and technology, but it could be usefully applied to domestic architecture, and more broadly as a strategy for managing change in the built environment. The idea that buildings and their uses change over time by organic processes contrasts with the concept of a fixed sculptural art piece that informs many in the architecture profession.

Although some of these ideas about organic change can be easily grasped and widely applied as simple common sense, there is great value in exploring more deeply the patterns of natural system change and the implications for systems that are designed and managed by humans.

Ecological Models of Succession

The concept of ecological succession, where there is a pattern to changes in vegetation composition and structure, has been a central idea in ecological theory for nearly 100 years. Although some of the more classical models of ecological succession have since been challenged by new evidence and models, the basic idea of ordered and directional change as a characteristic of ecosystems is still useful.

Classic succession

The classic model of ecological succession was based on observation of patterns of change in moist, fertile and temperate North American and European landscapes. Following disturbance, the colonisation of bare ground by herbaceous weeds leads to grassland and then pioneer shrubs and brambles, followed by fast-growing forest trees; finally, slow-growing, long-lived trees form a stable climax ecology. These stages might all be viable ecosystems with their own dynamics, but for a given climate and soils the climax vegetation prevails in the absence of disturbance.

In this and similar examples of classic succession there is generally change from:

- low biomass to high biomass
- low soil humus to high soil humus
- low resilience to disturbance to high resilience
- low diversity to high diversity
- predominance of competitive relationships to predominance of co-operative and symbiotic relationships.

The latter two points are contentious, even in the northern hemisphere temperate deciduous forests. This model has been useful in understanding agriculture and other human land uses.

- Most crop agriculture is analogous to the herbaceous weed phase; it involves the rapid growth of annual plants over bare (cultivated) ground.
- Pastures grazed by animals are a managed version of the grassland, which is dominated by perennial species, but still herbaceous rather than woody.
- Many of our marginal and abandoned patches of land are covered by vigorous pioneer shrubs and fast-growing forest trees.
- In traditional pre-industrial Europe, the great, managed, long-rotation royal forests of valuable long-lived timber trees were close to a climax state.

By recognising the stages of ecological succession which particular land uses represent, we are more able to see their systemic characteristics, strengths and weaknesses and, most importantly, predict the sort of succession forces that can be expected to operate within them.

- Bare ground is an unstable ecological void into which any possible plant life will flow to bind the surface against erosion and prevent leaching of nutrients.
- Pastures grazed by livestock tend to be invaded by spiny and unpalatable plants, especially shrubs and brambles.
- Forests of light-demanding pioneer trees provide an environment for succession to shade-tolerant trees spread by berry-eating birds, but they do not provide a good environment for regeneration of their own kind.

Initial floristic composition

This model of ecological succession suggests that disturbance leads to a simultaneous regeneration of all species; but over time, the short-lived species die out, leaving the long-lived ones. Within this model there is still a change from low biomass and humus to high biomass and humus. However, diversity is greatest following regeneration and lowest at maturity.

This model describes most Australian ecosystems from heathlands to wet sclerophyll forests, where disturbance, typically fire, results in the shortest-lived and longest-lived species regenerating together. For example, in a dry sclerophyll eucalypt forest, a hot fire or other major disturbance produces a thick regeneration of seedling eucalypts and wattles, along with many flowering understorey, herbaceous and grassy species; but after a few years some of the ephemerals have disappeared. After 30 years there may be no wattles present; the diversity of understorey has been reduced to the most shade-tolerant species and those adaptable to soil conditions where most of the nutrients are tied up in dominant trees or slowly decomposing organic matter.

Permaculture use of plant and animal succession

Both models, the classic and initial floristic composition, influenced the ideas and examples of designed succession proposed in *Permaculture One*, where fast-growing pioneer plants provide some yields and improve the environment for more long-lived and valuable plants.

More broadly, we recognised that if we could provide more of our needs from later successional stages dominated by perennial plants, especially trees, then our cultivated ecologies would be more ecologically balanced and resilient to seasonal variability than those based largely on annual crops.

Some of the ways we have made use of succession at Melliodora include:

- prevention of burning to allow accumulation of organic matter and nitrogen fertility
- rabbit-proof and stock-proof fencing to exclude animals in the early stages, thus allowing a wider range of more palatable pasture species to grow and improve the soil
- slashing grass and weeds to mulch tree plantings
- mixed shelter plantings of fast-growing, nitrogen-fixing acacias, tagasaste and casuarinas with slower-growing, longer-lived, shade-tolerant shrubs and trees
- interplanting of widely spaced walnut trees with hazelnut hedgerows to provide nut yields in the decades before the walnuts fully occupy the site
- planting of dense canopied deciduous and evergreen trees along the gully to shade out blackberry
- use of goats to eat blackberry, coarse pasture grasses and weeds, to encourage richer pasture species suitable for grazing geese.

An understanding of successional dynamics is essential to working with plants and land. The following examples illustrate some of the problems that can be attributed to a lack of understanding of succession.

- Most people tend to design and establish gardens (unconsciously following the initial floristic composition model) and are disappointed that there isn't enough space for everything and that some things die out or grow poorly as the system changes.
- Sometimes the response to understanding that some trees grow very large is to plant them at wide spacings in shelterbelts, where they struggle with competing grass and damaging wind (which actually increases in speed as it whistles between the slowly growing trees).

- In other situations, people give up growing some species because frost, wind and sun damage the plant, when a more mature system with more shade and shelter allows these delicate species to thrive.
- When new weeds appear, people frequently assume that large numbers herald an invasive takeover. Instead, it may be that disturbance such as earthworks or ploughing have allowed the growth of weeds like thistle which, in many situations, decline dramatically in abundance as perennial grass species become established in successive years.

The role of animals in preventing or steering succession in a particular direction is often ignored or misunderstood. In the case of the thistles, close grazing, especially by sheep or rabbits in autumn, is a recipe for maintaining the thistles. Even botanists sometimes fail to recognise the signs that grazing by wild animals is often the dominant factor determining what regenerates and persists in natural systems.

It could be argued that these problems arise from failure to understand the specific needs and characteristics of plants (and animals). However, if we expect change and are familiar with a variety of successional patterns from observing natural and cultivated ecosystems, we are in a better position to predict and apply succession to get better results from what we plan and plant.

Observation, the first principle of permaculture, is the key to making use of succession. Landscapes embody dynamic processes with a history and, to some extent, a future that can be read from the signs observable now. This ability is a critical observational skill that can be developed.[7] Dynamic change in nature and landscape is easily accepted as an intellectual concept, but our direct visual experience of landscape is most commonly a static image or picture in which the past and future are invisible.

Many modern urbanites experience seasonal and even daily weather as unique characteristics of a place without any clear picture of how that experience might be part of a larger cycle or process. Real estate agents know that spring is the best time for marketing rural land because the grass is green and the streams are flowing. As a consultant working with new rural land owners, an important part of my job is helping decode the signs in the landscape of past events, use and potentials which are not readily visible. Once we start to see all landscapes in this way, we understand that, although chance events are the cause of much of what we see, the patterns described by ecologists from study of natural landscapes are also useful in understanding landscapes radically changed by human intervention. These temporal patterns that we call succession are analogous to the spatial patterns of landscapes described as Land Systems in **Principle 7**: *Design from Patterns to Details*. The urban weedscape case studies later in this principle primarily rely on reading landscape skills to understand novel succession processes.

Plants Come Before Animals

Leaving aside the complex possibilities of the various models of vegetation succession, fundamental ecological successional rules can also be used in simple ways to ensure sound development of land. The basic fact of ecological succession is that plants provide the resources to support animals. This reminds us that we should establish pastures and fodder crops before we get animals. Imagine a small rural property which has too many livestock, fed out of purchased bags and bales, accumulating manure that cannot be absorbed by the overgrazed pasture — an example of the contradiction of succession order. This contradiction of ecological patterns may not be obvious to someone whose main contact with animals has been through pets, the zoo, the racecourse, or the bird fancier. The commercial feedlot illustrates this contradiction on a grand scale: it is financially viable only because of cheap energy, externalised environmental costs, and agricultural subsidies.

Collingwood Children's Farm

When I worked with Collingwood Children's Farm in Melbourne to refine and improve the farm's sustainability and environmental education value,[8] we proposed fodder trees and shrubs that could be lopped by children to feed the goats, as an alternative to the then current practice of providing stale bread. The bread, apart from being bad for the goats' digestion, was giving children the message that goats, like people, eat "sponge white" out of a plastic bag from the shop. The tagasaste and willow tree fodder has been a great success with the staff, the children and the goats. The possible ecological lessons that might be absorbed by osmosis from this activity could include the fact that ruminant animals get their sustenance from plants that we find inedible and, hopefully, that you need plants before you can sustain grazing animals.

Permaculture Design Course experience

In introducing systems ecology thinking in our Permaculture Design Courses, I have found that *the idea of energy and successional hierarchy flowing from the plants to herbivores and on to carnivores is, for many people, a new way of thinking about nature. The usual perspective focuses on the active role of animals in consuming food. This perspective is natural enough (for animals and humans) but it misses the point: it is not the acts of eating or living that create food and habitat. These resources are created for animals and humans by the slow-moving processes of plants that appear (superficially) to be doing nothing.*

To understand our place in the world, this idea is simple and yet quite profound. In an address to a permaculture conference[9] I suggested that permaculture could be seen (at one level) as "remedial wholistics" for citizens of affluent countries divorced from nature. As the rest of the world becomes rapidly urbanised (but not necessarily affluent), this remedial education has become a global necessity.

Pulsing Model of Ecological Succession

Ecological theory since the 1960s has begun to recognise that many ecosystems are evolved to go through periodic disturbance as part of an overall dynamic stability. The disturbance may be external to the ecosystem, in the case of bushfire or flood, or internal, in the case of defoliating insect plagues.[10] When viewed from the organism level, the disturbance may appear as a catastrophe, but for the ecosystem as a whole it beneficial. The pulse can contribute to efficient use of limited resources and maximise total energy capture. These pulsing ecosystems typically develop a pattern of long, slow accumulation of biomass (production) followed by a short intense pulse of consumption where total biomass falls rapidly and nutrients are recycled. Figure 32 shows the patterns of production by plants and consumption by animals (or fire) under the old climax concept and the more recent pulsing concept.

Landscape evolution

Before ecologists had begun to question the classic climax models of succession and stability, in another field of natural science, geomorphology, this pulsing model was already widely accepted. Called Catastrophism, this theory suggests that many, perhaps most, landforms are primarily the result of very powerful forces acting for relatively short periods of time. The obvious example is the relatively fast formation of mountains by volcanic activity, but more generally mountain-building happens at the edges of the earth's tectonic plates over relatively short periods of geological time.

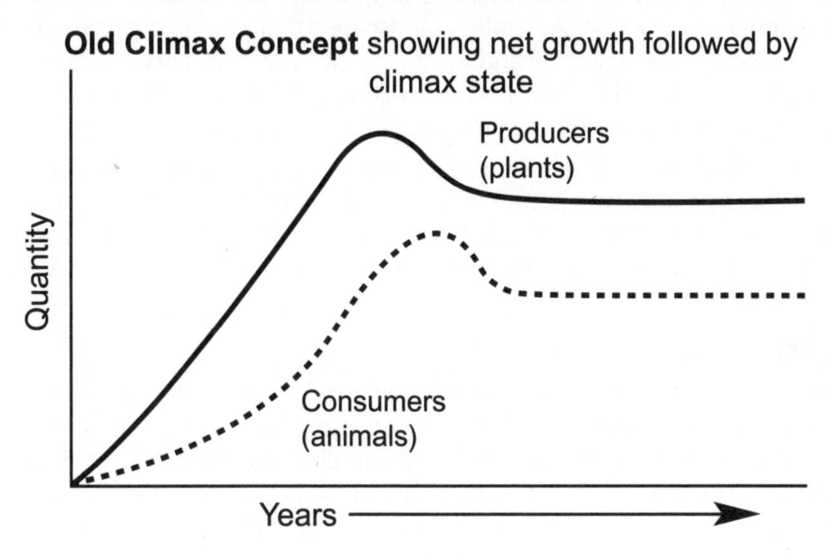

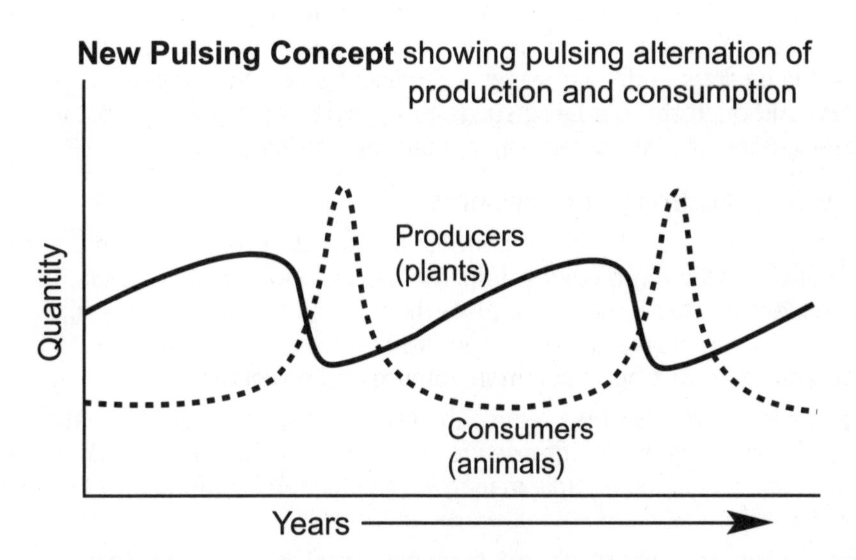

Figure 32: Comparison of two paradigms for growth and succession (after Odum 1987)

More surprising is the idea that river courses may be set by catastrophic flood events on an enormous scale, rather than by gradual, slow erosion. Under this model, the formation of a river is similar to what we can observe on a small scale with erosion gullies. These minor watercourses are generally stable for long periods of time, until extreme run-off and soil conditions, often combined with degradation of catchment vegetation, result in major extension of the gully.[11] Most of the fully vegetated U-shaped gullies in farm and bushland in central Victoria date from a series of weather events during the gold rush in the late 19th century, at a time when the catchments were largely cleared of vegetation.

Earthquakes in mountainous country result in multiple landslips, stream blockages and even lake formation; this is another dramatic example of catastrophic land formation. Even in old, comparatively stable Australian landscapes there are many examples of the fast formative processes. Dust storms and wind erosion may be minimal for decades, but a single event, such as the dust storm of 1983 that blanketed Melbourne in Victorian Mallee topsoil, can move more soil than decades of normal wind erosion.

The old theory of Gradualism in geomorphology still has a place in explaining the slow adjustment to the new forms created by catastrophic events. Examples are the slow erosion and occasional collapse of the sidewalls of an erosion gully until a new stable profile is established, and the gradual erosion of landslips damming a valley after an earthquake.

Chaos Theory

Chaos Theory has provided a general framework for understanding the role of change and stability across many scientific disciplines. It suggests that events (changes) are not randomly distributed in space or time, but occur in clusters. Further, this clustering of events operates at multiple scales. Thus there are clusters within clusters. The Big Bang theory of the origin of the universe is similar to the catastrophic geomorphological event followed by gradual adjustments (formation of galaxies of stars etc.). Galaxies can themselves be seen as intense points of activity (energy) and form (matter), which are very dense and separated by thousands of light years of almost empty space. Within galaxies, star clusters, binary stars and solar systems represent smaller-scale clusters within the larger clusters. Although the Big Bang suggests a single origin point, a recent theory[12] suggests the Big Bang was one expansion in a pulsing universe.

Fire as a pulse in Australian ecosystems

Most Australian forests are adapted to periodic fire, which recycles nutrients, results in plant regeneration, and often provides a flush of nectar and seed for insects, birds and animals dependent on these resources. Also, the effects of heat can release previously unavailable minerals in clays and rocks. The fire is a form of consumption of biomass analogous to animal grazing but much more intense and powerful.

By reducing fuel levels, smaller fires reduce the likelihood of larger and potentially more damaging fires. How frequent and intense fire should be to maintain ecosystem health is a contentious issue for national parks managers, as illustrated in **Principle 11**: *Use Edges and Value the Marginal*.

There is evidence that frequent fires result in an ecological drift to more vegetation.[13] The reasons for this are complex, but the loss of nitrogen and phosphorous following fires has been shown to be substantial, providing evidence for a drift to lower fertility. important chemical changes may have been missed in ecological studies due to misleading soil fertility theory.[14]

The details of the effects of fire on natural ecosystems are enormously complex and contentious. For any particular low-fertility ecosystem adapted to fire, a particular frequency and intensity will provide an optimal balance between system stimulation and stability. The pulse of fire appears to provide the action and the benefits, but the long slow phase of plant growth is just as essential for system health.

Pulsing Agricultural Ecosystems

Cell grazing

Allan Savory[15] and others put forward the idea that roaming herds of grazing animals maintained the health and vigour of rangelands by grazing hard and then moving on. This has been successfully applied to pastoral farming as cell grazing. There is now ample evidence that pastures grow better without the constant nibbling of animals, but a short burst of grazing maintains the vegetative growth of pastures and recycles nutrients. An occasional hard graze exposes and cuts the soil surface, providing favourable conditions for regeneration of short-lived legumes and other useful species.

In a systems ecology sense, this separating of the production aspect of the system (plant growth) from the consumption (animal grazing) increases the total energy processing and resilience of the system. Further, if the system is to remain healthy, the consumption phase is always brief and relatively intense. How brief and how intense are key management questions for pastoral farmers.

Pasture ley cropping

Another important agricultural example of pulsing systems is the rotation of fertility-demanding annual crops with longer rotations (leys) of soil-improving perennial pastures. In moderate rainfall temperate climates, mixtures of clovers, perennial grasses and other herbaceous plants improve soil organic matter, structure and fertility, especially if the grazing regime follows the pulsing patterns referred to previously. Ploughing under the pasture releases much of the accumulated fertility leading to high yields in annual crops. However, if the sequence of annual crops is maintained for too many years, fertility declines. Variations of this system prevailed in the Australian cropping zones until the pasture phase supporting sheep became uneconomic.

Today, the typical regime is direct drill seeding without cultivation, using herbicides to control weeds, and adding fertility (from the bag); this compensates for the loss of the pasture phase. Whether this more intensive land use will rebuild organic matter and mineral fertility in cropping soil or lead to another downward spiral of land degradation remains contentious, but the evidence is not encouraging.

Organic sources of bagged nitrogen fertiliser are too expensive for use on broadacre organic farms. Green manure and legume pasture leys used by organic farmers remain a more sustainable method of soil improvement that reflects the pulsing ecosystem model.

Bush regrowth as soil-conserving ley

In high-rainfall, low-fertility hill country in southern Australia, regrowth bush (especially of soil-improving wattles) has, by default, acted as a fertility rebuilding phase between more opportunistic clearing and sowing to pastures, which last for a decade or two before the bush reclaims the land. Inexperienced and enthusiastic landholders, spurred by observations of good pasture growth following clearing, have done this repeatedly over the last 100 years. In most cases, the clearing and pasture-growing phase probably caused more erosion and loss of organic matter and nutrients than 20 or more years of bush regrowth have been able to accumulate. Consequently, each pulse contributes to a downward spiral of land degradation.

Tropical slash and burn

Of the pulsing agricultural systems that have been well documented, the most evolved and sustainable example is the swidden or slash-and-burn agriculture of New Guinea and other tropical regions. Small areas of rainforest are cleared, burnt and planted to a great mixture of perennial crops, which form food forests thriving on the fertility released by the burning. The structure of the food garden mimics the rainforest, with multiple levels of tree, shrub, vine and root crops minimising the leaching of nutrients. Over time the garden becomes harder to harvest and there is more growth of woody non-food plants. It is then abandoned and returns to forest. Sometimes naturally regenerating forest trees are retained, known as "mother of the gardens" in recognition of the need for rainforest regeneration to rebuild soil fertility. In the New Guinea highlands before Europeans arrived in the 1930s, this system of agriculture sustainably supported dense populations. It is the most energy-efficient form of agriculture documented by scientists.[16]

The rotation time under regrowth forest is critical, with 20 years generally necessary to fully rebuild fertility before another cycle of gardening. Consequently the ratio of garden to forest is always small. In tropical regions this form of agriculture is now widely regarded as unsustainable because the exploitation phase has become more and more frequent due to pressure of population living from declining areas of forest.[17]

Lessons for Permaculture Design and Land Management

These examples emphasise the importance of the pulsing ecosystem model in understanding a wide range of land uses. *The use of fire, grazing and cultivation to provide a pulse of high yield between longer phases of biological rebuilding can be recognised as strategic use of change against a background of catching and storing energy. The discovery of this process was more fundamental to the success of* Homo sapiens *than the invention of annual crop agriculture.* The more difficult lesson is that the benefits from the pulse are dependent on the slow accumulation of the building phase; pulsing the system too frequently leads to a downward spiral of land degradation. This dynamic can be seen as a generalised model of land degradation, where addiction to a particular management pattern provides immediate benefits but produces long-term decline.

The use of grazing, fire or cultivation can be seen as an interruption of successional processes. This is the antithesis of the classic permaculture strategy of accelerating succession towards climax vegetation. To suggest that these productive uses of disturbance follow patterns observable in nature could be seen as dangerous, because they may encourage the over-exploitation and land degradation which permaculture is trying to reverse through a fundamental redesign of agriculture and land use.

In *Permaculture: Design for Cultivating Ecosystems*[18] I showed how easy it is to slip back into the pioneering mentality which always tries to stop nature's successional processes under the frequently mistaken idea that great yield or value will result. But I also gave examples of how pulsing the system with axe, fire and grazing animals can harvest the production that vegetation has slowly accumulated.

A friend of mine, who had been an enthusiastic no-dig permaculture gardener for 15 years, discovered the power and productivity of a rotary hoe in his extensive vegetable garden. But after several years of cultivation, his wife's new sheet-mulch garden on a piece of old pasture land out-produced his cultivated patch, which began to suffer from soil structure problems. Ironically, the sheet-mulch technique is itself a kind of exploitative pulse, where the decomposition of the mulch and smothered pasture results in a high-yielding garden for a few years before the underlying mineral imbalances tend to become limiting factors and reduce yields.

Sheet mulching and successional alternatives

It is annoying that permaculture has become equated with the sheet-mulch gardening technique as if it were the pinnacle of sustainable land use. For me the technique has always been an easy way to convert lawn and pasture into productive food gardens without the hard work of digging and weeding and by making use of locally abundant and wasted organic materials. While vast quantities of organic materials are available (from urban landscapes and farms), sheet mulching is good use of what others undervalue or waste but in a lower-energy future such materials will not be available cheaply. Used as a maintenance, rather than an establishment technique, sheet-mulch creates dependence on a constant supply of mulch. If that mulch is high-quality animal feed such as lucerne hay, this is hardly sustainable.

Second, there are doubts about the health and balance of garden soils continually covered by quantities of organic material at a density 10 to 100 times greater than that possible from litter in any natural ecosystem. From a biodynamic perspective, excessive and continuous mulching is said to smother the life in the soil; from an Albrechtian perspective, over time it typically results in excessively high levels of potassium.

Successional alternatives to continuous maintenance of a sheet-mulch vegetable garden include:

- reversion to a grass and weed fallow, and moving the garden to a fresh location
- gradual dominance by fruit trees and other shade-tolerant perennials, with natural litter fall and prunings providing adequate mulch
- more intensive methods of soil management, including application of high-quality compost, carefully selected and measured rock minerals, and modest use of mulching, combined with light surface cultivation and deep aeration forking to maintain a balanced and productive soil, hopefully in perpetuity.

The first alternative is the fate of many gardens for a variety of reasons, not the least of which is that the gardeners move house. The second creates the classic permaculture food forest, which many see as the ideal but still leaves the question about where to grow the tomatoes and lettuces

At Melliodora we have pursued a combination of the second and third strategies, but have also experienced the first in some locations where creeping bent grass has got the better of us.[19]

Food forest succession

Many intensive sheet-mulch permaculture gardens of mixed annuals and perennials are an application of the initial floristic composition model of succession. The succession to fruit trees, climbers, shade-tolerant herbs and a few perennial vegetables has resulted in many and varied food forests, which continue to provide a variety of yields for very little work. On the other hand, high-yielding production of the most valued annual vegetables requires an open sunny space.

After a decade or so of nurturing an expanding food forest, I have seen permaculturalists take to overgrown leguminous trees and excessively numerous fruiting species with the chainsaw and chipper, to release some fertility and sunlight for vegetable gardening or other valued plants. I gain an iconoclastic delight in seeing this reinvention of slash and burn agriculture. I plant soil-improving trees in the knowledge that the greatest benefits will come when I cut them down and release the fertility in their roots and tops for use by other plants. Perhaps less confronting examples of the use of nature's abundance in creating a food forest at Melliodora are provided by our living fences of wild cherry plum seedlings transplanted from the gardens, and countless seedling fruit trees for fruit and root stock given away to others establishing gardens.[20]

In *Natural Gardening and Intensive Biological Gardening*[21] I considered the broad range of strategies and techniques that might be relevant to garden agriculture as a continuum between intensive and extensive approaches. Many of the options considered relate to the issues of scale addressed in **Principle 9**: *Use Small and Slow Solutions* but this principle about the constructive and creative use of change is also central to an important understanding: what exists now will evolve into something else. Simple notions of stability and sustainability will not help us decide on the best actions.

Four-Phase Model of Ecosystem Change

The two-phase model of pulsing ecosystem change is useful, but a more elaborate four-phase model[22] is also instructive. The four phases are Conservation, Release, Reorganisation and Exploitation.

- Conservation: the long-lived, steady-state climax, where there is a high degree of interconnection between system elements, a large amount of stored biological capital, and little leakage of nutrients.
- Release: the pulse of disturbance, typically very short in duration.
- Reorganisation: the most unstable phase, when open niches, soluble nutrients and energy are available to be tapped or lost. During this phase there is the potential to flip to some system that is less, or more, productive and organised.
- Exploitation: fast-growing pioneer species colonise the opportunities, catch and store energy quickly, and cement the patterns for the more gradual build-up of biomass and greater connectedness, leading toward a new conservation phase.

This model focuses attention on the conditions within the climax that make the system vulnerable to disturbance, and on the complex and uncertain dynamics that follow the release phase, before a new vigorous exploitation phase is established. Most importantly, the reorganisation phase incorporates:

- the high risk of ecological drift to lower productivity (generally due to loss of mineral nutrients)
- the lower risk of major degradation (often due to coincidence of severe weather events)[23]
- the potential for ecological invasion and possible shift to more productive systems better able to use available resources.

Economic and Social Succession

Models of ecological succession are also useful in understanding micro-economic and social change as well as large-scale human cultural patterns.

Back-to-the-land pioneers

The term "pioneer plants" derives from the original meaning of "pioneer" as one who first settles a region, thus opening it for occupation and development by others. Since the 1970s back-to-the-land migrants left the cities for a more autonomous rural life. They have acted as pioneers in changing the social and economic environment of the depressed rural areas into which they moved. In the process, they laid the foundations for more mainstream rural resettlement and for the tourist industries that make these regions stand out from the ongoing decline of rural Australia. The restored old buildings, arts and craft workers, cafes and health-food shops, permaculture properties and alternative buildings, small-scale organic farms, conserved patches of forest and wetland, alternative schools and health practitioners were the achievements of the first generation of back-to-the-land settlers, and they attract a second phase of new settlers and tourists. The social infrastructure created by the pioneers enabled these regions to attract people and money, countering the intensifying hold of the big cities over national culture and economy.[24]

The irony in this social succession process is that the pioneers rarely view the arrival of the new migrants and tourists with equanimity. They see the newcomers as unappreciative consumers of the rural cultural and resources, who bring with them dysfunctional urban attitudes. There is much truth in this view, but if we understood more about social succession we would not be so disappointed by this process. As pioneers we make the social environment favourable for like-minded people, but over time we will collectively create conditions that suit another class or sector of society. This rural phenomenon is a variation of the recognised social succession process in inner-city areas called "gentrification".

Gentrification

The process begins when students, artists and other trendsetters are attracted to working-class inner-city areas that are physically rundown but socially dynamic. The pioneers make the area more attractive to owner-renovators. The renovators raise rental and resale values, and this pushes out many of the original working-class occupants. Over the last 30 years this gentrification process has transformed the old inner suburbs of most cities of the Western world.

There is some evidence that there are further stages to this succession process. Investor owners rent out the fancy townhouses to upwardly mobile young professionals, but by this stage the dynamic community, which attracted the original bohemians in the first place, is dead. Absent and disinterested owners, increasing crime and, eventually, declining values create a new opportunity for the poor to obtain shelter and create the possibility of community. This looks like a pulsing or even a four-phase system.

Nouveau riche succession

Cultural succession occurs within families. The shift over several generations from traditional rural peasants connected to place, to migrant urban workers, to successful small business persons, to educated professionals, is a classic one that rides the wave of fossil-fuel-based affluence. This pattern can be seen in all cultures and is now a truly global social process. The first generation that makes the break from home is often driven away by grinding poverty, oppression or dispossession, but may also be lured by the carrot of hoped-for riches, like Dick Whittington who believed that the streets of London were paved with gold. This successional pattern, and variations on it, have been repeated in the histories of hundreds of millions of families worldwide over the last century.

Like the rapid succession of gentrification, there is evidence that the transition to affluence is not a final stable climax state but a pulse of consumption, which is not sustained. The maxim about the first generation creating the wealth, which the second generation spends and third loses, shows a common understanding of the limits to affluence. It is true that the aristocracy and the rich often maintain an affluent lineage for many generations by careful nurturing of specific characteristics in their children. However, this maintenance of wealth has been achieved in a context where their very distinction from the masses and small numbers help to maintain their culture. If everyone were equally rich, it would be very hard for the elites to maintain any distinctive culture.

Affluent Western industrial societies have to some extent managed to maintain generations of moderately functional affluence for the majority, but only in a global context where the masses remain poor. Further, many of the problems of mounting health and welfare costs, the loss of a work culture, addictive behaviours, crisis in the education system and so on, are indicators that the costs of affluence are starting to catch up with us. It is hard to imagine a world where this pulse of affluent consumption can persist for more than a few generations, even if there were not severe constraints on materials and energy resources.

Socio-political cycles as a pulsing system

Neo-classical economic theory and its attendant political beliefs, which dominated mainstream public policy over the last two decades of the 20th century, is the subject of an increasing body of critical analysis from both social and environmental perspectives. Almost everyone, except the true believers, has some story about the follies of so-called economic rationality. Rather than adding to that litany, I think it is useful to understand the last 20 years as an intense pulse of economic activity based on consuming the social and economic capital built up over the preceding four decades.

This current phase of laissez-faire capitalism began with Margaret Thatcher and Ronald Regan in the early 1980s. It signalled the demise of Keynesian economics and social democratic politics that had prevailed in the industrialised world since the disasters of the Great Depression and World War II.

During this time a mix of factors contributed to an accumulation of public infrastructure and assets, education, health, and other social wealth. They include:

- the work ethic and culture of frugality
- faith in societal structures and values
- rising wages and working conditions
- strong public policy to control the worst excesses of capitalism and to direct societal wealth for the common good.

By the 1970s, the oil crises, the Cold War and other factors impacted on the capacity to support a continually rising standard of living and public wealth. The agents of the elites sold the message that the inefficiency of public bureaucracy and control of the economy inhibited the market from efficiently distributing resources and creating wealth. The privatisation, income tax cuts, downsizing and corporate buyouts since the controls on capital have been relaxed, have certainly increased measurable economic wealth. Most of this new wealth has fallen into corporate and private hands in rich countries.

After 20 years we can now see that much of the so-called productivity of this period was built by cashing in the public assets created over previous decades. Like the farmer marvelling at his crops growing on the ashes of a burnt forest, many mistakenly believe that this phase of intense economic consumption can go on indefinitely and falsely identify it as productive. Public physical assets have been sold off and dismantled. In addition, the knowledge and skills of people in large organisations, which represent system intelligence and foresight created by a public education system and long-term visions, have been discarded in favour of fast-moving animal cunning driven by the bottom line. The commonwealth of shared knowledge in science has become the intellectual property goldmine to be plundered, traded and controlled. Even the goodwill and credibility of non-government organisations, built up by decades of voluntary action and commitment, are valued in terms of millions of dollars.[25]

Can the rich countries avoid another economic tragedy on the scale of the Great Depression before there is widespread recognition that we need to rebuild the collective wealth? It is tempting to see this in terms of a holy crusade of good against evil, but most would acknowledge that some good has come from the pulse of rampant capitalism. The structures of public wealth that are rebuilt will of necessity be different from those that have been swept away. Those structures arose from the social democratic politics of the 1930s, 1940s and 1950s, at a time of rapidly rising net energy. In the 1930s tapping of conventional oil for transport and coal for electrification was beginning to accelerate. Thus, the building of social and economic capital was based on the rapid consumption of natural capital.

The Large-scale Pulse of Fossil Fuels

Today we have consumed the easy half of the total likely reserves of conventional oil, and opportunities for major expansion of electricity networks and consumption are contracting. At the same time the world's population has more than tripled. Thus the rebuilding of social and cultural capital must occur within a context of declining net energy availability, and in ways that reinforce the simultaneous rebuilding of natural capital in our landscapes.

Grasping this energetic reality leads many to despair at the magnitude of the task and the impediments to success. While global capitalism has been like a fire converting green forests to ashes, it has likewise released potential and information from the constraints of cultural norms and institutions that were hopelessly inappropriate for dealing with a world of declining energy. The ashes of the consumed forest provide opportunities for the seeds of pioneering species to reform the forest in a way that better reflects large-scale realities, such as fertility or climate change. Similarly, globalisation provides the opportunities for social seeding to create new bioregional cultures adapted to energetic realities.

The cultural and ecological destruction created by global capitalism is providing new opportunities to setting the parameters of very large-scale and long-term cultural systems adapted to declining energy. Thus the permaculture aphorism "the problem is the solution" is not some naive optimism in the face of terrible prospects, or the delusion of those with all the opportunities, but a simple idea with powerful relevance to our time. If we view global capitalism as releasing the earth's accumulation of renewable and non-renewable resources according to Holling's Four-Phase Cycle,[22] then permaculture is the new potential of the Reorganisation phase.

In developing a post-affluent culture it is not necessary to denigrate what our parents, grandparents or ancestors did as ignorant, shortsighted or anti-nature. Instead, we recognise that the ground on which we stand has been prepared for us by those who have gone before. A more positive view of generational change enables us to see how to act constructively in the face of the immense opportunities from energy descent.

Evolutionary Change

Plant and animal selection

Although the use of ecological succession has been emphasised in permaculture design, other uses of biological changes are equally important. The application of deliberate selection pressure to create and change cultivated plant varieties and domestic animal breeds is perhaps the most ancient and important use of genetic change for building a sustainable future.

Principle 10: *Use and Value Diversity* explained some of the factors of plant and animal selection in terms of a balance between productivity and diversity. As human-induced change in environmental conditions becomes faster and more erratic, so predicting which selected forms will prove adaptive becomes more uncertain. Hence the argument for the primacy of diversity, including the chaotic wild diversity of weeds.

The spectrum of plant, animal and human diversity allows enormous scope both to conserve and to create new, selected diversity to suit emerging conditions. In the same way that industrialisation led to a surge in plant and animal breeding to suit higher energy availability, a decline in energy availability will lead to wider use of traditional varieties and breeds (where they still exist) and the creation of new ones. Permaculture, and organic agriculture are at the vanguard of this trend.

While plant and animal breeding has greatly contributed to humanity's expansion over the last 10,000 years, we tend to forget that plant and animal breeding have failed to create any new species. Until the advent of genetic engineering, selection was confined to working within the pool of natural diversity of each species.

Species evolution and extinction

The vast majority of species that have ever existed are now extinct: Gaia is savage, as well as a nurturing mother. The 3800-million-year history of life on earth etched in the fossil record is a sobering reminder of the finality of extinction. The process of extinction is well understood, especially because humanity gains the dubious distinction of being one of the most powerful agents of species extinction in the history of the planet.

Conversely, the evolution of new species and larger families of life by Gaia over that same time is the astonishing reality exposed by the fossil record. The processes involved in this explosion of life remain a mystery 150 years after the publication of Darwin's Origin of Species.

Post-Darwinian evolution

Darwin's theory of evolution changed thinking fundamentally toward an acceptance of deep and persistent change at the heart of the natural world. In evolutionary theory, the idea that evolution proceeds by gradual small increments has now been overturned by the idea of Punctuated Equilibrium,[26] where evolution proceeds by occasional jumps. This view is in keeping with the shift from Gradualism to Catastrophism in geomorphology, from Climax to Pulsing ecosystems in systems ecology, and Chaos Theory more generally.

The clustering of otherwise random mutation might be a more realistic model than classic Darwinian gradualism in explaining the pulses of species emergence in the fossil record. However, there is also widespread doubt among evolutionists that natural selection, even operating on clustered genetic mutations, is adequate as a sole mechanism to explain the evolution of species.[27]

Natural selection might be the editor of life, but it is not the author. The possibility that randomly generated genetic mutations are the lone creator of life's variety has been problematic since Darwin first proposed it. After a century and a half, it seems even less adequate to explain the evidence of nature's creativity, especially because of the great difficulty in observing signs of classical speciation under any conditions.[28]

This idea is uncomfortable for most biologists because it is the same argument that Creationists use. Even Punctuated Equilibrium can be seen as acknowledging that creative events are the real drivers in evolution. Although these processes might move the science of life closer to concerns of theology with creation, there are many mechanisms that may explain how the self-organising characteristics of life lead to evolutionary change without resorting to a paternalistic god.

Some of the processes that may be generating variation in addition to mutation include:

- gene transfer between bacteria
- symbiosis between previously free-living organisms to create a new organism
- internally directed mutations within organisms in response to environmental pressures
- internal selection operating on mutations.

Recent evidence about the widespread nature of gene transfer between bacteria suggests it may be a major mechanism of evolution.[29] Lynn Margulis, who co-wrote *The Gaia Hypothesis* with James Lovelock in the 1970s, is well known in biological science for her 1965 hypothesis: that the nucleated cell, which is the basis of all complex organisms, originated from a novel symbiosis. This hypothesis, now widely accepted, suggested that membraned cells incorporated free-living bacteria along with their own millions of years of evolutionary history. The evidence about gene transfer between bacteria shows one mechanism by which symbiotic unions between species may be a continuing source of evolutionary innovation.

The billiard table analogy

Each species can be thought of as representing a design solution; minor variations are possible, but over time an archetypal form (called an Attractor in Chaos Theory) holds sway. Whenever selection pressure is released, the archetypal form re-emerges in the way described in animal breeding. (see "Genetic Diversity" in **Principle 10** *Use and Value Diversity*)

These design solutions can be thought of as depressions in a flat table. Evolving species are balls that roll around on the table until they are captured in a design solution. The range of possibly viable design solutions is limited. For example, an elephant that has legs as thin as a horse's is just not possible. In many ways, the order in nature is these possible viable design solutions as much as it is in the species which represent them.

Plant and animal breeding moves the balls around the depressions. Some large force or different mechanism is needed to get a ball (a species) out of a depression (a possible design solution) and onto the table surface. Once out on the table, the slightest force can cause the ball to go anywhere. Sooner rather than later, the ball will be captured by another depression.

As well as the physical and metabolic limits to possible design solutions, there are organisational and ecological limits. Ecosystem self-organisation provides niches into which species are more or less constrained to fit. The patterns of niches and relationships follow more general system design rules determined by energy quality and flow. For example, in most temperate forests, rodents occupy the forest floor. In the relatively recent creation of New Zealand's ecosystems,[30] there were no rodents or even mammals, but there was a need and opportunity for a small animal that scurries around in the forest litter eating seeds and insects. A series of flightless wrens were the evolutionary solution. The arrival of rodents with the Maoris and Europeans put paid to the wrens, which are now all extinct.

The extensive extinction of specialised species and their replacement by hardy generalist species has led many biologists to believe we are witnessing the "McDonaldisation of the world's biota".[31] Although it is difficult to refute the evidence for this process, the catastrophic changes that lead to species extinctions also create new and novel conditions; these may stimulate or accelerate mechanisms which lead to the emergence of new species.

If evolution does proceed by diverse processes in jumps analogous to the pulsing ecosystems model, it is hard to imagine a greater stimulation for evolutionary jumps than the current planetary situation. The same factors that are the source of species extinction may generate new ones and possibly a flood of new species. Those factors include:

- mining and use of almost half of the fossil fuels accumulated over the last 1000 million years
- 6000 million people highly connected and concentrated to form one global population
- world trade, leading to unintentional transfer of lifeforms on a scale never possible in the history of life, on top of the energetic intentional introductions of the last few hundred years
- risk of widespread nuclear contamination from weapons, power production and waste, as well as ongoing chemical contamination from industrial processes
- the rapid expansion in release of genetically engineered organisms.

We need to keep this unpredictable state in mind when thinking about designs that depend on very long-term strategies, in the same way that we now need to consider climate change as a factor in design. The thought of novel lifeforms suggests new animal and human disease organisms or rampant weeds taking over the landscape; in a similar way, climate change raises many negative possibilities. While it is important in a public policy sense to attempt to minimise the human contribution to destabilising global environmental change, it is clear that we also have to be prepared, to the extent possible, for large-scale changes already beyond human control. Like the onset of the last ice age 18,000 years ago, any change also presents opportunities, but it is very hard to speak of opportunities without appearing flippant about negative effects. The permaculture slogan that "the problem is the solution" provides a good discipline to look for opportunities in what may be the bleakest of situations.

I am arguing that an eruption of new species is both a positive expression of nature's abundance and a potential mortal threat to humanity. To simultaneously hold both concepts is difficult: the focus on one tends to dilute or diffuse the power of the other.

System Evolution

Although the potential for evolution of new species is an important threat and opportunity for humanity, the concept of evolution has a much wider currency when applied to much more fluid self-organised systems of natural ecosystems and landscapes, as well as businesses, organisations, communities, and cultures. This wider use of the concept of evolution has been generally understood in terms analogous to Darwinian evolution of species.

Just as the concept of species evolution has itself changed from gradual imperceptible change to one of novel and transformative change, so too is it appropriate to understand ecosystem and social evolution as transformative change. The idea of constant change in technology, economics, society, politics and culture is now the norm as a result of the continuous, multi-dimensional and radical changes of recent decades. It is now appropriate to think of system evolution as resulting from "emergent" behaviour or characteristics that transform the system to a degree where, like a new species, we would call it something fundamentally different with its own dynamic.

In Holling's Four-Phase Model of ecological change, the highly connected and co-evolved structures of the Conservation phase maintain current diversity but prevent the development of new possibilities. The Release (disturbance) breaks the connectivity; it allows the risky and uncertain development of new connections during the Reorganisation phase, which exhibit their real power in the Exploitation phase before settling into a new long Conservation phase. If that Conservation phase lasts for a long time and survives several cycles of disturbance then we could accept it having achieved some sort of evolutionary success.

Ecosynthesis

The focus of conservation biology on the ancient co-evolved nature of some ecological relationships has led to a distorted view that most ecosystems are fixed and ancient systems, almost like species themselves. Although species may be remarkably durable and persistent systems, the relationships between species and the physical environment factors that make up an ecosystem are remarkably plastic, with new relationships and adaptations leading to rapid ecosystem evolution.

For example:

- palaeobotanical evidence[32] shows that many of today's remnant and predominantly natural ecosystems did not survive the fluctuations of ice age intact in climatic refuges, as previously thought. Instead, they appear to be novel and contingent combinations of species that have come together during this interglacial (the last 10,000 years). These species survived the ice ages in many different places and systems

- further, there is increasing evidence that Australia's current fire-prone, low-fertility, eucalypt-dominated vegetation is decidedly atypical of other interglacial periods, when rainforest vegetation dominated. The human factor seems the most likely cause of this anomaly, but it also means that some of our indigenous systems are not climatically optimal and are ripe for invasive domination by rainforest-type vegetation
- the recognition that indigenous land management has for tens of thousands of years shaped the rapid evolution of ecosystems and landscapes is a relatively new idea in science
- the co-evolution of European and Mediterranean ecosystems to cultural management systems spanning a thousand years or less is now acknowledged as providing some of the greatest reservoirs of biodiversity in Europe
- at the very time that concern about spread of non-native plants and animals has captured the imagination of the scientific and wider community, there is mounting evidence of continuous and eruptive evolution and adaptation in ecosystems on time-scales as short as a human generation.

European trade and expansion across the globe over 500 years to 1900 was the source of what will prove to be the furthest-reaching and most permanent changes.[33] There is even evidence that European wooden ships seeded the coastal waters of the whole world, but especially the temperate zones, with European coastal lifeforms, so that what marine ecologists are recording for the first time are already radically changed ecosystems.[34] These changes are typically described in terms of ecosystem degradation. I, and others have defended this "ecosynthesis", the evolution of new ecosystems of native and exotic species responding to novel conditions. Ecosynthesis has had, and continues to have, beneficial effects in moderating and repairing the environmental impact of human expansion, as well as providing new resources as plants and animals naturalise. Some of the more thoughtful research[35] in conservation biology has recognised that mutualism (co-operative relationships) is a major factor in assisting the spread of exotic species, but it generally fails to make the connection to the idea of ecosystem evolution.

Urban weedscapes as ecological succession and ecosynthesis

Abandoned gardens and farms were for me some of the earliest inspiration for the permaculture concept. A trip to New Zealand in 1979 intensified my interest in the emergent ecosystem qualities of weedscapes. The article "*Impressions of New Zealand*"[36] written at the time includes a description of a large-scale weedscape that was inspirational for later work.

Riparian Weedscapes of Melbourne

In 1982 a personal quest to improve my skills in landscape pattern recognition and my understanding of succession dynamics of weedscapes led me, along with a colleague, to an observational study of the Yarra River floodplain in the heart of metropolitan Melbourne.[37] We observed the composition and structure of vegetation and land use at a wide range of alluvial flats, riparian edges and wetland sites on the Chandler floodplain and correlated this with recorded and anecdotal information about vegetation and land use in the past.

Despite the wide disparity and severity of land use impacts from grazing and farming to golf courses and stormwater drainage works, patterns of vegetation succession were evident. Of greatest value were areas where there had been no substantial management for decades: these places showed slowly emerging stages of vegetation succession that appeared to have some characteristics of a classic climax.

For example, in the absence of fire, grazing and clearing, the floodplain vegetation changed towards a forest, dominated by an overstorey of indigenous river red gums, with a closed canopy understorey of tall shrubs and small trees, including indigenous, Australian native and exotic species. The ground layer was dominated by wandering jew (*Tradescantia albiflora*) that loves fertility and is very shade-tolerant. The few mature (80-year-old)[38] examples of this forest type were exciting and beautiful places. Our recognition that the earlier stages of this forest structure and composition were emerging at many other sites was the evidence that the old sites were more than unique chance mixtures of vegetation with no systemic pattern. The robustness of this "successional climax" was given further weight by the fact that the vegetation pattern was emerging from diverse starting points — grazed red gum woodland, cultivated fields, and even freshly bulldozed areas.

At the time that we were identifying successional patterns and other systemic characteristics in these new urban ecosystems, indigenous revegetation advocates, under a new management regime for urban bushland, were identifying all the non-indigenous species as environmental weeds to be removed. After nearly two decades of this management, many of these areas have been turned into a purely indigenous version of the same forest type with massive inputs of herbicides and labour. It is ironic that, despite the continued dominance of the ubiquitous red gum, the emerging wild systems (and the intensively managed revegetation alternative) bear little functional or structural relationship to the pre-European ecosystem: an open grassy woodland, much drier, more dependent on fire and grazing, and lower in fertility.

Four factors important in driving the evolution of novel successional patterns in this urban ecosystem were the abundance of:

- water, from both natural and stormwater sources
- nutrients, from farming, stormwater and other sources
- seed and vegetative propagules, from diverse suburban garden, agricultural and indigenous forest sources
- absence of grazing animals and fire.

The absence of grazing animals and fire began the breakdown of the pre-European ecosystem, while the water and nutrients provided the energy for vigorous growth and decay cycles, as well as abundant food for birds and animals. The seed sources ensured a wide diversity of potential species from which systemic characteristics could develop.

Riparian Weedscapes of Hepburn Springs

The extensive riparian weedscape of Spring Creek and its tributary gullies attracted my attention when I moved to Hepburn Springs in 1985. Over the following decade we (an informal group of local residents) created tracks and planted trees over several kilometres of blackberry (*Rubus fruticosus*) and willow (*Salix albaXfragilis*) dominated stream corridor.[39] Our management decisions were constrained by the great scale and power of the system and our limited resources. We used reading landscape skills and ecological theory to help guide a strategy for accelerating the succession to a low fire hazard, high amenity mixed corridor forest. For some of us predicting the ecosynthesis processes at work, it was as interesting as any interventions we made. A succession from primary weedscape of blackberry, gorse (*Ulex europaeus*) and cape broom (*Genista monospessulana*) to deciduous forest of willow, hawthorn (*Crataegus monogyna*) and native blackwood (*Acacia melanoxylon*) was evident. This reflected similar processes to those observed on the Yarra. Other deciduous species, most notably sycamore (*Acer pseudoplatanus*) and European ash (*Fraxinus excelsior*) were also present and we predicted that over time they would overtake the willow canopy. The role of animals and birds was critical in both spreading the invading trees and preventing their establishment by predation.

Two formal university research projects by members of the group were able to confirm and quantify some of our more general observations and predictions. A PhD study[40] comparing the streambed ecology of this creek with a parallel stream dominated by eucalypts showed the willows were capturing 40 times more sediment and 10 times more phosphorous than the eucalypts. Most of the sediment was trapped in the extraordinary willow root mattresses that clothe the streambed. This and many other measures confirmed that these willow corridor forests provided a near optimum water filtering system in a steep catchment receiving runoff and sediment from both native forest (subject to sheet erosion), and suburban storm water. We were working with a powerful example of nature catching and storing energy.

The other study[41] aged the willows, sycamores and ashes along the creek and in an elegant use of statistical analysis and successional theory predicted that sycamore and ash would begin to dominate as the first generation of willows died. This hypothesis agreed with our more wholistic observations and predictions about the likely successional patterns in this novel ecosythesised forest. These may take a lifetime to confirm but they show the potential for more predictive ecological science. These examples suggest that *the development of systemic properties in novel combinations of plants and animals, is not so much the final outcome of slow bottom up evolutionary processes. Instead they are the design rules that guide the cyclical and evolving dance of species towards a new nature.* The ecological holocaust of the Central Victorian goldrush in the late 19th century was the release phase which allowed the piecemeal settlement processes since to lay down the energetic regimes and biological resources for this ecosynthesis.

Humanity as part of nature

The idea that complex weedscapes, especially in our urban landscapes, represent new evolving ecosystems of great importance in an energy descent future is a major theme of my work in permaculture over the last 20 years.[42] A broad understanding of successional models and evolutionary dynamics has helped me to see ecological pattern and order where others see meaningless and destructive chaos. However, the following quote shows that others with more official endorsement agree:

> there is no fundamental difference between natural, wild or modified, semi-natural or developed, domesticated or purely artificial vegetations. The laws governing these ecosystems are identical.
>
> Use and Conservation of the Biosphere, UNESCO, Paris.

The recognition of the reality of ecosynthesis is important at a number of levels.

Knowing that ecosystems can evolve at remarkably fast rates in response to human and other influences, we hope that the self-organisational power of nature can be enhanced and augmented to create functional human ecosystems in relatively few generations. Thus permaculture and similar concepts such as Analogue Forestry, which design agricultural and other land use systems modelled on natural systems, are not using some vague analogy but making a direct and conscious application of processes that are continuously at work everywhere in nature.

Because of this ubiquity of self-organisation and system evolution, even the most carefully planned and controlled biological system will change and evolve. Conventional agriculture, master planning, ecological preservation and other rigid methods are of limited use in dealing with the complexity of biological and human systems. Use of excessive energy to simplify and control the system is the typical solution. Simplification and control might be a brief description of what we tend to do in all areas — using natural systems, organising scientific research, managing organisations, and designing our living environment. If we can develop more open, flexible and interactive processes for planning, design and management, we are more likely to see the benefits from wild nature and human complexity.

To recognise our actions and ourselves as a part of nature is a cultural transformation, begun but not completed. The change from seeing human impacts on nature as improvements to seeing them as destructive may be a necessary step in that evolution, but the transformation comes when we no longer see ourselves as outside nature. To make this final step, we must first set aside the judgments about our actions being good or bad. Then we can see the examples of ecosynthesis all around us as nature at work.

Observation skills are necessary to perceive the subtle signs of changes over time-scales much greater (slower) than the observation period. This is a critical issue in making sense of change on all scales and for making sense of the broader issue of sustainability. **Principle 9**: *Use Small and Slow Solutions* referred to the impediments that modern life creates to developing these skills. In the article "Do Media Technologies Scramble Young Minds?",[43] I suggest that ways of thinking built into very young minds through TV and other media technology are perhaps the greatest impediment to pattern understanding involving the temporal dimension.

But these more recent deficiencies are only compounding factors. The real Achilles heel of modern thinking comes from the European Enlightenment separation between humanity and nature, mind and body, good and evil, which blinds us to more wholistic integrated understandings. *The science of ecology provided the overwhelming evidence that everything is connected, so it is a great irony that conservation biology is now dominated by an orthodoxy that is blind to ecosynthesis as nature's way of weaving a new tapestry of life.*

Emergence

One of the key ideas from systems theory, which has gained a wider currency, is that of "emergence". This suggests that self-organisation within complex systems results in activity, structures and behaviours that clearly emerge from within the system but have the effect of either transforming it or producing some completely new system. This is analogous to the Reorganisation phase of Holling's model of ecosystem succession, or the creation of new species in biological evolution.

An obvious example is the global network of personal computers which increasingly dominates world communication and economy; it arose out of the information technology industry quite unpredicted by the mainstream players or commentators. Most experts instead predicted a world controlled through a hierarchical structure from a few centrally located mainframe computers. It is important to understand how strongly committed the corporations and governments were to this centralised technology, and how fast they have had to adapt to the anarchic emergence of the networked world of personal computers and the Internet.

Of course everything has moved so fast that we can now see huge corporate and government interests riding the wave of network anarchy (to varying degrees of success). But it is important to remember that, not much more than 15 years ago, the advocates and entrepreneurs for the networked world were small, fringe actors with apparently little power to influence anything. The phrase "having a life of its own" comes to mind. Many of the proponents of this networked world see global computer networks as the beginning of a global human consciousness, a hive mind (as in beehives), with either utopian or sinister connotations.[44]

Emergent possibilities are all around us as a natural outcome of the extraordinarily fast co-evolutionary growth of human systems based on fossil fuels. Many of them, like information technology and genetic engineering, show the potential to overturn many of our assumptions.

As evidence mounts for the climax and rapid energy descent, the particular and local nature of that future is filled with uncertainty and emergent possibilities. The choices we take in responding to large-scale forces — fight to the death, uneasy compromise, going with the flow — may or may not make any difference.

Even the grand assumption of permaculture that the future of descending energy is inevitable may be mistaken. There is the possibility that some combination of computer networks, human culture and biotechnology will create a runaway world of accelerating energetic and informational growth (at least for another few hundred years).

It is clear that new evolutionary jumps emerge out of the chaos of high-energy systems, but these will not necessarily lead to human salvation. In any case, the forces at work are clearly beyond human control, so we might as well get on doing what we can to create a world that reflects human values and ethics within the constraints of nature's laws.

Genetic engineering: explosive evolution or techno dream?

Much has been written about the hazards of genetic engineering. The rapid emergence of consumer resistance to food from genetically modified organisms (GMOs), despite the grand plans of the corporations, has made it one of the most controversial environmental issues of our time.

Of all of the emergent factors being generated by human systems, genetic engineering of novel lifeforms seems most interesting because it feeds back into nature's well-established biological foundations for evolutionary change. Unlike information technology, it does not depend on complex infrastructure and energy flows to have impact once released into nature. Because gene transfer between bacteria in nature has been identified as common, the idea that the genetics of novel lifeforms will stay put in GMOs released on farms, or even used as medicines, is laughable. New microbes will spread to the limits set by ecological constraints, not those set by human management or edict, much faster than the spread of introduced plants and animals.

If, on the other hand, like so many favoured and pampered introduced species, genetic engineering turns out to be a failure, then the resources devoted to it will have been just another lost opportunity for creative and graceful adaptation to an energy descent.

The vast and pervasive nature of genetic promiscuity may mean genetically engineered organisms are lost in the sea of much more robust genetics responding to real factors in the environment, rather than the cocoon of human illusions created in the laboratory and the boardroom. The grand illusion on which virtually all genetic engineering is founded is the faith in an ability to manipulate our environment using high energy and technology. Whether the biotechnology industries can continue to attract massive investment capital once the high cost of energy becomes entrenched is also uncertain. In these conditions competition for capital will be intense from the established energy-harvesting (fossil-fuel) industries as well as the proven quick returns on investment from conservation and some renewable energy sources. Even if the biotech industry can continue to attract investment capital, its ability to create lifeforms that are very powerful without the support of a high-energy economy is dubious.

If genetic engineering does manage to mine the world's genetic diversity to generate real wealth rather than more economic illusions, then the Maximum Power Law would predict a rapid economic and technological reorganisation around genetic engineering (probably combined with computer technology) to create a massive acceleration in change. This would include the transformation of humanity into some new recombinant species, as suggested by Jeremy Rifkin.[45]

The scientists who enthusiastically look forward to such evolutionary transformation or replacement of humanity by self-organising genetic and/or information technology might seem like Dr Strangelove. But I believe they are more realistic than the larger group of enthusiasts within the scientific community who believe that genetic engineering can allow steady economic growth, save us from ecological disaster, but remain as a technological tool in our hands rather than becoming the proverbial monster of Dr Frankenstein in Mary Shelley's portentous novel.

We are unable to stop the accumulation of risk of catastrophic change resulting from nuclear radiation, the greenhouse effect or genetic engineering. Much of our frustration is the residual feeling that we should be able to prevent these things because they are of human origin. On the other hand, these out-of-control forces have simply replaced in the human psyche some of the high-order, large-scale forces of nature and/or gods that previously precipitated catastrophic change as well as bestowing gifts of abundance.

Gender Balance and Sustainability

I started this principle by saying that we need to integrate a new balance between stability and change into our daily lives and designs, and that we need an understanding of the multi-scale nature of change and stability to overcome the limitations we inherit from industrial culture.

Although we know that change is continuous, the balance and quality of that change may be one of the most important characteristics of a sustainable culture. In the Introduction, I characterised the bias of industrial culture toward episodic change, while sustainable culture is biased towards rhythmic change. The classical view of history as a series of episodic crises and conflicts, with great men as the leading actors, can be seen as an elitist and patriarchal view, which ignores the less dramatic, rhythmic, cyclical nature of ordinary life which is the greater part of human experience across time and space.

Similarly in our use of nature, environmental concepts, including permaculture, emphasise working with the rhythmic cycles of change in nature, rather than excessive reliance on the episodic intervention that kicks the system into some hopefully preferable state. It is reasonable to see this view of nature as more in tune with feminine rather than masculine culture.

The pulsing ecosystem model and the land use examples that I discuss show how episodic, even chaotic, change becomes integrated into ecological systems. But this type of change is a two-edged sword: it can lead either to a spiral of degradation, or to novel and often unintended transformative change.

Rather than simply rejecting episodic change as destructive, I have sought to show how we can use this type of change creatively rather than being captive to its dynamics. In terms of gender cultures, it is a process of beginning to see the limitations of patriarchal culture, while recognising that the masculine ways of action that have built that culture are potent agents of change which do have a place.

The value of the pulsing model of ecosystems is that it clearly demonstrates that in any system, a small amount of this type of change goes a long way. Any system that persists (relatively) is dominated by small-scale rhythmic patterns that maintain and conserve a larger stability.

Bringing this all down to earth, it is the patterns of traditional life focused on the home and a domestic connection to nature, the cycles of the seasons, and even the mundane, supposedly boring aspects of childcare and education, housework and building maintenance, plant and animal husbandry, community support and maintenance, which must dominate any notions of sustainable culture. It is hard to avoid the conclusion that women might be leaders in this transformation.

Ivan Illich argued that industrialisation has transformed patriarchy into a genderless culture that has absorbed both men and women but curiously leaves men structurally dominant.[46] Despite the achievements of feminism, modern industrial culture has condensed around a core of hard-to-budge ways of thinking, doing and being, which most men and women take for granted as normal but which are, in essence, masculine in origin. One of the most important of these is the bias towards seeing all systems as passive and inactive until a powerful actor or force pushes the system somewhere.

To me it seems inevitable that society eventually evolves a new structure of "ambiguous complementarity"[47] between the genders, if only because it reflects a fundamental energetic efficiency for organisation of households. Before this is likely in any enduring form, restoring the asymmetrical balances explored in this and the other principles is a task for all of us. As a man, I recognise the deep masculine roots of my own ways of thinking, acting and being. Although the obvious influences on the development of my ideas may have been men, the wisdom and guidance that have tempered and tested those ideas have come from the women in my life, especially at critical times of decisive action or change.

My inability to be concrete and explicit about that process is a reflection of the structural bias in our language and culture, in which feminine ways of being remain apparently passive and invisible. The Indian eco-feminist Vandana Shiva has identified the cultures of women, indigenous and peasant peoples as having a common invisibility that is consumed by the dominant culture; for me this is very relevant. That Vandana Shiva, like all of us, uses the education and tools of modernity to convey these understandings is itself a paradox.

Clearly the progressive rebalancing of feminine and masculine characteristics within a culture of sustainability will be different to that in any traditional sustainable culture. The feminist movement has done much to break down the hegemony of men in industrial society, but I believe much feminist orthodoxy, like that of the environment movement, unconsciously accepts the high-energy genderless society as some sort of natural enduring state within which we must develop new values and ideas. Like with the environment movement, the diversity and vitality of feminism are good signs for its continued contribution as we move into the uncharted territory of gender politics in descent.

Long-term Thinking, Large-scale Cycles

Making sense of our era demands that we get past the limited notion of change driven by the inexorable arrow of progress to see that nature and humanity are governed by cycles, both large and small, nested within each other like Russian nesting dolls (matryoshka) and overlapping as depicted by Venn diagrams.

We need to integrate into our thinking multiple cycles on a larger scale than those suggested by current fashion or business or political cycles.

Historians explaining the lessons of history have never been popular, especially since they have lost the excitement of the "great men" theory of history. When geologists, palaeo-biologists and archaeologists try to paint the really big pictures that provide a context for the present, this is perceived as academic musings of little consequence, or at best as exciting snapshots of other times which are compared to our own in a one-dimensional way. On the other hand, apocalyptic predictions of either a scientific or spiritual nature lead some to follow very specific plans and preparations rather than displaying a wholistic and flexible attitude to unpredictable change.

Geology and biology long ago overturned the Biblical notion that the world is only 6000 years old. The expansion of human power based on consumption of a substantial portion of the most accessible and potent fossil fuels of the planet requires us to think simultaneously on multiple time-scales if we are to develop the wisdom necessary for a culture of sustainability to emerge during energy descent. We must rebuild and surpass the long-term thinking that was natural to indigenous cultures of place. Learning how to see and feel the multi-dimensional, patterned nature of the earth's living history is central to the ability to use and creatively respond to the changes we face.

The Pulsing Patterns of Life on Earth

The history of life on earth includes vast periods of tens of millions of years when all the earth's land masses were worn down, shallow seas covered most of the earth's surface, and the climate was almost uniformly mild and moderately wet. Biodiversity was low and ecosystems changed little over vast areas and millions of years. This stability has been punctuated by shorter periods of great change. The last 1.6 million years, known as the Quaternary, has been characterised by tectonic uplift and volcanism, global and montane ice caps, more extensive continents with vastly varied climates, ice ages and warmer interglacials, massive extinctions and prolific speciation. Much of the wondrous biological and geographic diversity of the earth, which we accept as natural only exists in these periods of great change. It is unlikely that a species such as Homo sapiens could emerge, or even survive, the long, slow, stable periods that dominate the earth's geological history. On that scale, humanity is a short-lived microbial explosion.

Within these ages of great geologic, climatic and biotic change, there are small-scale patterns of stability, followed by brief pulses of activity. Ice ages lasting hundreds of thousands of years represent the stable norm of the Quaternary, with interglacials lasting roughly ten thousand years representing the explosive pulse of life based on new fertility, benign climates, and the growth of biodiversity.

At a smaller scale still, we can see both the glacial and the interglacial as relatively stable states with rapid transition in between: a temporal edge between two stable systems. For example, as sea levels rose at the end of the last ice age 12,000 to 8000 years ago, ancestors of Australian Aborigines would have seen the Bassian Plain between Tasmania and the mainland shrink each year as winter storms breached the coastal dunes, flooding thousands of hectares which would never return to dry land. More dramatic is the evidence that continental glaciers shed icebergs of such magnitude that sea levels may have risen to flood vast areas, as in the Atlantis story.

Mammoths were frozen with grass in their mouths, suggesting an early and severe winter precipitating the onset of last ice ages in the northern hemisphere.[48] This is more evidence that far-reaching change can be precipitated with great rapidity.

Within this current benign and remarkably stable interglacial climate, humanity has developed agriculture, what we call civilisation, and finally global industrial culture. All of them have accelerated the pace and power of human cultural and environmental change, but all are dependent on the fragile stability of the interglacial paradise. While the adverse effects of global warming have created a reason once more to contemplate our dependence on larger forces and circumstances, any serious consideration of the next ice age (due any time in the next thousand or so years) would have us designing a modest, generalist, flexible culture, carried by a small global population able to make the transition into and through the long slow years of the ice age.[49] We need to break out of the delusion of apparently linear acceleration of human material and numerical progress to a world view in which everything is contained by cycles, waves and pulses that flow between polarities of great stability and intense change, all nested one within another.

This cyclical view of time is not some new idea, but simply a rediscovery of ways of understanding that are embedded in human cultural history and our collective unconscious.

For example, the ancient Hindu cosmology gives us a breathtaking perspective on the cyclical nature of time and proportional pulsing balance between change and stability (Figure 33).[50] A 60,000-year cultural cycle is dominated by a long matriarchy phase composed of three great enduring ages; it collapses in a vortex of chaos before the birth of the patriarchy phase, the Kali Yuga. The Kali Yuga (or Age of Conflict) lasts only 6000 years before another vortex of change leads to the next cycle. By the very precise reckoning of the Hindu numerologists, we still have over 440 years of chaotic change within the Kali Yuga before the new matriarchal age emerges. No time at all in the scheme of things!

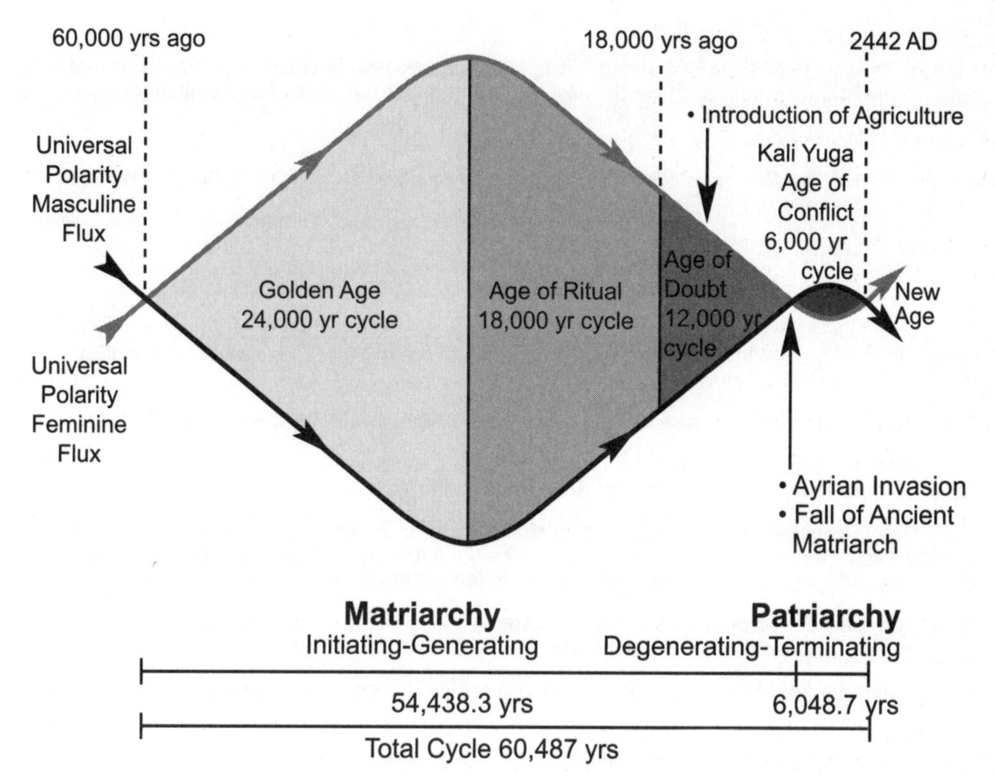

Figure 33: The traditional Hindu cultural cycle calendar (after Lawlor 1991)

Without granting any particular credence to this or any other predictions, it is reasonable to conclude that life for our children, grandchildren and their descendants is hardly likely to be characterised either by a steady state of illusory continuous growth or by a quiet, predictable, sustainable culture. We must prepare them for uncertainty and rapid change, while instilling in them a living appreciation of the enduring rhythms of nature along with the value of long-term thinking, persistent ethical behaviour, and the importance of the simple, ordinary, even mundane, aspects of life.

Permaculture is a dynamic interplay between two phases: on the one hand, sustaining life within the cycle of the seasons, and on the other, conceptual abstraction and emotional intensity of creativity and design. I see the relationship between these two as like the pulsing relationship between stability and change. *It is the steady, cyclical and humble engagement with nature that provides the sustenance for the spark of insight and integration (integrity), which, in turn, informs and transforms the practice. The first is harmonious and enduring; the second is episodic and powerful. The joyful asymmetric balance between the two expresses our humanity.*

1 J. Ralston-Saul, *The Unconscious Civilisation*.

2 See S. Bright, *The Line Ahead* Catalyst Communication Consultants 1996. A series of short stories that trace the possible future of Queensland Rail over the next 50 years was one outcome of a consultancy project.

3 Ricardo Semler *Maverick* 1995.

4 Paraphrasing Bill Mollison from the video *Bill Mollison The Permaculture Concept: In Grave Danger of Falling Food*.

5 For an outline of this hypothesis, see Article 6 "Historical Precedents for Permaculture" in *David Holmgren: Collected Writings* 1978-2000.

6 Expanded from a concept by British architect Frank Duffy.

7 See Article 3 "An Eclectic Approach to the Skills of Reading Landscape and their Application to Permaculture Consultancy" and Article 9 "Whole-Farm and Landscape Planning" in *David Holmgren: Collected Writings* 1978-2000.

8 M. Wilson and D. Holmgren, *Collingwood Children's Farm Property Management Plan Consultation* April 1996.

9 See Article 13 "Development Aid for the Industrialised North: Turning an Idea on Its Head, or The Problem Is the Solution" in *David Holmgren Collected Writings*: 1978-2000.

10 Studies of insect plagues in forests suggest that the defoliation of trees may recycle nutrients, while the death of some trees removes weaker individuals and thins the forest. See references to Ludwig et al. 1978 in H. T. Odum, *Living with Complexity* Crafoord Prize in the Biosciences, Royal Swedish Academy of Science 1987.

11 This extension of an erosion gully always travels upstream from a nick point in the gully course, where water falling over a vertical face undermines the cliff of soil leading to a collapse.

12 Neil Turok. The Chair of Mathematical Physics at Cambridge University interviewed on "AM" ABC Radio National, April 26 2002.

13 See W. Jackson, "Ecological Drift" in *The South West Book* Australian Conservation Foundation 1978.

14 From the Albrechtian perspective referred to in **Principle 2:** *Catch and Store Energy* the ratio of calcium to potassium is the critical factor in determining the quality of plant biomass and the resulting litter. The higher the calcium–potassium ratio, the softer and more nitrogen-rich will be the vegetation, resulting in rapid decomposition and low fire hazard. The lower the calcium–potassium ratio, the harder, more fibrous and unpalatable to mammal grazers will be the plants. This results in accumulation of fuels. I am not aware of any fire ecology study that included measurement of levels of calcium, let alone calcium–potassium ratios. Given the poor record of agricultural science in maintaining the health of farm soils, let alone natural ecosystems, it is surprising, even shocking, that most ecologists seem to accept conventional agricultural soil science as a foundation for designing studies to monitor ecosystems.

15 A. Savory, *Holistic Resource Management* Island Press 1988.

16 For a classic study done in the early 1960s, before contact with industrial society had changed social and agricultural systems, see R. A. Rappaport, "The Flow of Energy in an Agricultural Society" in *Biology and Culture in Modern Perspective: Readings from Scientific America* W.H. Feeeman & Co. 1972.

17 Due more to forest logging and conventional agricultural development by governments, corporations and migrants than to rises in indigenous populations.

18 Article 1 in *David Holmgren: Collected Writings* 1978-2000.

19 Creeping bent grass (*Agrostis stolonifera*) is a creeping rhizomous grass (similar to couch), which is active in both summer and winter in our climate.

20 In our area, mulched gardens turn into cherry plum forests with an assortment of co-dominant trees, unless they are regularly weeded of fruit trees.

21 Article 8 in *David Holmgren: Collected Writings* 1978-2000.

22 See C. S. Holling, "The Renewal, Growth Birth, and Death of Ecological Communities" *Whole Earth Review* Summer 1998. Holling is one of the pioneers of this new model of ecosystem change and its wider application to human institution and societal change.

23 For example, the heavy rains and floods that followed the 1983 Ash Wednesday fires in Victoria removed large amounts of minerals (as ash) from whole regions.

24 For expansion of this theme of rural resettlement see Article 5 "Submission In Response to a Review of Rural Land Use in Victoria" and Article 27 "The Counterculture as Dynamic Margin" in *David Holmgren: Collected Writings* 1978-2000.

25 The Greenpeace "label" has been valued at over $US410 million worldwide: *The Economist*, 1998, quoted by S. Beder, *Global Spin* Scribe Publications 2000.

26 Richard Goldschmidt, a geneticist, was the first to show that gradualism was not a plausible explanation of evolutionary change, but his ideas were derided until revived by Steven Jay Gould in the 1970s.

27 Kevin Kelly, "Post Darwinian Evolution" in *Whole Earth Review* no. 76, Fall 1992.

28 Ironically, studies with fruit flies, which have shown new characteristics attributed to symbiotic virus, are defined as non-evolutionary by neo-Darwinian theory.

29 See news article "Recent revelation about bacteria's evolution could affect thinking on how higher organisms evolved" in *University Times* (University of Pittsburgh Faculty & Staff Newspaper) June 8th 2000 http://www.pitt.edu/utimes/issues/32/000608/12.html

30 New geomorphological evidence suggests that New Zealand was completely under the sea 20 million years ago, so its total flora and fauna has evolved in that time from what managed to cross the ocean.

31 B. Holmes, "Day of the Sparrow" *New Scientist* June 1998.

32 Kale Sniderman, pers. comm.

33 For an excellent analysis of the ecological foundations of European expansion and the role of introduced weeds and pests in stabilising landscapes degraded by European exploitation, see A. W. Crosby, *Ecological Imperialism: European Expansion 900 to 1900* Cambridge University Press 1986

34 See T. Low, *Feral Future* Penguin Books 1999, for reference to this issue. He reviews the degree to which exotic lifeforms are moving around the world and changing ecosystems. Low's factual research and original observations and anecdotes are excellent, but the doom and gloom analysis of consequences, and the conclusions drawn reveal little systemic understanding.

35 David M. Richardson et al., "Plant Invasions: The Role of Mutualisms" *Biological Review* no. 75, 2000 provides an interesting review of examples.

36 Article 2 in *David Holmgren: Collected Writings* 1978-2000.

37 This work was documented in D. Holmgren and P. Morgan, *The Yarra Floodplain: The Study of an Urban Ecosystem* Environment Studies Association of Victoria 1982.

38 Most notably Wilson's Reserve.

39 D. Holmgren *Melliodora*, includes a brief description of the strategy and species used in the "gully project".

40 Michael Wilson *Post gold rush stream regeneration: implications for managing exotic and native vegetation* Centre for Environmental Management, University of Ballarat (presented at the Second Australian Stream Management Conference in February 1999)

41 J. M. (Kale) Sniderman, *Successional Dynamics In A Mixed Native/Introduced Riparian Forest In Central Victoria* University of Ballarat 1998.

42 I have a book in preparation, entitled "Weeds or Wild Nature?" challenging the environmental orthodoxy about naturalised plants and animals.

43 See Article 24 in *David Holmgren: Collected Writings* 1978-2000.

44 See Kevin Kelly, "Hive Mind" *Whole Earth Review* no. 82, 1994. *Whole Earth Review* over the last 20 years has provided an insight into many of the emergent phenomena of information technology, especially when it was edited by Kevin Kelly in the late 1980s.

45 J. Rifkin, *Entropy: A New World View* Viking 1980.

46 See I. Illich, *Gender* Pantheon 1982.

47 Illich uses this term to describe the relationship between the genders in pre-industrial societies, where men and women existed in separate but complementary cultures, which could never be fully understood or controlled by each other.

48 Once heavy snows accumulate and fail to melt over the summer at high latitudes, whole regions become devoid of life. Thus one bad winter triggers the ice age.

49 Suggested by Rhys Jones on "The Science Show" ABC Radio National, 4 November 2000 as a way of putting concern about global warming in context.

50 See R. Lawlor, *Voices of the First Day: Awakening in the Aboriginal Dreamtime* Inner Traditions 1991.

Postscript: After September 11

Under the heading "Top-down Thinking, Bottom-up Action" in **Principle** 4: *Apply Self-regulation and Accept Feedback* I observed:

> Although I am articulating top-down thinking and bottom-up action as a restatement of the environmentalist slogan "thinking globally acting locally", there is plenty of evidence that global power elites have learnt this new mode of understanding and action. A shift from evident, formal and managerial exercise of power to invisible, informal and collaborative modes allows elites to bypass the democratic and bureaucratic controls on their power demanded by the public. More fundamentally, when elites overcome their own egos to see they are just small players in these vast human systems of global energy peak, then they will recognise that collaborative and subtle exercise of power is more effective than threats and brute force. The ways in which this revolution in understanding and action must have permeated the highest levels are not yet clearly visible.

If the truly powerful global elites have retained and reinforced their wholistic systemic understandings with the recent insights from Chaos theory, cybernetics and systems ecology and developed alternatives to top-down managerial control, then they wield a power few of us can imagine. When systemic thinking is a skill in the hands of a few, it can and will be used for good and ill. For example, in the 1980s I realised that the EMERGY evaluations of Howard Odum can be used, and probably have been, by global corporations to identify free and undervalued resources, which they can exploit.

Political and economic analysts frequently misunderstand the impact and consequences of this new mode of behaviour. Events are generally explained in terms of the almost random aggregations of market and political forces, which represent some sort of balanced and democratic outcome. At the other extreme, conspiracy theories explain events in terms of fully controlled outcomes by some particular grouping of elites using managerial methods.

I believe that top-down thinking, bottom-up action provides a way of comprehending the powerful synergistic forces to which the majority of the intelligentsia seems blind. Or perhaps they are afraid to speak, for fear of being labelled a conspiracy theorist (which is akin to certification of insanity).

A barrage of propaganda from the mainstream media denigrates any systemic understandings of political processes as conspiracy theories. This ensures that all evidence is considered and evaluated through reductionist logic which separates causes and effects, processes and products, system and element. Conspiracy theories provide a simple and useful way of explaining this power, but they are inadequate because they assume that exposure of the persons apparently near the centre of the particular conspiracy will remedy the adverse effects of these new abuses of power.

A more sophisticated understanding would see that world events tend to be driven by loose coalitions of economic, political and military interests, which function like guilds of species in an ecosystem. These guilds generate patterns of events that meet the interests of those coalitions, without there being any unity of purpose or clear plan. When powerful players accept they are not all-powerful, they increase their effectiveness but are also able to deny and cover any responsibility for the adverse outcomes of those actions.

During the Cold War, the CIA and the KGB functioned as complementary and self-reinforcing institutions rather than representing the interests of their respective nations. In the process, they exposed us all to the risk of nuclear annihilation. Today, new, more powerful ecosystems of global institutions and invisible emergent "organisms" are dancing humanity toward energy Armageddon, with the nuclear scenario still as real as ever.

There is abundant evidence[1] that September 11 was an outcome of these shadowy coalitions, which link global energy corporations, US foreign policy, the global "intelligence community", Islamic fundamentalists, arms dealers and the illegal drug trade. Discussion of this bizarre symbiosis remains beyond the pale of mainstream media in Western nations partly because of the blanket agreement that conspiracy theories that could affect the highest levels of global affairs are, ipso facto, invalid. This is the best example of the paralysis of public discourse due to an absence of language to comprehend top-down thinking and bottom-up action as a new mode of power.

Any study of the history of covert action by the US government around the world over the last 30 years (primarily by the CIA) shows that systemic understandings have informed the application of relatively limited resources to provide massive political and economic leverage. The CIA "community" uses the term "blowback" to describe the adverse consequences of agents of influence who turn and bite the hand that feeds them. Liberal-left analysts have generally interpreted this as evidence of the incompetence of the CIA and the need for reforms (more top-down managerial control). A more wholistic understanding sees these eruptions of evil by drug barons, dictators like Manuel Noriega and Saddam Hussein, and movements like the Taliban as essential to sustain the power of CIA as a "virtual" private army of the global energy corporations. September 11 highlights that this new locus of global power may be consolidating around an apparent fight to the death between fundamentalist religion and rational materialism. Most people around the world find themselves having to take sides.

Dr Sherif Youssef Hetata, a leading Egyptian activist against religious fundamentalism, has pointed out[2] that globalisation, as delivered by hegemonic Western power, not only drives the rise of religious (especially Islamic) fundamentalism but also shares many characteristics with religious fundamentalism. Figure 4 in Ethical Principles allows us to make sense of this apparent conflict as the destructive pathway to the emergent union of our materialist and spiritual aspects.

Puns like "Bush Laden" are light hearted attempts to recognise the locus of power behind this dark union, but removal of one set of unsavoury characters just leads to their replacement with another. Just as important as the decadent power elite is the complicity of the billion or so middle-class recipients of most of the fossil-fuel wealth that flows from this system of power.

During the 1990s social and environmental movements lost ground through the declining effect of hard won top-down bureaucratic control over excesses of corporate power. These new excesses of power are increasingly exercised through proxies, lobbyists, agents provocateurs, think tanks, public relations, identification of invisible allies in grassroots movements, and other methods.[3] On the other hand, those same movements have had extraordinary success in using limited resources to influence society through small but significant consumer and shareholder boycotts, culture jamming,[4] empowering and celebrating sustainable alternatives. More fundamentally, the creation of sustainable alternatives exposes the big lie that we and our descendants are all beholden to this global hegemonic power for our very existence.

Social democratic movements of the late 19th and early 20th centuries mastered the then novel model of bottom-up thinking and top-down action in order to control some of the excesses of corporatist and fascist power. Today the so-called anti-globalisation movement is the tip of the sustainable alternatives iceberg, which has the potential to stop those same titanic forces precipitating global Armageddon after the oil peak. Understanding and applying top-down thinking and bottom-up action is the key to success.

1 Readily available on the Internet by dedicated private researchers and activists.
For example: Mike Rupert http://www.copvcia.com/
Jared Israel & Michel Chossudovsky http://emperors-clothes.com/index.html

2 Interviewed on Radio National, 27 May 2002

3 See S. Beda, *Global Spin: The Corporate Assault On Environmentalism*.

4 For a sample of creative and entertaining culture jamming see the following websites:
http://rtmark.com/home.html
http://futurefeedforward.com/
http://www.guerrillanews.com/redux/

Selected Bibliography

Alexander, Christopher et al A *Pattern Language: Towns, Buildings, Construction* Oxford University Press 1977. This classic design text is being used by some permaculture designers as a model framework for developing a permaculture pattern language for productive landscapes.

Beda, S. *Global Spin: The Corporate Assault On Environmentalism* Scribe Publications 2000. Documents the scope and depth of methods used by corporations to disarm and deflect environmental activism.

Berry, W. *Culture and Agriculture: The Unsettling of America* Sierra Club Books 1977.A classic and articulate advocacy of the role of small traditional farming as a foundation for modern environmental concerns and the destructive forces of big agriculture pushed by corporations and governments.

Campbell, C. *The Coming Oil Crisis* Multi-Science Publishing 1997. One of the most authoritative books on the global energy peak phenomenon and its consequences. For further sources see http://www.hubbertpeak.com/

Crosby, A. W. *Ecological Imperialism: European Expansion 900 to 1900* Cambridge University Press 1986. An excellent analysis of the ecological foundations of successful colonisation by Europe of other global regions and the role of introduced weeds and pests in stabilising landscapes degraded by European exploitation.

Flannery, T. *The Future Eaters* Reed Books 1994. A palaeoecological overview of the nature and constraints of the Australian environment. Flannery's description of the process and outcomes of human exploitation of new free resources can be used to understand fossil fuel peak.

Fukuoka, M. *The One Straw Revolution* Rodale Press 1978. The classic book about agriculture and philosophy from Japan's foremost exponent of "natural farming" published the same year as Permaculture One.

Furuno, T. *The Power of Duck: Integrated Rice and Duck Farming* Tagari 2001. Excellent documentation of one of the best integrated polyculture systems from south east Asia.

Gall, John *General Systematics* Harper & Row 1977. An accessible guide to systems theory.

Hall, C. ed., *Quantifying Sustainable Development: The Future of Tropical Economies*, Academic Press, 2000. An overview with case studies of how the field of environmental accounting is being applied to third world development issues.

Hawkin, P Lovins. A. and Lovins, H. *Natural Capitalism: Creating the Next Industrial Revolution*, Rocky Mountain Institute, 1999. The arguments and evidence for environmentally benign industries modelled on principles of natural design.

Holmgren, D. *Melliodora (Hepburn Permaculture Gardens): Ten Years of Sustainable Living* Holmgren Design Services 1996. Documents the design and development of our own small rural property in central Victoria.

Holmgren, D. *David Holmgren: Collected Writings 1978-2000* (CD) Holmgren Design Services 2002. Articles on a diverse range of subjects related to permaculture which illustrate and expand on some of the concepts and examples in this book. Also available in screen readable format at www.holmgren.com.au

Jackson, W. *New Roots for Agriculture* University of Nebraska Press 1980. Classic book about using the prairie ecosystems as a model in the development of perennial grain crops as a major element of sustainable agriculture. For more recent information see website http://www.landinstitute.org

Levins, R. and Lewontin, R. *The Dialectical Biologist* Harvard University Press 1985. A useful introduction to the Marxist dialectical perspective on many issues directly relevant to the permaculture agenda.

Lovelock, J. *The Ages of Gaia: A Biography of Our Living Planet* Oxford University Press 1989. An overview of the dynamics of the earth as a self regulating system by the co-originator of the Gaia hypothesis (in the 1970's).

Low, T. *Feral Future* Penguin Books 1999. A readable review of the spread of exotic lifeforms both in Australia and around the world. Low's doom and gloom perspective reveals little systemic understanding although he does provide plenty of evidence to justify a more balanced view.

McCamant K. and Durrett, C. *Co-Housing: A Contemporary Approach to Housing Ourselves* Ten Speed Press 1994. An overview of both Danish co-housing and the application of the concept in America.

Mollison, B. & Holmgren, D. *Permaculture One* Corgi 1978 and since published in 5 languages (now out of print). Outlines the concept and its initial applications.

Mollison, B. *Permaculture: A Designers Manual* Tagari 1988. Mollison's most complete work on the concepts and design systems of permaculture with extensive graphical interpretation by permaculture designer Andrew Jeeves. Widely uses as the definitive text for permaculture teaching.

Newman, P. and Kenworth, J. *Sustainability and Cities: Overcoming Automobile Dependence* Island Press 1999. A comprehensive analysis of the dysfunctional nature of urban transport infrastructure as well as some examples of more enlightened solutions to urban development.

Odum, H.T. *Environmental Accounting: EMERGY and Environmental Decision Making* Wiley 1996. The definitive text explaining the principles and methods of EMERGY evalution and their diverse applications.

Odum, H.T. & Odum, E.C. *A Prosperous Way Down: Principles and Policies* Wiley 2001. A readable and timely explanation for the lay reader of the EMERGY concepts and implications of energy transition for the economy, society and culture. It updates their much earlier text for the lay reader *Energy Basis for Man and Nature* McGraw-Hill 1979.

Kropotkin, P. *Mutual Aid* Heinemann 1902. The classic answer to the social Darwinists by the Russian naturalist and anarchist.

Ralston-Saul, John *Voltaire's Bastards: The Dictatorship of Reason in the West* Penguin 1993. A novel and integrated history of the conceptual flaws in our civilization and an alternative view to the defunct left-right political framework. Although the environmental crisis is marginal to Ralston-Saul's analysis, he provides a new way of understanding how enlightenment thinking has been a seed for the environmental crisis.

Savory, A. *Holistic Resource Management*, Island Press, 1988. The conceptual text behind an agricultural training system and network of pastoral farmers using ecological principles to maintain and improve pastoral ecosystems.

Schumacher, E. F. *Small Is Beautiful: A study of economics as if people mattered* Blond and Briggs 1973. One of the classic books envisioning human scale economic systems and processes connected to decades of alternative economic success stories before and since.

Shiva, V. *Monocultures of the Mind: Perspective on Biodiversity and Biotechnology* Zed Books 1993. Goes behind the dysfunctional simplicity of Third World agricultural development to show the conceptual simplicity and naked exploitation driving these systems.

Smith, J. Russell *Tree Crops: A Permanent Agriculture* Devin Adair 1953. The classic book on the under-developed potential of tree crops to substitute for grain agriculture

Walters, Charles Jr, ed., *The Albrecht Papers: Volume One Foundation Concepts* Acres USA 1975. A collection of readable articles about the nature of living soil by an eminent soil scientist. William Albrecht's work and research encompassed and digested the successes and failures of chemical agriculture and laid the foundations for the scientific understanding of organic agriculture.

Yeomans, P.A. *Water For Every Farm* Murray Books 1965. Technical details and results of the Keyline system of land design, possibly Australia's greatest contribution to broadacre sustainable land use. Revised and reprinted by Ken Yeomans, see http//www.keyline.com.au

Index